Introductory Solid Mechanics

Introductory Solid Mechanics

James McD. Baxter Brown

Department of Mechanical Engineering
University of Dundee

JOHN WILEY & SONS   LONDON · NEW YORK · SYDNEY · TORONTO

Library of Congress Catalog Card No. 73–2777

ISBN 0 471 05985 4 Cloth bound
ISBN 0 471 05986 2 Paper bound

Printed in Great Britain by
Adlard & Son Ltd., Bartholomew Press, Dorking

To Isobel, Morven and Iona

Preface

Structural analysis and stress analysis have tended to exist as largely independent branches of the subject of the Mechanics of Deformable Solids. This dichotomy has been principally due to the great mathematical difficulties presented by the complex shapes of engineering structures and components. However with the general availability of high-speed digital computers it has become possible to integrate stress and structural analysis using a matrix stiffness process based on representation of a system as a combination of discrete elements of finite size. The extent to which these methods are now generally adopted suggests the reorientation of first degree teaching to provide a more suitable preparation than the traditional emphasis on particular results and formulae. This is the aim of my text. Much of the traditional material is retained but is presented in such a form and sequence as to permit the concurrent development of the matrix stiffness procedure. The latter is first introduced for collinear assemblies of bars and then extended to the treatment of plane bodies and plane frames.

Chapters 1–4 contain the basic background development of stress, strain, stress/strain relationships and strain energy. Chapters 5–9 treat the behaviour of bars formed into collinear-type assemblies and subject to direct loads, bending, torsion, and combinations thereof. Chapters 10, 11 and 15 deal with topics which are necessary for completeness in coverage of the subject but which space does not permit to be incorporated into the matrix procedure. Chapter 12 provides further development of the fundamental basis of the subject as a preliminary to the extension of matrix procedures to plane bodies in Chapter 13. In Chapter 14 the application of earlier methods to frames is considered.

References to specialized literature are given and the notation has as far as practicable been harmonized with the more widely available advanced texts. SI units are used throughout except for the occasional use for convenience of cm in relation to beam properties. It may be useful to note that in handwork bold-face type can be conveniently indicated by wavy underline, e.g. T, the standard proof correction mark for this purpose.

Contents

Principal Notation

Superscripts * virtual quantity
T transpose of a matrix

Overlaid bar (Chapter 14)—local coordinates

A	Cross-sectional area of bar
$\mathbf{b}, \mathbf{B}$	Strain/displacement matrices
$\mathbf{c}, \mathbf{c_j}$	(Chapter 5) Nodal loads due to misfit—for assembly, for jth bar (typical component $c_{i;\,j}$)
$c, \mathbf{c}$	(Chapter 14) Numbers of internal pin-joints, support constraints
d	Diameter
$D; D_f$	Flexural rigidity of plate; number of nodal degrees of freedom
$\mathbf{D}, \mathbf{D_T}$	Stress/strain matrices
e_{xx}, e_{yy}, e_{xy}	Total strain components
E	Young's Modulus
f_s	Factor of safety
$\mathbf{f}, \mathbf{f_j}, \mathbf{f}_{i;\,j}$	Nodal loads due to intermediate loading of assembly, of jth bar, at ith node of jth bar
$\mathbf{F}, \mathbf{F_j}, \mathbf{F}_{i;\,j}$	Nodal loads of assembly, of jth bar, at ith node of jth bar
$\mathbf{F_R}, \mathbf{F_P}$	Vectors of known, reactive, nodal loads of assembly
g	Gravitational constant
G	Shear modulus
h	Depth of beam
$\mathbf{H}$	Displacement/amplitude matrix
$I, I_y, I_z; I_{yz}$	Second moments of area; product moment of area
$\mathbf{I}$	Unit matrix
J	Polar moment of area
$k; k_{IJ}, k_J$	Stiffness; (Chapter 14) rotational stiffness, of bar IJ, of joint J
$\mathbf{k_j}; \mathbf{k}_{ij;\,j}$	Stiffness matrix of jth bar (element); sub-matrix of $\mathbf{k_j}$
K, K_f	Bulk modulus; bulk modulus of fluid
$\mathbf{K}; \mathbf{K}_{ij}$	Assembly stiffness matrix; sub-matrix of $\mathbf{K}$
$L; L_e$	Length of bar; effective length of bar (Chapter 11)
$m; m_x, m_y, m_z$	Moment; moments about x, y, z axes
$M, M_y, M_z; M_0, \hat{M}$	Bending moments; yield, fully plastic bending moments
$M_r, M_\theta, M_x, M_y; M_{xy}$	(Chapter 15) Plate bending moments; twisting moments
$n, \mathbf{n}$	Number of nodes, bars, in an assembly
$p, p_0, \hat{p}$	Pressure, yield pressure, fully plastic pressure
p_x, p_y	Body force intensities (Chapter 12)
q	Distributed load intensity on plate
Q_r, Q_θ, Q_x, Q_y	(Chapter 15) Plate shearing forces

Q, Q_j, $Q_{i;j}$	Nodal thermal loads of assembly, of jth bar, at ith node of jth bar
r; r_0; R, R_y, R_z	Radius, radius of gyration (Chapter 11); shaft outer radius; radii of curvature of beam
R_x, R_y	(Chapter 15) Radii of curvature of plate surface
R	State of strain
s	Length of mean perimeter of hollow section (Chapter 8)
t	Thickness of plate, of tube wall
T	Temperature; torsion moment (Chapters 8 and 9)
T_x, T_y	Surface tractions on boundary
T	State of stress
u, v, w	Components of linear displacement in x, y, z directions
U, U_0	Strain energy, strain energy density
$\bar{U}$, $\bar{U}_0$	Complementary energy and density
V; $\bar{V}$	Shearing force, volume; potential of external loads (Chapter 12)
w	Distributed load intensity on beam
W, $\bar{W}$	Work, complementary work
X, Y, Z	Components of force in x, y, z directions
X_C, X_{CE}	Critical load, Euler buckling load
Z	Section modulus
α	Coefficient of linear expansion
α	Vector of displacement function amplitudes
γ_{xy}, γ_{yz}, γ_{xz}; $\gamma_{x\theta}$	Shear strain components in rectangular; polar coordinates
δ, δ_j, δ_i	Nodal displacements of assembly, of jth bar, at ith node
δ_R, δ_P	Vectors of unknown, prescribed displacements of assembly
Δ	Volume strain (Chapters 2 and 3); area of triangle (Chapter 13)
ϵ; ϵ_I	Vector of element strain components; initial strain
ϵ_x, ϵ_y, ϵ_z; ϵ_x, ϵ_θ, ϵ_r	Normal strain components in rectangular; polar coordinates
ϵ_1, ϵ_2, ϵ_3	Principal strains
λ; λ	Lamé's constant; coordinate transformation matrix
μ	Direction normal to a surface; number of deflexion curve half waves (Chapter 11)
ν	Poisson's ratio
ρ	Mass density
σ	Vector of element stress components
σ_x, σ_y, σ_z; σ_x, σ_θ, σ_r	Normal stress components in rectangular; polar coordinates
σ_1, σ_2, σ_3; σ_I, σ_{II}, σ_{III}	Principal stresses in arbitrary order; algebraic order
τ_{xy}, τ_{yz}, τ_{xz}; $\tau_{x\theta}$	Shear stress components in rectangular; polar coordinates
Φ	Airy stress function
θ; θ, ϕ, ψ	Slope of beam; rotations in xy, xz, yz planes
ω	Angular velocity
σ_0, τ_0; τ_A, τ_{max}	Yield stress in tensile test, in pure shear; allowable shear stress, maximum shear stress

The State of Stress

1.1 External and Internal Forces

The external forces which may act on a body are of two types viz. body forces and surface forces. Body forces (represented notionally in Figure 1.1 by dashed arrows F_i) are those due for example to gravitation and magnetism, and act on the mass of matter present but without the need of any physical contact with the body. Surface forces are exerted over the bounding surface of a body (the arrows P_i in Figure 1.1), and are due to contacts with other bodies.

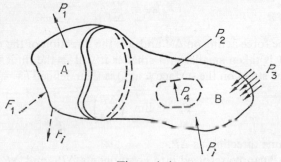

Figure 1.1

The external forces cause forces distributed through the material of the body. These *internal forces* transmit the effects of the external forces between their points of application, and a knowledge of their distribution is necessary in order that the body may be so designed that the forces do not exceed its strength. These internal forces are specified in terms of a quantity called *stress* whose nature and properties will next be described.

1.2 Stress on a Surface

Consider a thin slice of material lying between two imaginary parallel planes as in Figure 1.1. Parts A and B of the body exert internal forces distributed over the surfaces of the slice. If the body as a whole is in equilibrium then so is the slice. Furthermore if the slice is now taken to be of infinitesimal thickness then the lines of action of the forces on the opposite surfaces at any position Q will coincide (Figure 1.2a), since Q becomes effectively a particle subject to two forces in equilibrium.

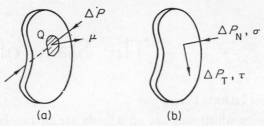

(a)   (b)

Figure 1.2

Since the internal forces are continuously distributed over the surfaces of the thin slice it is reasonable to relate the forces to the magnitude of the area they act on. If ΔA is a small part of the surface of the slice then the average force per unit area is

$$(T_\mu)_{\text{av}} = \frac{\Delta P}{\Delta A}$$

where ΔP is the force acting on ΔA and μ is the direction of the normal to the surface. If ΔA is taken smaller and smaller it will in the limit tend to some point Q and the stress on the surface μ at Q is then defined as

$$T_\mu = \lim_{\Delta A \to 0} \left(\frac{\Delta P}{\Delta A} \right)$$

and has the same direction as ΔP.

The force ΔP can be resolved into components ΔP_N and ΔP_T normal and tangential to ΔA as in Figure 1.2b. Then the stress on the surface μ at Q may be described in terms of

Normal (or direct) stress

$$\sigma_\mu = \lim_{\Delta A \to 0} \left(\frac{\Delta P_N}{\Delta A} \right) \tag{1.1a}$$

and *Shear* (or tangential) stress

$$\tau_\mu = \lim_{\Delta A \to 0} \left(\frac{\Delta P_T}{\Delta A} \right) \tag{1.1b}$$

The resolved stress components are shown in Figure 1.2b acting at Q on the surfaces of the slice.

1.3 State of Stress

If the parallel planes isolating the slice in Figure 1.1 had another inclination but still contained the point Q as in Figure 1.3, then the force distribution on the slice would be different in general and so would be the stress on the

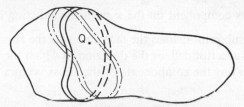

Figure 1.3

surface at Q. It might therefore seem that the *state of stress* could only be fully defined if slices were taken with all possible orientations. Were this necessary, stress would not be a very useful quantity. Fortunately it is found that complete definition of the state of stress at Q requires only the use of any three mutually perpendicular sets of planes around Q which then isolate a rectangular parallelepiped of infinitesimal dimensions around Q as in Figure 1.4a.

Choosing coordinate axes parallel to the edges, the three sets of opposite faces are named after the directions of the outward normals to their surfaces. Thus of the two faces hatched in Figure 1.4a one has its normal in the positive

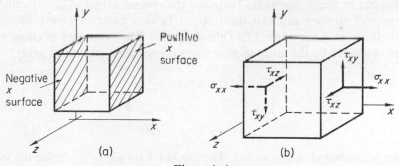

Figure 1.4

direction of the x axis and is known as the positive x surface, while the other has its normal in the negative direction of the x axis and is called the negative x surface.

The stresses on the three sets of surfaces are now resolved into their components in the directions of the x, y, z axes. Figure 1.4b shows the stress components

$$\sigma_{xx} \qquad \tau_{xy} \qquad \tau_{xz}$$

on the x faces. The meaning of the double subscripts is as follows:

σ_{xx} is the stress component on the x surface and acting in the x direction.

τ_{xy} is the stress component on the x surface and acting in the y direction.

τ_{xz} is the stress component on the x surface and acting in the z direction.

Thus the first subscript defines the face on which the stress component acts and the second subscript defines the direction in which the stress component acts. In the same way the components of the stress vectors on the y surfaces would be

$$\tau_{yx} \qquad \sigma_{yy} \qquad \tau_{yz}$$

For the z surfaces the stress components are

$$\tau_{zx} \qquad \tau_{zy} \qquad \sigma_{zz}$$

The rotation of the double subscripts suggests that the nine stress components might be conveniently written as an array

$$\sigma_{xx} \qquad \tau_{xy} \qquad \tau_{xz}$$
$$\tau_{yx} \qquad \sigma_{yy} \qquad \tau_{yz}$$
$$\tau_{zx} \qquad \tau_{zy} \qquad \sigma_{zz}$$

Now the normal stress component σ_{xx} acting on the x surface by its nature as a *normal stress* must act in the x direction, since the x surface is that with its normal in the x direction. Therefore the second subscript is superfluous for normal stresses and can be dropped in this notation where different symbols are used for normal and shear stresses. Thus the state of stress at a point is defined by the array of nine stress components (the stress array)

$$\mathbf{T} = \begin{pmatrix} \sigma_x & \tau_{xy} & \tau_{xz} \\ \tau_{yx} & \sigma_y & \tau_{yz} \\ \tau_{zx} & \tau_{zy} & \sigma_z \end{pmatrix} \qquad (A1.1)$$

which is illustrated in Figure 1.5. The symbol $\mathbf{T}$ is used to represent the state of stress at a point in a material.

Reference to Figure 1.5 will show that each row of the stress array (A1.1) consists of the stress components acting on a particular set of surfaces. For example the components τ_{yx}, σ_y, τ_{yz} are components of stress on the y surfaces. On the other hand, each column of the stress array represents the stress components acting in a particular coordinate direction, e.g. those in the third column all act in the z direction though on different surfaces. It should be

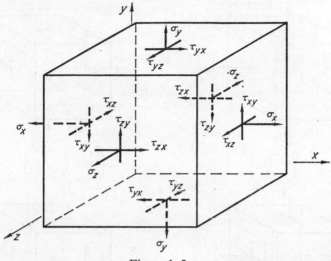

Figure 1.5

noted that the normal stress components occupy a diagonal of the stress array. This diagonal is called the principal diagonal.

The following *sign convention* is now adopted. A stress component is considered positive if, when acting on a positive surface, it is parallel to the positive direction of an axis; when acting on a negative surface the stress component is positive if it acts parallel to the negative direction of a co-ordinate axis. It will be seen that in Figure 1.5 all the stress components have in fact been given their positive directions. A useful simplification of the procedure is first to apply the sign convention to the stress components on the positive surfaces of the elemental block. The components on the negative surfaces are then drawn acting in the opposite direction to the corresponding component on a positive face. It may be further noted that a positive normal stress is termed *tensile* and a negative normal stress *compressive*.

Example 1.1 Examples of a stress array and the corresponding illustrated state of stress are given in Figure 1.6, and should be studied carefully, referring to array (A1.1) and the sign convention given above.

Problems for Solution 1.1, 1.2.

1.4 Symmetry of the Stress Array

The state of stress has been represented by the array (A1.1) of nine stress components comprising three normal stresses and six shear stresses. However, as will now be shown, only three of the shear stress components are independent. Without loss of generality the cube of side a in Figure 1.6d can be

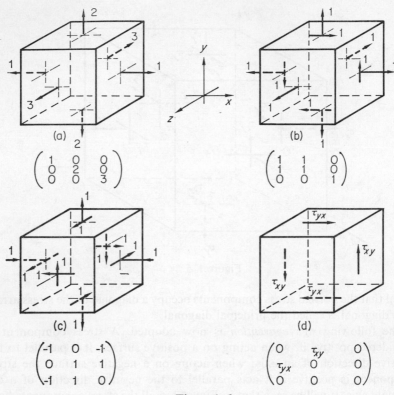

Figure 1.6

considered as an infinitesimal region of a body in equilibrium subject to the state of stress represented by the array

$$\mathbf{T} = \begin{pmatrix} 0 & \tau_{xy} & 0 \\ \tau_{yx} & 0 & 0 \\ 0 & 0 & 0 \end{pmatrix} \qquad (A1.2)$$

Comparison with Figure 1.5 shows that this array contains only those stress components capable of exerting moments in the xy plane. The two faces perpendicular to the x axis (i.e. the x faces) bear forces of magnitude, $\tau_{xy}a^2$ acting parallel to the y axis but in opposite directions and separated by a distance a. These forces therefore exert a couple of magnitude $((\tau_{xy}a^2)a)$ on the cube in the xy plane. This moment is anticlockwise when viewed in the negative direction of the z axis. Now the stress components τ_{yx} acting on the y faces represent forces $\tau_{yx}a^2$ parallel to the x axis and distance a apart, so exerting a clockwise couple $(\tau_{yx}a^2)a$. Therefore for equilibrium in the xy plane

$$\tau_{xy}a^3 = \tau_{yx}a^3$$

i.e.

$$\tau_{xy} = \tau_{yx}$$

A similar argument could be carried out for the yz and xz planes to prove that

$$\tau_{yz} = \tau_{zy} \quad \text{and} \quad \tau_{xz} = \tau_{zx}$$

respectively. Thus if i and j represent any of x, y, z and if $i \neq j$, then

$$\tau_{ij} = \tau_{ji} \qquad (1.2)$$

This result is sometimes called the Principle of Complementary Shear Stress.

Examination of the stress array (A1.1) shows that the positions of the equal shear stress components are such that the array is symmetrical about the diagonal line occupied by the normal stress components (the principal diagonal). This fact has been assumed in the previous worked examples. Another consequence of (1.2) is that $\tau_{ij} \neq 0$ unless $\tau_{ji} = 0$. In other words if, say, τ_{xy} is present, then so is τ_{yx}.

1.5 Superposition of Stress Arrays

By definition any one of the stress components on a surface is proportional to the corresponding resolved component of force (equation 1.1). Therefore if several increments of a force component are applied the corresponding stress component is the sum of the stresses due to each force increment acting separately. It follows that if a body is subjected to several sets of external forces the combined state of stress due to all the sets of forces acting together is the algebraic sum of the separate states of stress, provided that each stress array is measured with respect to the same coordinate axes. Superposition of stress arrays is thus performed by taking the algebraic sums of corresponding elements of the separate arrays.

Example 1.2 At a point in a plate stress arrays due to two separate load systems A and B have been measured as follows with respect to the same axes x, y, z

Set A
$$\begin{pmatrix} 2 & 1 & -3 \\ 1 & 4 & -2 \\ -3 & -2 & 6 \end{pmatrix} \text{MN/m}^2$$
Set B
$$\begin{pmatrix} 0 & 1 & 3 \\ 1 & 0 & 2 \\ 3 & 2 & 0 \end{pmatrix} \text{MN/m}^2$$

Then the combined effect of the load systems A and B, would result in a state of stress

$$\begin{pmatrix} (2+0) & (1+1) & (-3+3) \\ (1+1) & (4+0) & (-2+2) \\ (-3+3) & (-2+2) & (6+0) \end{pmatrix} = \begin{pmatrix} 2 & 2 & 0 \\ 2 & 4 & 0 \\ 0 & 0 & 6 \end{pmatrix} \text{MN/m}^2$$

Problems for Solution 1.3, 1.4

1.6 Transformation of Stress Arrays by Rotation of the Coordinate Axes

The state of stress at a point was described in Article 1.3 by the array (A1.1) measured with respect to coordinate axes x, y, z. However, since the choice of axes was arbitrary the state of stress could equally well be defined for any other set of axes x', y', z' at the point, giving the array

$$\mathbf{T} = \begin{pmatrix} \sigma_{x'} & \tau_{x'y'} & \tau_{x'z'} \\ \tau_{y'x'} & \sigma_{y'} & \tau_{y'z'} \\ \tau_{z'x'} & \tau_{z'y'} & \sigma_{z'} \end{pmatrix} \tag{A1.3}$$

The stress components in the two arrays would in general be different since they act on surfaces having different inclinations to the internal force system. However the internal force system and, therefore, the state of stress, depends only on the external force system and is therefore independent of the orientation of coordinate axes. Therefore since the arrays (A1.1) and (A1.3) represent the same state of stress there must be some relationship between their components and the angles between the sets of axes x, y, z and x', y', z'. If such a relationship can be found it would be possible to compute from any one array the components of the stress array for any other set of coordinate axes.

The establishment of these relationships for a completely general state of stress is beyond the scope of this work. The derivation may be found in any text on the Theory of Elasticity.[1] However the significant features of the general analysis can be outlined from a study of the more simple state of stress defined by the array

$$\begin{pmatrix} \sigma_x & \tau_{xy} & 0 \\ \tau_{yx} & \sigma_y & 0 \\ 0 & 0 & \sigma_z \end{pmatrix} \tag{A1.4}$$

If $\sigma_z = 0$ then the array represents the system shown in Figure 1.7a which is often called a state of *plane stress* since all the stress components associated with one coordinate axis (here the z axis) are zero. However in what follows it is only convenient, not *necessary*, for σ_z to be zero. If Figure 1.7a is viewed

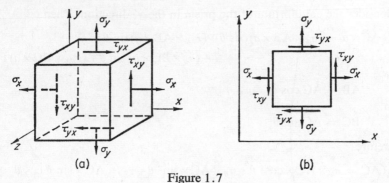

Figure 1.7

along the z axis then Figure 1.7b is obtained in which the shear stress components are shown at a distance from the edges of the element for clarity (the actual components being coincident with the edges of the element). Let the x, y axes rotate through $+\theta$ to positions x', y' (Figure 1.8a) and consider the

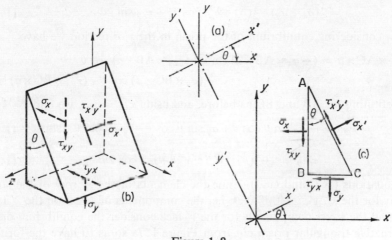

Figure 1.8

equilibrium of a triangular prism cut from the original cube (side a) by a plane parallel to the y' axis, as shown in Figure 1.8b. The stress components acting on the prism are then

$$\sigma_x, \ \tau_{xy}, \ \sigma_y, \ \tau_{yx}, \ \sigma_{x'}, \ \tau_{x'y'}$$

acting as shown in Figures 1.8b and c, and the stress array for the x', y', z axes becomes

$$\begin{pmatrix} \sigma_{x'} & \tau_{x'y'} & 0 \\ \tau_{y'x'} & \sigma_{y'} & 0 \\ 0 & 0 & \sigma_z = 0 \end{pmatrix} \qquad (A1.5)$$

Consider the equilibrium of the prism in the x' direction. Then

$$\sigma_{x'} \times AC \times a = (\sigma_x \times AB \times a)\cos\theta + (\tau_{xy} \times AB \times a)\sin\theta$$
$$+ (\sigma_y \times BC \times a)\sin\theta + (\tau_{yx} \times BC \times a)\cos\theta$$

Now

$$AB = AC\cos\theta$$

and

$$BC = AC\sin\theta$$

Therefore

$$\sigma_{x'} \times AC = \sigma_x \times AC \times \cos^2\theta + \sigma_y \times AC \times \sin^2\theta + \tau_{xy} \times AC \times \sin\theta\cos\theta$$
$$+ \tau_{yx} \times AC \times \sin\theta\cos\theta$$

But

$$\tau_{xy} = \tau_{yx}$$

so that

$$\sigma_{x'} = \sigma_x\cos^2\theta + \sigma_y\sin^2\theta + 2\tau_{xy}\sin\theta\cos\theta \tag{1.3a}$$

or

$$\sigma_{x'} = \tfrac{1}{2}(\sigma_x + \sigma_y) + \tfrac{1}{2}(\sigma_x - \sigma_y)\cos 2\theta + \tau_{xy}\sin 2\theta \tag{1.3b}$$

Now considering equilibrium of the prism in the y' direction we have

$$\tau_{x'y'} \times AC \times a = (-\sigma_x \times AB \times a)\sin\theta + (\tau_{xy} \times AB \times a)\cos\theta$$
$$+ (\sigma_y \times BC \times a)\cos\theta - (\tau_{yx} \times BC \times a)\sin\theta$$

Substituting for AB and BC as before, and using $\tau_{xy} = \tau_{yx}$, we find

$$\tau_{x'y'} = -\sigma_x\sin\theta\cos\theta + \sigma_y\sin\theta\cos\theta + \tau_{xy}(\cos^2\theta - \sin^2\theta) \tag{1.4a}$$

or

$$\tau_{x'y'} = -\tfrac{1}{2}(\sigma_x - \sigma_y)\sin 2\theta + \tau_{xy}\cos 2\theta \tag{1.4b}$$

Equations (1.3) and (1.4) define the elements of the x' row of the stress array for the x', y', z set of axes, i.e. the components of stress on the x' face. To find the stress components for the y' face consider the equilibrium of the alternative triangular prism cut from Figure 1.7a so as to have the form of Figure 1.9 when viewed along the z axis. For equilibrium in the y' direction

$$\sigma_{y'} \times AC \times a = (\sigma_x \times BC \times a)\sin\theta - (\tau_{xy} \times BC \times a)\cos\theta + (\sigma_y \times AB \times a)\cos\theta$$
$$- (\tau_{yx} \times AB \times a)\sin\theta$$

and so

$$\sigma_{y'} = \sigma_x\sin^2\theta + \sigma_y\cos^2\theta - 2\tau_{xy}\sin\theta\cos\theta \tag{1.5a}$$

or

$$\sigma_{y'} = \tfrac{1}{2}(\sigma_x + \sigma_y) - \tfrac{1}{2}(\sigma_x - \sigma_y)\cos 2\theta - \tau_{xy}\sin 2\theta \tag{1.5b}$$

From consideration of equilibrium in the x' direction it is found that $\tau_{y'x'}$ is, as would be expected from array symmetry, equal to $\tau_{x'y'}$ as expressed by equations (1.4).

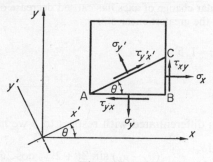

Figure 1.9

Thus if the state of stress at a point is given by the stress array (A1.4) then by application of equations (1.3) to (1.5) it is possible to calculate the components of the array (A1.5) which define the same state of stress but with respect to another set of axes x', y', z where x', y' are inclined at θ to x, y as in Figure 1.8a. It will be noted that equation (1.5a) might have been obtained from (1.3a) by substituting $\theta + \pi/2$ for θ and $\sigma_{y'}$ for $\sigma_{x'}$. In conclusion it may be noted that an analogous procedure may be carried out for a completely general state of stress.[1]

Example 1.3 The state of stress at a point is

$$\begin{pmatrix} 100 & 100 & 0 \\ 100 & 300 & 0 \\ 0 & 0 & -100 \end{pmatrix} \text{ (units MN/m}^2)$$

with respect to axes x, y, z. Find the state of stress which would have been obtained with axes x', y', z where the relative orientation is $\theta = 30°$.

Solution Applying equation (1.3b)

$$\sigma_{x'} = \tfrac{1}{2}(100+300) + \tfrac{1}{2}(100-300)\cos 60° + 100 \sin 60°$$
$$= 200 - 50 + 86·6 = 236·6 \text{ MN/m}^2$$

Putting $\theta = 30°+90°$ in (1.3) or $\theta = 30°$ in (1.5) we obtain $\sigma_{y'} = 163·4$ MN/m². From (1.4b)

$$\tau_{x'y'} = -\tfrac{1}{2}(100-300)\sin 60° + 100 \cos 60° = 136·6 \text{ MN/m}^2$$
$$= \tau_{y'x'} \text{ (symmetry of the stress array)}$$

Thus the measured stress array would have been

$$\begin{pmatrix} 236·6 & 136·6 & 0 \\ 136·6 & 163·4 & 0 \\ 0 & 0 & -100 \end{pmatrix} \text{ MN/m}^2$$

Note how this particular change of axes has caused decrease of the greatest normal stress and increase of the greatest shear stress.

Problem for Solution   1.5

1.7   Principal Axes

If equation (1.3b) is differentiated with respect to θ we have

$$\frac{d\sigma_{x'}}{d\theta} = -(\sigma_x - \sigma_y)\sin 2\theta + 2\tau_{xy}\cos 2\theta \qquad (1.6)$$

Comparing this with (1.4b) shows that

$$\frac{d\sigma_{x'}}{d\theta} = 2\tau_{x'y'}$$

For a stationary value of $\sigma_{x'}$

$$\frac{d\sigma_{x'}}{d\theta} = 0$$

and therefore from (1.6)

$$(\sigma_x - \sigma_y)\sin 2\theta = 2\tau_{xy}\cos 2\theta$$

i.e.

$$\tan 2\theta = \frac{2\tau_{xy}}{\sigma_x - \sigma_y} \qquad (1.7)$$

Since there are two values of 2θ differing by $180°$ which satisfy this equation, it follows that two values of θ differing by $90°$ are obtained from it. However $90°$ is just the angle between the x' and y' axes. Therefore equation (1.7) defines one particular orientation of the axes x', y' for which stationary values of the normal stresses are obtained and for which the shear stress components are zero. When the axes have this orientation they are called the *principal axes*, and the associated normal stress components $\sigma_{x'}$ and $\sigma_{y'}$ are called *principal stresses*. The stress array then takes the diagonal form in (A1.6) which may be termed the principal stress array.

$$\begin{pmatrix} \sigma_{x'} & 0 & 0 \\ 0 & \sigma_{y'} & 0 \\ 0 & 0 & \sigma_z \end{pmatrix} \qquad (A1.6)$$

The principal stress components $\sigma_{x'}$ and $\sigma_{y'}$ acting on the x' and y' surfaces have no associated shear stress components. It will be recalled that in the stress array (A1.4) there were no shearing stresses on the z surfaces leaving only σ_z in the z row of the array. Thus in terms of the definition given above the z axis was in fact a principal axis and σ_z a principal stress.

Therefore if one principal axis is known it is possible to find the others by using equation (1.7). If all three principal axes are unknown it is possible to deduce their directions using more general but analogous procedures. However work of this type is beyond the scope of this text, see reference 1. Fortunately in a great many problems of practical interest one principal axis is known.

Thus in conclusion it may be said that at any point in a material there is always one orientation of the three coordinate axes for which the measured stress array comprises only normal stress components. The axes are then the principal axes at the point, the normal stress components are the principal stresses, usually called σ_1, σ_2, σ_3 and the stress array is the principal stress array

$$\begin{pmatrix} \sigma_1 & 0 & 0 \\ 0 & \sigma_2 & 0 \\ 0 & 0 & \sigma_3 \end{pmatrix} \tag{A1.7}$$

By considering second derivatives of $\sigma_{x'}$ (equation 1.3) it is possible to show that, in the plane stress situation, one of the principal stresses is the greatest, and one the least normal stress (algebraically) which can be measured at the point for the given loading and any choice of the axes x', y'. Explicit expressions for these extreme values of normal stress can be obtained from substitution of equation (1.7) in equations (1.3) Thus

$$\sigma_1 = \tfrac{1}{2}(\sigma_x + \sigma_y) + \sqrt{(\tfrac{1}{2}(\sigma_x - \sigma_y))^2 + \tau_{xy}^2} \tag{1.8a}$$

$$\sigma_2 = \tfrac{1}{2}(\sigma_x + \sigma_y) - \sqrt{(\tfrac{1}{2}(\sigma_x - \sigma_y))^2 + \tau_{xy}^2} \tag{1.8b}$$

The above discussion can be generalized to three dimensions[1] to show that one of the principal stresses is the greatest, and one the least (algebraic) normal stress which can be measured for any choice of axes. Some writers use σ_1, σ_2, σ_3 to represent the principal stresses in algebraic order. Here when this meaning is required the principal stresses will be denoted σ_I, σ_{II}, σ_{III} where $\sigma_I \geqslant \sigma_{II} \geqslant \sigma_{III}$.

Example 1.4 In Example 1.3 it will be noted that the z axis is in fact a principal axis. Find the orientation of the other two principal axes and the other principal stresses.

Solution From equation (1.7) $\tan 2\theta = 2 \times 100/(100 - 300) = -1$ and so $2\theta = 135°$. Hence the x', y' axes are inclined at $67\tfrac{1}{2}°$ to the x, y axes.

From equations (1.8)

$$\sigma_1 = \tfrac{1}{2}(100 + 300) + \sqrt{(\tfrac{1}{2}(100 - 300))^2 + 100^2} = 341 \cdot 4 \text{ MN/m}^2$$

and
$$\sigma_2 = 58 \cdot 6 \text{ MN/m}^2$$

The principal stress array is then

$$\begin{pmatrix} 341 \cdot 4 & 0 & 0 \\ 0 & 58 \cdot 6 & 0 \\ 0 & 0 & -100 \end{pmatrix} \text{MN/m}^2$$

The reader may confirm that the same result can be obtained by starting from the array calculated in Example 1.3.

Problem for Solution 1.6

1.8 Maximum Shear Stresses

If equation (1.4b) is differentiated with respect to θ we have

$$\frac{d\tau_{x'y'}}{d\theta} = -(\sigma_x - \sigma_y)\cos 2\theta - 2\tau_{xy}\sin 2\theta$$

and for a stationary value

$$\tan 2\theta = -\frac{\sigma_x - \sigma_y}{2\tau_{xy}} \tag{1.9}$$

Equation (1.9) defines an orientation of the axes for which it can be shown that the shear stress has extreme values. Comparison of equations (1.9) and (1.7) shows that they define two sets of axes oriented at 45° to each other. Thus extreme values of shear stress are associated with a set of axes making 45° angles to the principal axes.

If the result expressed by equation (1.9) is substituted in equation (1.4a) then the extreme values of shear stress are found to be

$$(\tau_{x'y'})_{\max} = \pm \sqrt{\left(\frac{\sigma_x - \sigma_y}{2}\right)^2 + \tau_{xy}^2} \tag{1.10}$$

It may be noted that, from equations (1.8)

$$\tfrac{1}{2}(\sigma_1 - \sigma_2) = \sqrt{\left(\frac{\sigma_x - \sigma_y}{2}\right)^2 + \tau_{xy}^2}$$

Therefore comparing with equation (1.10) it can be said that *numerically*

$$(\tau_{x'y'})_{\max} = \tfrac{1}{2}(\sigma_1 - \sigma_2) \tag{1.11}$$

In the full three-dimensional theory of elasticity it is shown that this result can be generalized. The greatest shear stress in the material $\tau_{\max}$ is then obtained from the principal stresses as half the greatest algebraic difference between any two of them, i.e. using σ_1, σ_2, σ_3 to represent the principal stresses

$$\tau_{max} = \text{greatest of } \tfrac{1}{2}|\sigma_1 - \sigma_2| \text{ or } \tfrac{1}{2}|\sigma_2 - \sigma_3| \text{ or } \tfrac{1}{2}|\sigma_3 - \sigma_1| \qquad (1.12)$$

or, using the alternative notation

$$\tau_{max} = \tfrac{1}{2}(\sigma_I - \sigma_{III}) \qquad (1.13)$$

For instance, referring to the state of stress of Example 1.4 and applying result (1.12), the greatest shear stress in the material is

$$\tau_{max} = \text{greatest of } \tfrac{1}{2}|341{\cdot}4 - 58{\cdot}6| \text{ or } \tfrac{1}{2}|58{\cdot}6 - (-100)| \text{ or } \tfrac{1}{2}|-100 - 341{\cdot}4|$$

$$= \text{greatest of } |141{\cdot}4| \text{ or } |79{\cdot}3| \text{ or } |-220{\cdot}7|$$

$$= 220{\cdot}7 \text{ MN/m}^2$$

Thus the greatest shear stress for this plane stress problem is associated with the z axis and this would have been true even if σ_z had been zero, when τ_{max} would have been $\tfrac{1}{2}|0 - 341{\cdot}4| = 170{\cdot}7$ MN/m^2.

1.9 Specification of the State of Stress

It was shown in Article 1.6 that a particular state of stress may be represented by an infinite number of different stress arrays according to the infinite number of sets of coordinate axes which may be selected. While each of these arrays gives a complete description of the state of stress, as shown in Articles 1.7 and 1.8 there are two particular sets of axes which are associated with the extreme values of normal and shear stress. Since the purpose of stress analysis is usually to design against the extreme stresses, it follows that the stress arrays for these axes are the most useful. In fact only the stress array for the principal axes need be determined since, as shown by equation (1.13) the extreme shear stress τ_{max} can be deduced from the principal stresses.

Thus the specification of a state of stress is most concise when the coordinate axes are the principal axes. If the measuring coordinate axes are not the principal axes the measured array is first transformed to the principal axes and for plane stress systems this can be carried out graphically as described in the next article.

1.10 Mohr Circle Construction

If a state of stress is measured with respect to sets of axes x, y, z and x', y', z then the corresponding stress arrays are

$$\begin{pmatrix} \sigma_x & \tau_{xy} & 0 \\ \tau_{yx} & \sigma_y & 0 \\ 0 & 0 & \sigma_z \end{pmatrix} \qquad \begin{pmatrix} \sigma_{x'} & \tau_{x'y'} & 0 \\ \tau_{y'x'} & \sigma_{y'} & 0 \\ 0 & 0 & \sigma_z \end{pmatrix} \qquad \text{(A1.8, 9)}$$

The transformation equations (1.3) and (1.4) relating these arrays can be written as

$$\sigma = \sigma_{x'} = \tfrac{1}{2}(\sigma_x + \sigma_y) + \tfrac{1}{2}(\sigma_x - \sigma_y)\cos 2\theta + \tau_{xy}\sin 2\theta \qquad (1.14a)$$

$$\tau = \tau_{x'y'} = -\tfrac{1}{2}(\sigma_x - \sigma_y)\sin 2\theta + \tau_{xy}\cos 2\theta \qquad (1.14b)$$

where θ is the angle between the x', y' and x, y axes. When equation (1.14a) is rearranged and squared and then added to the square of (1.14b) there results, after simplification

$$(\sigma - \tfrac{1}{2}(\sigma_x + \sigma_y))^2 + \tau^2 = (\tfrac{1}{2}(\sigma_x - \sigma_y))^2 + \tau_{xy}^2$$

which is the equation of a circle in σ, τ coordinates having its centre at $(\tfrac{1}{2}(\sigma_x + \sigma_y), 0)$ and radius $\sqrt{(\tfrac{1}{2}(\sigma_x - \sigma_y))^2 + \tau_{xy}^2}$.

Therefore for given values of σ_x, σ_y, τ_{xy} it is possible to construct a circle in σ, τ space for which the coordinates of a point on the circumference are the values of $\sigma = \sigma_{x'}$ and $\tau = \tau_{x'y'}$ for a specific value of θ. This circle therefore provides graphically the information given by equations (1.14).

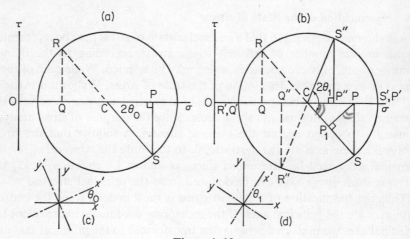

Figure 1.10

To construct this circle axes σ and τ are drawn as in Figure 1.10a, and the same scale chosen for each axis. Point P is plotted on the σ axis with OP representing, to scale, σ_x. Similarly Q is plotted with OQ representing σ_y. A perpendicular is erected at P and point S plotted on it with PS representing to the same scale τ_{xy}. PS is measured downwards if τ_{xy} is positive, upwards if τ_{xy} is negative. Then the centre of the circle has coordinates

$$(\tfrac{1}{2}(\sigma_x + \sigma_y), 0) \qquad \text{i.e.} \qquad (\tfrac{1}{2}(OP + OQ), 0)$$

which correspond to C the midpoint of QP. Similarly, substituting $\sigma_x = OP$

etc., in the expression obtained above for the radius of the circle we find

$$\text{Radius} = \sqrt{(\tfrac{1}{2}(OP-OQ))^2 + PS^2}$$
$$= \sqrt{CP^2 + PS^2} = CS$$

With C as centre and CS as radius, the circle is drawn and the construction completed by producing SC to meet the circle at R. We note that from the method of construction the coordinates OP, PS of the end S (henceforth called the leading end) of the diameter SCR represent to scale the stress components σ_x, τ_{xy}; also the coordinate OQ of R represents σ_y. Thus the coordinates of the ends of the diameter contain the information needed to write the stress array (A1.8) for the axes x, y, z.

Consider now another diameter, say S'CR' in Figure 1.10b. Dropping a perpendicular PP_1 as shown the angles $S\hat{P}P_1$ and $S\hat{C}P$ are equal and are denoted $2\theta_0$. Then substituting in equation (1.14a) $OP = \sigma_x$, $OQ = \sigma_y$, $PS = \tau_{xy}$ we find

$$\sigma = \sigma_{x'} = \tfrac{1}{2}(OP+OQ) + \tfrac{1}{2}(OP-OQ)\cos 2\theta_0 + PS \sin 2\theta_0$$
$$= OC + CP_1 + P_1S$$
$$= OC + CS' = OS'$$

Substituting in (1.14b),
$$\tau = \tau_{x'y'} = 0$$

and so S' and P' can be considered to be coincident. Similarly, by setting $2\theta_0 = 180° + 2\theta_0$ in equation (1.14a) we obtain

$$\sigma = \sigma_{y'} = OQ'$$

where R' and Q' are coincident. Thus the coordinates of the ends of the diameter S'CR' would enable the stress array (A1.9) to be written for axes x', y' inclined at θ_0 to x, y as in Figure 1.10c. This conclusion can be extended to any other diameter such as S"CR" (Figure 1.10b). There the coordinates OP", P"S", OQ" represent $\sigma_{x'}$, $\tau_{x'y'}$, $\sigma_{y'}$ when the angle between the axes is as in Figure 1.10d.

Therefore the stress array (A1.9) for any set of axes x', y', z is obtained from that (A1.8) for axes x, y, z by plotting the Mohr Stress Circle as it is called, and reading the coordinates of a diameter S'CR', where the angle $S\hat{C}S'$ is twice that between x and x' and is measured in the same sense. When the diameter S'CR' is horizontal $\tau_{x'y'}$ is zero and so the principal stress array is obtained.

Summary

Given array (A1.8): to find (A1.9).

A. *Construction* (i) Plot OP and OQ to scale on the σ axis to represent σ_x and σ_y respectively. C is the midpoint of PQ.

(ii) At P erect a perpendicular and plot PS $= \tau_{xy}$ to the same scale, down if positive, up if negative.

(iii) With centre C and radius CS draw the Mohr Stress Circle; produce SC to meet the circle at R.

B. *Interpretation* (i) The coordinates of the ends of any diameter S'CR' making an angle $2\theta_0$ to SCR at C give the stress components $\sigma_{x'}$, $\tau_{x'y'}$ $(= \tau_{y'x'})$, $\sigma_{y'}$ in the stress array for axes x', y' inclined at θ_0 to the axes x, y. The angles $2\theta_0$ and θ_0 are measured in the same sense.

(ii) When the diameter S'CR' is horizontal it defines the principal stress array.

Example 1.5 Given the state of stress for axes x, y, z as

$$\begin{pmatrix} 8 & 4 & 0 \\ 4 & 2 & 0 \\ 0 & 0 & -4 \end{pmatrix} \text{MN/m}^2$$

(a) Find the principal axes and stress array and state the maximum shear stress.

(b) Find the stress array for axes x', y' inclined at $-22\tfrac{1}{2}°$ to x, y.

Solution Following the procedure described above, we set out to scale along the σ axis OP = 8 units, OQ = 2 units. Since τ_{xy} is positive PS is 4 units down as shown in Figure 1.11a. C is at the midpoint (5, 0) of QP. With centre C and radius CS the Mohr Stress Circle is drawn.

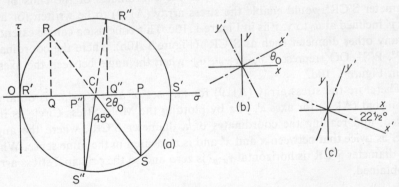

Figure 1.11

By measurement from a scale drawing, or by calculation,

$$\text{CS} = \sqrt{\text{CP}^2 + \text{PS}^2} = \sqrt{3^2 + 4^2} = 5 \text{ units}$$

(a) To find the principal axes draw diameter S'CR' in the horizontal position.

Horizontal coordinate of leading end is $OS' = \sigma_{x'} = OC + CS' = 10$ MN/m².
Horizontal coordinate of R' is $OR' = \sigma_{y'} = OC - CS' = 0$.
Vertical coordinate of S' is zero so that $\tau_{x'y'} = 0 = \tau_{y'x'}$.

Hence the stress array measured with respect to the principal axes is

$$\begin{pmatrix} 10 & 0 & 0 \\ 0 & 0 & 0 \\ 0 & 0 & -4 \end{pmatrix} \text{MN/m}^2$$

Here

$$S\hat{C}S' = 2\theta_0 = 53°8'$$

Therefore $\theta_0 = 26°34'$ and the principal axes are inclined at $+26°34'$ to the original x, y axes as shown in Figure 1.11b.

The maximum shear stress in the material is, from result (1.12), the greatest of $\frac{1}{2}|10-0|$ or $\frac{1}{2}|0-(-4)|$ or $\frac{1}{2}|-4-10|$, i.e. 7 MN/m².

(b) Draw diameter S"CR" so that angle $S\hat{C}S''$ is $2 \times (-22\frac{1}{2}°) = -45°$ (Figures 1.11a and c). Then by measurement or calculation

$$\sigma_{x'} = OP'' = 4·14 \text{ MN/m}^2$$
$$\sigma_{y'} = OQ'' = 5·86 \text{ MN/m}^2$$
$$\tau_{x'y'} = \tau_{y'x'} = P''S'' = 4·92 \text{ MN/m}^2$$

The stress array is

$$\begin{pmatrix} 4·14 & 4·92 & 0 \\ 4·92 & 5·86 & 0 \\ 0 & 0 & -4 \end{pmatrix} \text{MN/m}^2$$

Example 1.6   Find the principal stress array for the state of stress described by the array

$$\begin{pmatrix} -10 & -10 & 0 \\ -10 & 20 & 0 \\ 0 & 0 & 0 \end{pmatrix} \text{MN/m}^2$$

Show the orientation of the principal axes to the given axes and state the maximum shear stress in the material.

Solution   Here $OP = -10$ units, $OQ = 20$ units, PS is 10 units up, C is at (5, 0), and CS is found to be 18·02 MN/m².

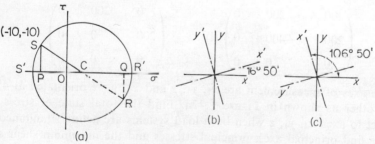

Figure 1.12

To find the principal axes consider the horizontal diameter S'CR', in Figure 1.12a.

$$\sigma_{x'} = \text{horizontal coordinate of S'} = -13 \cdot 02 \text{ MN/m}^2$$
$$\sigma_{y'} = \text{horizontal coordinate of R'} = 23 \cdot 02 \text{ MN/m}^2$$
$$\tau_{x'y'} = \text{vertical coordinate of S'} = 0 = \tau_{y'x'}$$

Therefore the required array is

$$\begin{pmatrix} -13 \cdot 02 & 0 & 0 \\ 0 & 23 \cdot 02 & 0 \\ 0 & 0 & 0 \end{pmatrix} \text{MN/m}^2$$

The angle between the diameters is $S\hat{C}S' = 2\theta_0 = 33°40'$. Thus the principal axes are inclined at $16°50'$ to the original axes in Figure 1.12b.

In result (1.13) $\sigma_I = 23 \cdot 02$, $\sigma_{III} = -13 \cdot 02$, MN/m², and so the maximum shear stress in the material is

$$\tau_{max} = \tfrac{1}{2}(23 \cdot 02 - (-13 \cdot 02)) = 18 \cdot 02 \text{ MN/m}^2$$

There is of course no reason why the diameter SCR may not be rotated to the alternative horizontal position in which the leading end is made to occupy the right-hand end, i.e. S' and R' are interchanged in Figure 1.12a. The horizontal coordinate of the leading end is now $OS'' = 23 \cdot 02 \text{ MN/m}^2$ while the vertical coordinate is still zero. The horizontal coordinate of R'' is now $\sigma_{y'} = -13 \cdot 02$ MN/m² and $S\hat{C}S'' = 2\theta_0 = 180° + 33°40'$.

The stress array is

$$\begin{pmatrix} 23 \cdot 02 & 0 & 0 \\ 0 & -13 \cdot 02 & 0 \\ 0 & 0 & 0 \end{pmatrix} \text{MN/m}^2$$

and the axes are as shown in Figure 1.12c. It will be seen that the effect has been simply to interchange the names of the principal axes and the principal stresses. It is important to note that the top row stress components are taken from the co-ordinates of the *leading end* of the diameter and the corresponding inclination of the axes must also be obtained from the angle of rotation measured from the leading ends.

Example 1.7   At a particular point in a structure two separate load systems acting on different occasions result in the measurement of states of stress described by the arrays (i) and (ii) below, (units of MN/m²),

$$\begin{pmatrix} 200 & 200 & 0 \\ 200 & -400 & 0 \\ 0 & 0 & 0 \end{pmatrix} \text{(i)} \qquad \begin{pmatrix} 0 & 200 & 0 \\ 200 & 0 & 0 \\ 0 & 0 & 0 \end{pmatrix} \text{(ii)}$$

The axes of measurement are x_1, y_1, z and x_2, y_2, z orientated at 45° to each other as shown in Figure 1.13a. Find the total state of stress with respect to axes x_1, y_1, z when both load systems are acting simultaneously. Hence find principal axes, principal stresses and the maximum shear stress in the material.

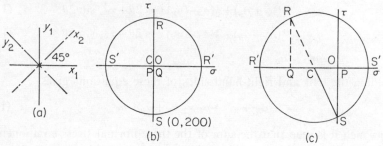

Figure 1.13

Solution The Mohr Circle for stress array (ii) is shown in Figure 1.13b with $OP = OQ = 0$, $PS = 200$ down, and centre at (0, 0). To transform this stress array to correspond to measuring axes x_1, y_1, z it will be necessary to rotate the axes by $-45°$, i.e. to draw a diameter $S'CR'$ at $2 \times (-45°) = -90°$ to SCR as shown. Then $CS = 200$ units and the state of stress due to the second loading system is

$$\begin{pmatrix} -200 & 0 & 0 \\ 0 & 200 & 0 \\ 0 & 0 & 0 \end{pmatrix} \text{MN/m}^2 \qquad \text{(iii)}$$

for axes x_1, y_1, z.

When both load systems act simultaneously the combined state of stress will (see Article 1.5) be the algebraic sum of arrays (i) and (iii) which are both measured with respect to axes x_1, y_1, z.

i.e. $\begin{pmatrix} 200 & 200 & 0 \\ 200 & -400 & 0 \\ 0 & 0 & 0 \end{pmatrix} + \begin{pmatrix} -200 & 0 & 0 \\ 0 & 200 & 0 \\ 0 & 0 & 0 \end{pmatrix} = \begin{pmatrix} 0 & 200 & 0 \\ 200 & -200 & 0 \\ 0 & 0 & 0 \end{pmatrix} \text{MN/m}^2$

To find the principal axes and stresses a Mohr Circle is drawn with $OP = 0$, $OQ = -200$, $PS = 200$ down, and centre C at $(-100, 0)$, (Figure 1.13c). Then since $CS = \sqrt{100^2 + 200^2} = 223·5 \text{ MN/m}^2$ and $SCS' = 2\theta_0 = 63°28'$ the principal stress array is

$$\begin{pmatrix} 123·5 & 0 & 0 \\ 0 & -323·5 & 0 \\ 0 & 0 & 0 \end{pmatrix} \text{MN/m}^2$$

and the inclination of the principal axes is similar to Figure 1.10c. The maximum shear stress in the material is, from (1.13),

$$\tau_{max} = 223·5 \text{ MN/m}^2$$

Problems for Solution 1.7–1.10

1.11 Sum of Normal Stresses

In Article 1.6 equations were derived as follows to allow a state of stress expressed for axes x, y, z by an array (A1.4) to be restated for alternative coordinate axes x', y', z by the array (A1.5), viz.

$$\sigma_{x'} = \tfrac{1}{2}(\sigma_x + \sigma_y) + \tfrac{1}{2}(\sigma_x - \sigma_y)\cos 2\theta + \tau_{xy}\sin 2\theta \qquad (1.3b)$$

$$\sigma_{y'} = \tfrac{1}{2}(\sigma_x + \sigma_y) - \tfrac{1}{2}(\sigma_x - \sigma_y)\cos 2\theta - \tau_{xy}\sin 2\theta \qquad (1.5b)$$

$$\sigma_z = \sigma_z$$

Adding the left- and right-hand sides of these equations gives

$$\sigma_{x'} + \sigma_{y'} + \sigma_z = \sigma_x + \sigma_y + \sigma_z \qquad (1.15)$$

from which it is seen that the sum of the three normal stress components is unaffected by a change of two coordinate axes at a point. This result can be generalized for rotation of all three axes as is proved in, for example, reference 1, and is usually described as an *invariant* property of the state of stress. A physical interpretation of this result will be found in Section 3.3.2.

1.12 Stress Failure

In the previous articles of this chapter it has been shown that the internal forces at any point of a structure are most concisely described by the principal stress array

$$\begin{pmatrix} \sigma_1 & 0 & 0 \\ 0 & \sigma_2 & 0 \\ 0 & 0 & \sigma_3 \end{pmatrix} \qquad (A1.10)$$

From this stress array the greatest magnitudes of normal and shear stress at the point can be derived if required.

Having described the expected state of stress the question arises as to what materials, if any, may be able to withstand the stresses. This might be investigated by actually constructing the article from a given material and testing it under the expected working loads. Alternatively we might try to subject a sample of the material to the expected state of stress. Both these procedures are likely to be rather expensive and time consuming.

A third possibility which proves to be useful in preliminary design, is to see whether there is any one characteristic property of a material which determines whether a state of stress will cause failure. For instance it might be supposed that a material would fail when the greatest magnitude of normal stress attained a characteristic value. Denoting this value by σ_0, and recalling that the principal stresses include the extremes of normal stress, this hypothesis predicts failure when

$$\sigma_0 = \text{greatest of } |\sigma_1| \text{ or } |\sigma_2| \text{ or } |\sigma_3| \qquad (1.16)$$

The property σ_0 would usually be measured in a simple tensile test in which a rod of material is subject to uniaxial tension as in Figure 1.14a. The state of stress at failure is then

$$\begin{pmatrix} \sigma_0 & 0 & 0 \\ 0 & 0 & 0 \\ 0 & 0 & 0 \end{pmatrix} \tag{A1.11}$$

However, this hypothesis, which was advanced by Rankine, does not correspond well with experiment.

An alternative hypothesis, associated with the name of Tresca, states that failure occurs at a characteristic value of shear stress denoted τ_0. From Article 1.8 it is known that the greatest shear stress at a point may be deduced from the principal stress array (A1.10) by application of results (1.12) or (1.13). Therefore failure should occur when

$$\tau_0 = \text{greatest of } \tfrac{1}{2}|\sigma_1 - \sigma_2| \text{ or } \tfrac{1}{2}|\sigma_2 - \sigma_3| \text{ or } \tfrac{1}{2}|\sigma_3 - \sigma_1| \tag{1.17}$$

i.e. when

$$\tau_0 = \tfrac{1}{2}(\sigma_\mathrm{I} - \sigma_\mathrm{III}) \tag{1.18}$$

This hypothesis, which is in reasonable agreement with experiment, is known as the *Maximum shear stress* or *Tresca Theory of Failure* and will be used henceforth unless otherwise stated.

Example 1.9 Find the Rankine and Tresca prediction of failure for a material in states of stress given by the following arrays

$$\text{(a)} \begin{pmatrix} \sigma & 0 & 0 \\ 0 & 0 & 0 \\ 0 & 0 & 0 \end{pmatrix} \qquad \text{(b)} \begin{pmatrix} \sigma & 0 & 0 \\ 0 & \sigma & 0 \\ 0 & 0 & \sigma \end{pmatrix}$$

Solution Applying conditions (1.16) and (1.18) failure is predicted at

	Rankine	*Tresca*
(a)	$\sigma_0 = \sigma$	$\tau_0 = \tfrac{1}{2}\sigma$
(b)	$\sigma_0 = \sigma$	No failure

The mutual inconsistency of these two hypotheses is evident.

If the validity of the Tresca hypothesis is accepted, then the value of τ_0 can be obtained conveniently from the results of the simple tensile test described above. At failure the maximum shear stress is obtained from array (A1.11) as $0.5 \, \sigma_0$ and so

$$\tau_0 = 0.5 \, \sigma_0 \tag{1.19}$$

The Tresca failure conditions (1.17) and (1.18) could therefore be restated as

$$\sigma_0 = \text{greatest of } |\sigma_1 - \sigma_2| \text{ or } |\sigma_2 - \sigma_3| \text{ or } |\sigma_3 - \sigma_1| \tag{1.20}$$

$$\sigma_0 = \sigma_\mathrm{I} - \sigma_\mathrm{III} \tag{1.21}$$

1.13 Factor of Safety

There is almost always some degree of uncertainty in predicting the stresses which will be developed in a proposed article when in use. There may be inadequate knowledge of the load system; the method of analysis probably depends on simplifying assumptions; deficiencies may arise due to production methods, and so on. Furthermore there is always the possibility of overloading, or of accidental damage to part of the structure. To provide some margin against such uncertainties it is widespread practice to allow a *factor of safety*. Thus a component or structure would be designed to have an anticipated maximum stress less than that to cause failure. The ratio of the material failure stress to the maximum designed stress is called the factor of safety, i.e.

$$f_\mathrm{s} = \frac{\tau_0}{\tau_\mathrm{max}} = \frac{0 \cdot 5 \, \sigma_0}{\tau_\mathrm{max}} \tag{1.22}$$

The size of the safety factor used depends very strongly on previous experience for any particular type of component or structure. However it is clear that the value will tend to increase with the degree of uncertainty. Nevertheless it must be borne in mind that a generous safety factor may mean waste of material. For instance an aeroplane with an excessive safety factor has unnecessary dead weight to support in the air. The decision in this case must depend on the relative cost of extra design and testing compared with the extra materials to be used and the loss of payload.

It is convenient here to define one additional term, viz. allowable stress. This is obtained by dividing the material failure stress by the safety factor. Thus the allowable shear stress is

$$\tau_\mathrm{A} = \frac{\tau_0}{f_\mathrm{s}} = \frac{\sigma_0}{2 f_\mathrm{s}} \tag{1.23}$$

Example 1.9   At a critical point in a structure the state of stress is found to be given by the array

$$\begin{pmatrix} 100 & 200 & 0 \\ 200 & -300 & 0 \\ 0 & 0 & 100 \end{pmatrix} \mathrm{MN/m^2}$$

A sample of the material failed in a tensile test at 700 MN/m². Find the factor of safety.

Solution The first step is to apply the Mohr Circle construction to obtain the principal stress array

$$\begin{pmatrix} 182 \cdot 8 & 0 & 0 \\ 0 & -382 \cdot 8 & 0 \\ 0 & 0 & 100 \end{pmatrix} MN/m^2$$

Here $\sigma_0 = 700$ MN/m² and from the stress array $\tau_{max} = 282 \cdot 8$ MN/m². Hence applying equation (1.22)

$$f_s = \frac{0 \cdot 5 \times 700}{282 \cdot 8} = 1 \cdot 24$$

Problem for Solution 1.11

1.14 Simple Stress Systems

Calculation of the state of stress throughout a body from a knowledge of the applied forces and the configuration alone is not, in general, possible. The reason lies in the fact that six independent quantities are needed to specify the state of stress, viz.

$$\sigma_x \ \sigma_y \ \sigma_z \ \tau_{xy} \ \tau_{yz} \ \tau_{xz}$$

All these quantities are intensities of force. The conditions for static equilibrium can provide only three equations of force equilibrium—one for each coordinate direction. (It might be thought that the three equations for moment equilibrium would provide the extra three equations. However these were used to prove the symmetry of the stress array (see Article 1.4), i.e. to reduce the number of independent stress components from nine to six.) Only by considering the deformation of the material can sufficient additional equations be obtained to permit solution in principle for the distribution of the state of stress.

There are, nevertheless, certain very simple situations for which the state of stress can be *calculated* without recourse to the deformations, though *proof* of the validity of these calculations is still only possible through consideration of deformation, or by experiment.

(a) *Uniform Uniaxial Stress*

As an example consider a bar with uniform cross-section as in Figure 1.14a with end faces perpendicular to the *x* axis and subject to uniformly distributed

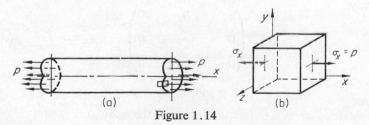

Figure 1.14

normal outward forces. A block of material intersecting one end (Fig. 1.14b) is clearly subject only to a tensile normal stress $\sigma_x = p$ and so the state of stress is given by the array

$$\begin{pmatrix} \sigma_x & 0 & 0 \\ 0 & 0 & 0 \\ 0 & 0 & 0 \end{pmatrix} \qquad (A1.12)$$

The uniformity of both geometry and loading makes plausible the assumption that the state of stress is the same throughout the bar, and this can be verified experimentally. As mentioned in Article 1.12 this state of stress corresponds to that in a simple tensile test which is one of the most common means of assessing material properties.

When the load is such that failure occurs $\sigma_x = \sigma_0$ and from this value using equation (1.18) the Tresca theory predicts critical shear stress $\tau_0 = \frac{1}{2}\sigma_0$.

(b) *Uniform Stress in Two Perpendicular Directions*

An analogous situation can be described for a thin plate subjected to uniformly distributed normal forces over its edges. Taking axes as in Figure 1.15a, a

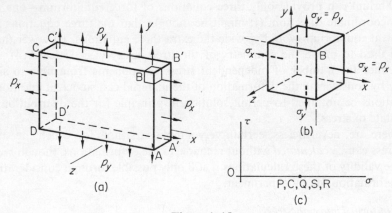

Figure 1.15

block from a corner (Figure 1.15b) is subject to biaxial tension described by the array

$$\begin{pmatrix} \sigma_x & 0 & 0 \\ 0 & \sigma_y & 0 \\ 0 & 0 & 0 \end{pmatrix} \qquad (A1.13)$$

where

$$\sigma_x = p_x; \qquad \sigma_y = p_y$$

This state of stress is in fact uniform over the body.

It is interesting to examine the limit behaviour for some particular cases. When $\sigma_x = \sigma_y$ the Mohr Circle reduces to a point as in Figure 1.15c since OP = OQ. Here applying result (1.18) $\tau_{max} = \frac{1}{2}\sigma_x = \frac{1}{2}\sigma_y = \tau_0$ at the limit which is determined by the third (zero) principal stress. When σ_x and σ_y are not equal but have the same sign, as in Figure 1.10a, the limit conditions are again controlled by the zero principal stress.

When σ_x and σ_y have opposite signs the Mohr Circle is of the type Figure 1.12a and $\tau_{max} = \frac{1}{2}|\sigma_x - \sigma_y| = \tau_0$ at the limit. In this case only, are the limit conditions controlled by the stresses in the xy plane.

(c) *Three-dimensional Uniform Stress*

If the surfaces of a rectangular parallelepiped are subjected to uniformly distributed outward normal forces then the state of stress is

$$\begin{pmatrix} \sigma_x & 0 & 0 \\ 0 & \sigma_y & 0 \\ 0 & 0 & \sigma_z \end{pmatrix} \tag{A1.14}$$

where $\sigma_x = p_x$; $\sigma_y = p_y$; $\sigma_z = p_z$ and p_x, p_y, p_z are the normal force intensities on the surfaces of the block.

It will be noted that if $\sigma_x = \sigma_y = \sigma_z = p$ a state of *isotropic* stress exists and from result (1.13) τ_{max} is zero. There is no limit to the magnitude of p as long as the maximum shear stress theory remains valid. A body under hydrostatic pressure is in this condition.

The stress systems (a), (b), (c) enable a surprising variety of situations to be investigated. This gives the opportunity to acquire a feel for magnitudes of quantities involved in stress calculations. A number of worked examples and problems for solution now follow.

Example 1.10 A conduit of square cross-section, as shown in Figure 1.16, has ends closed by strong rigid plates with frictionless surfaces. The plates

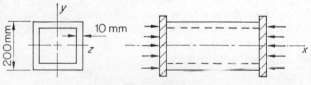

Figure 1.16

are subjected to uniformly distributed normal forces of total magnitude P. Find the maximum permissible value for P if the conduit material has $\sigma_0 = 180$ MN/m², and a factor of safety of 1·5 is required.

Solution   The normal forces are exerted on the plates. Since these are rigid the forces are transmitted uniformly to the end faces of the conduit. Since the plates are frictionless there are no forces parallel to the end faces. The area of conduit normal to the forces is

$$0 \cdot 2^2 - 0 \cdot 18^2 = 0 \cdot 0076 \text{ m}^2$$

Therefore the normal stress is

$$\sigma_x = -(P/0 \cdot 0076) \text{ N/m}^2$$

where P is measured in Newtons, and is compressive. The state of stress is the same throughout having the form of array (A1.12), and $\tau_{max} = P/0 \cdot 0152 \text{ N/m}^2$. From equation (1.23) the allowable shear stress is $(\sigma_0/2f_s)$, i.e. 60 MN/m².
    Therefore

$$\frac{P}{0 \cdot 0152} \leqslant 60 \times 10^6$$

and so

$$P \leqslant 912 \text{ kN}$$

Example 1.11   A plate of 20 mm thickness carries uniformly distributed normal forces of 1 MN acting outwards on its edge faces as in Figure 1.15a, the lateral faces ABCD and A'B'C'D' are not subject to forces. If AB = 1 m and BC = 2 m determine the state of stress, and the maximum shear stress. Find the limits within which the forces in the y direction could vary without affecting the magnitude of τ_{max}.

Solution   Areas of ABB'A' and CC'D'D are each 0·02 m² and so the intensity of force on these surfaces is

$$\sigma_x = p_x = 10^6/0 \cdot 02 = 50 \times 10^6 \text{ N/m}^2$$

Similarly

$$\sigma_y = p_y = 25 \times 10^6 \text{ N/m}^2$$

Since there are no lateral forces $\sigma_z = 0$ and the state of stress is

$$\begin{pmatrix} 50 & 0 & 0 \\ 0 & 25 & 0 \\ 0 & 0 & 0 \end{pmatrix} \text{MN/m}^2$$

Applying result (1.13) $\tau_{max} = \frac{1}{2}(50 - 0) = 25 \text{ MN/m}^2$.

If τ_{max} is not to vary then σ_y must remain the intermediate principal stress. Therefore the limits are $0 \leqslant \sigma_y \leqslant 50$ in units of MN/m². Then the forces on the y surfaces can vary within limits of zero and $(50 \times 10^6 \times 0 \cdot 04)$ N, i.e. 2 MN.

Problems for Solution   1.12–1.15

1.15   References

(a) *Cited*

1. Timoshenko, S. and Goodier, J. N., *Theory of Elasticity*, 3rd ed., McGraw-Hill, New York, 1970.

(b) *General*

1. Yokobori, T., *An Interdisciplinary Approach to Fracture and Strength of Solids*, Wolters-Noordhof, Groningen, 1968.
2. Dally, J. W. and Riley, W. F., *Experimental Stress Analysis*, McGraw-Hill, New York, 1965.

1.16 Problems for Solution

1.1. Illustrate the following stress arrays (units MN/m²).

(a) $\begin{pmatrix} 0 & 2 & 4 \\ 2 & 4 & 0 \\ 4 & 0 & 2 \end{pmatrix}$ (b) $\begin{pmatrix} 2 & 1 & 2 \\ 1 & -1 & 0 \\ 2 & 0 & -3 \end{pmatrix}$ (c) $\begin{pmatrix} 2 & -1 & 0 \\ -1 & 1 & -4 \\ 0 & -4 & 0 \end{pmatrix}$ (d) $\begin{pmatrix} 2 & -4 & 0 \\ -4 & -4 & 0 \\ 0 & 0 & 0 \end{pmatrix}$

1.2. Write down the stress arrays for the states of stress illustrated in Figure 1.17. Take the *numerical* magnitude of each non-zero stress component to be unity. (Note that only the stress components on the positive surfaces are illustrated in the figure.)

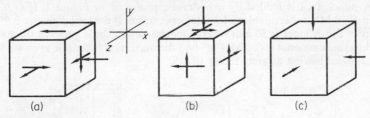

(a)   (b)   (c)

Figure 1.17

1.3. Obtain the combined effect of two loadings which separately induce the states of stress described by the given arrays. (Both arrays are measured at the same point with respect to the same axes.) Illustrate the combined state of stress

$$\begin{pmatrix} 2 & -4 & 0 \\ -4 & 6 & 0 \\ 0 & 0 & 3 \end{pmatrix} \qquad \begin{pmatrix} 0 & 6 & -2 \\ 6 & 2 & -2 \\ -2 & -2 & 2 \end{pmatrix}$$

1.4. Deduct from the result of Problem 1.3 the array

$$\begin{pmatrix} \bar{\sigma} & 0 & 0 \\ 0 & \bar{\sigma} & 0 \\ 0 & 0 & \bar{\sigma} \end{pmatrix}$$

where $\bar{\sigma}$ is the mean of the normal stresses obtained in Problem 1.3. Illustrate the resulting state of stress.

1.5. Transform the given stress arrays for changes of axes of (a) 45°, (b) 45°, (c) 30° (units MN/m²).

(a) $\begin{pmatrix} 100 & 0 & 0 \\ 0 & -100 & 0 \\ 0 & 0 & 50 \end{pmatrix}$ (b) $\begin{pmatrix} 100 & 0 & 0 \\ 0 & 0 & 0 \\ 0 & 0 & 50 \end{pmatrix}$ (c) $\begin{pmatrix} -40 & -20 & 0 \\ -20 & 60 & 0 \\ 0 & 0 & -50 \end{pmatrix}$

1.6. Determine the orientation of the principal axes, and deduce the principal stress array for the state of stress described in Problem 1.5c.

1.7. Use the Mohr Stress Circle to solve Problems 1.5 and 1.6.

1.8. Find the maximum shear stress for the following states of stress

(a) $\begin{pmatrix} 120 & 40 & 0 \\ 40 & 80 & 0 \\ 0 & 0 & 0 \end{pmatrix}$ (b) $\begin{pmatrix} 200 & 200 & 0 \\ 200 & -400 & 0 \\ 0 & 0 & 100 \end{pmatrix}$ (c) $\begin{pmatrix} 100 & 0 & 0 \\ 0 & 100 & 0 \\ 0 & 0 & 0 \end{pmatrix}$

1.9. If arrays (b) and (c) of Problem 1.5 were measured for different loads at the same point and with respect to the same axes find the principal axes and stress array when both sets of loads act together.

1.10. Find the stress arrays for axes inclined at $-30°$ to the original axes for the states of stress in Problem 1.8b and c.

1.11. If a sample of steel failed in a tensile test at 400 MN/m² examine whether it could be used for any of the duties expressed by the arrays in Problem 1.8, and if so calculate the safety factors.

1.12. A punch tool is loaded by a hydraulic piston as in Figure 1.18a. A fluid pressure of 10 MN/m² is unable to initiate penetration. If the tool diameter is 10 mm and that of the cylinder is 50 mm what is the maximum shear stress in the tool? If σ_0 for the punch material is 600 MN/m² find the maximum hydraulic pressure which can be allowed leaving a safety factor of 1·5.

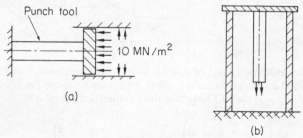

Figure 1.18

1.13. Figure 1.18b shows a half-sectional elevation of a tube resting on a plane, and closed by a rigid plate from which a rod is suspended along the tube axis. The lower end of this rod is subject to a uniformly distributed normal force. Find the maximum permissible value of this load if the shear and compressive stresses in both rod and tube are not to exceed 100 MN/m² and 80 MN/m² respectively. The tube and rod have cross-sectional areas of 2000 mm² and 1000 mm² respectively.

1.14. A rectangular plate is subject to uniformly distributed edge forces as in Figure 1.15a. Plot a graph of τ_{max} as p_y is varied within the limits $400 \geqslant p_y \geqslant -400$ (units MN/m²) when $p_x = 400$ MN/m².

1.15. The edges of a rectangular plate of thickness t are to be subjected to inward normal forces on ABB'A' and DCC'D' (see Figure 1.15a) and outward normal forces on the other edges. If AB = 0·5 m, BC = 1 m and the total force on each edge is 1 MN find the minimum value of t if the maximum shear stress is not to exceed 75 MN/m². If, as a further condition, the maximum compressive stress is not to exceed 80 MN/m², find the value of t.

2

Strain

2.1 Displacement and Deformation

Changes of shape and size of engineering structures and components under
load are studied not only for their intrinsic interest but also as a stage in the
development of additional relations between the stress components, the need
for which was pointed out in Article 1.14.

To distinguish between displacement and deformation consider a plane area
ABCD (Figure 2.1) which represents part of a body in the unloaded condition.
Suppose that after loading points A, B, C, D are at A_1, B_1, C_1, D_1. Then the

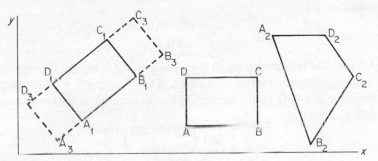

Figure 2.1

points have undergone a linear displacement, i.e. translation. In addition an
original line such as AB has experienced angular displacement, i.e. rotation.
If $A_1B_1C_1D_1$ has the same shape and size as ABCD then it is said that a rigid
body displacement has occurred.

Now suppose that after another loading the points A, B, C, D are at
A_2, B_2, C_2, D_2. As before all the points have experienced translation, and all
the sides rotation. However, it is clear that in addition ABCD has changed
both size and shape. The length of AB is different from that of A_2B_2 and so
linear deformation is said to have occurred. The angle $B_2\hat{A}_2D_2$ between lines
A_2B_2 and A_2D_2 is not the same as the angle $B\hat{A}D$ between the original lines
AB and AD; this is termed an angular deformation.

Lastly, consider displacement of A, B, C, D to A_3, B_3, C_3, D_3 such that $A_3A_1 = B_1B_3$ and $D_3D_1 = C_1C_3$. Clearly there has been deformation since $A_1B_1C_1D_1$ was the same size and shape as ABCD, but the centre of area of $A_3B_3C_3D_3$ is the same as that of $A_1B_1C_1D_1$. Therefore we could say that $A_3B_3C_3D_3$ was attained as the combination of a rigid body displacement (to $A_1B_1C_1D_1$) followed by a deformation.

Summarizing, we can say that when a body is displaced without change of shape or size it has undergone rigid body displacement; if as a result of displacement there has been any change of size and/or shape then deformation has occurred. It is deformation with which we are primarily concerned in this text.

2.2   Linear Strain

As noted above, when ABCD of Figure 2.1 was displaced to $A_2B_2C_2D_2$ the original line AB experienced a change in length, or *linear deformation* of $(A_2B_2 - AB)$. However a particular linear deformation of say 5 mm is likely to be more serious with respect to an original length of 1 m rather than to a length of 100 m. It is therefore reasonable to relate the linear deformation to the original length and so we define the average linear strain of line AB as the dimensionless quantity

$$e = \frac{A_2B_2 - AB}{AB}$$

If AB is made indefinitely small then the average intensity becomes in the limit a quantity representative of a point. This quantity is termed the *linear* (or *direct*, or *normal*) strain at the point, associated with some specified direction η. Then

$$\epsilon_\eta = \lim_{AB \to 0} \left(\frac{A_2B_2 - AB}{AB} \right) \tag{2.1}$$

where η is the direction of AB.

It should be noted that if any finite line is divided into a number of small segments then the strain will generally be different for each segment. That is to say that the linear strain will in general vary from point to point through the body.

2.3   Shear Strain

Angular deformation, as was discussed in Article 2.1, is detected by comparing the angle contained between two lines before and after loading. In one of the displacements considered in Figure 2.1 the angle $B\hat{A}D$ for example decreased to $B_2\hat{A}_2D_2$ and we say that the *angular deformation* associated with the original angle $B\hat{A}D$ is $(B\hat{A}D - B_2\hat{A}_2D_2)$.

If these angles are measured in radians then the angular deformation is dimensionless. In principle the deformation of any original angle might be considered. However it is customary to consider the deformation experienced by an original right angle in the material. By convention the *decrease* of an original right angle such as $B\hat{A}D$ is termed the *average angular deformation*. The qualification average is used above since the angle $B\hat{A}D$ was defined by the finite lines AB and AD so that the change in $B\hat{A}D$ is an average of the experience of a large region of material. If the defining lines AB and AD are taken indefinitely small then in the limit the average angular deformation tends to a quantity specific to a point. This limit quantity is termed the *shear strain* at the point and is associated with two directions α and β

i.e.
$$\gamma_{\alpha\beta} = \lim_{AB, AD \to 0} (B\hat{A}D - B_2\hat{A}_2D_2) \tag{2.2}$$

In general the shear strain will vary from point to point through the material.

2.4 Infinitesimal Strain

In the discussion of Article 2.2 no mention was made of the magnitude of the strains. For all purposes of structural or mechanical design these are in fact exceedingly small. There are two reasons for this. First, sagging bridges or swaying high flats would be socially unacceptable even if mechanically safe. Secondly, most of the materials used in engineering can only withstand small strains without undergoing, at best, irreversible changes, at worst, total collapse.

Figure 2.2 shows typical examples of stress/strain relationships from which

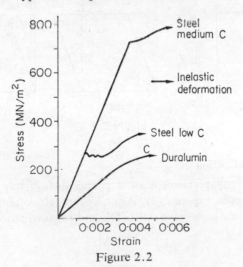

Figure 2.2

it is seen that the maximum strains in materials of interest in structural design rarely exceed 0·004 and are mostly of the order of 0·002 or even less. Thus in strain analysis the quantities involved are even at their greatest very small quantities indeed. (In plotting the stress/strain relation diagram for steel, increments of strain of less than 0·0001 may sometimes be measured.) It is therefore reasonable to describe structural strains as *infinitesimal*.

Infinitesimal strains have the convenient property that all increments of strain may be related to the unloaded (or free) state. Consider, for example, the linear strain of a body of length L subjected to two successive increments δL of linear deformation. Using equation (2.1) the linear strain for the first increment is

$$\lim_{L \to 0} \left(\frac{\delta L}{L} \right)$$

Similarly the second increment causes a strain which can be written as

$$\lim_{(L + \delta L) \to 0} \left(\frac{\delta L}{L + \delta L} \right) = \lim_{(L + \delta L) \to 0} \left(\frac{\delta L}{L} \right) \left(1 - \left(\frac{\delta L}{L} \right) + \left(\frac{\delta L}{L} \right)^2 - \ldots \right)$$

Thus if $(\delta L / L)$ is infinitesimal the second increment is indistinguishable from the first and so the strain for each increment of deformation can be evaluated by reference to the unloaded dimension. Strains outwith this range are of interest in the analysis of material-forming processes, which is a more specialized topic outside the scope of this text. In what follows it will be understood that strains are infinitesimal unless otherwise stated.

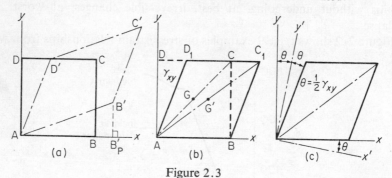

Figure 2.3

2.5   The State of Strain

An infinitesimal square element of a plane surface has sides parallel to coordinate axes as in Figure 2.3a. After loading the original square ABCD is deformed to some new shape AB'C'D', and from equation (2.1) the linear strain of AB is

$$\frac{AB' - AB}{AB}$$

since AB is already infinitesimal. Now $B'_P\hat{A}B$ is, by definition, part of the shear strain, which is itself infinitesimal, so that $\cos(B'_P\hat{A}B')\doteq\cos(0) = 1$. Hence AB′ can be replaced by its projection and the linear strain of AB is

$$\frac{AB'_P - AB}{AB}$$

It is therefore reasonable to associate the strain of AB with its original direction parallel to the x axis, so that it can be denoted as ϵ_x. In the same way ϵ_y is the linear strain of the line AD originally parallel to the y axis.

The shear strain is defined as the decrease of an original right angle, such as $B\hat{A}D$, for which it is $(B\hat{A}D - B'AD')$. This quantity is denoted γ_{xy} the subscripts indicating that this component is the shear strain in the xy plane.

Now suppose that the same total change of angle occurred in the manner shown in Figure 2.3b so that shear strain is at first sight due only to rotation of one defining line AD. However here there has been a rigid body rotation as can be seen by comparing lines joining A to the centres of area of ABCD and ABC_1D_1. Noting that γ_{xy} is infinitesimal we can write

$$B\hat{A}C = 45°; \; B\hat{A}C_1 = \tfrac{1}{2}B\hat{A}D_1 = \tfrac{1}{2}(90° - \gamma_{xy}) = 45° - \tfrac{1}{2}\gamma_{xy}$$

Hence the rigid body rotation is $C_1\hat{A}C = \tfrac{1}{2}\gamma_{xy}$. If the axes are rotated through this angle, as in Figure 2.3c, the effect of the rigid body rotation is eliminated and it is seen that half of the shear strain, i.e. $\tfrac{1}{2}\gamma_{xy}$, is associated with each axis. Thus when rigid body displacements are excluded the deformation of an area in the xy plane can be described by strain components ϵ_x and $\tfrac{1}{2}\gamma_{xy}$ associated with the x direction and by strains ϵ_y and $\tfrac{1}{2}\gamma_{xy}$ associated with the y direction.

The above argument could equally well have been applied to an area in the yz or the xz plane. For the yz plane the defining strain components would be ϵ_y, ϵ_z and γ_{yz} and for the xz plane ϵ_x, ϵ_z, and γ_{xz}. Thus the deformation of an original cube could be specified by examining the deformation of three planes as shown in Figure 2.4, which would then introduce three linear strains ϵ_x, ϵ_y, ϵ_z and three shear strains γ_{xy}, γ_{yz}, γ_{xz}, the latter being considered as associated half with each axis named in the subscript. Thus the strains

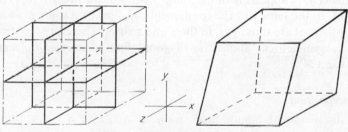

Figure 2.4

associated with the x axis are:

$$\epsilon_x, \quad \tfrac{1}{2}\gamma_{xy}, \quad \tfrac{1}{2}\gamma_{xz}$$

with the y axis:

$$\epsilon_y, \quad \tfrac{1}{2}\gamma_{xy}, \quad \tfrac{1}{2}\gamma_{yz}$$

with the z axis:

$$\epsilon_z, \quad \tfrac{1}{2}\gamma_{xz}, \quad \tfrac{1}{2}\gamma_{yz}$$

By analogy with the description of stress (see Article 1.3, array A1.1) these components are arranged in an array with the linear strains on the diagonal, viz.

$$\begin{pmatrix} \epsilon_x & \tfrac{1}{2}\gamma_{xy} & \tfrac{1}{2}\gamma_{xz} \\ \tfrac{1}{2}\gamma_{xy} & \epsilon_y & \tfrac{1}{2}\gamma_{yz} \\ \tfrac{1}{2}\gamma_{xz} & \tfrac{1}{2}\gamma_{yz} & \epsilon_z \end{pmatrix} \tag{A2.1}$$

This *strain array* completely describes the *state of strain* **R** at a point in a material.

N.B. The obvious correspondence between linear strain and normal stress suggests the adoption of the term normal strain (previously listed as an alternative) to describe the linear strain. Henceforth the term *normal strain* will be used in place of linear strain.

2.6 Properties of the Strain Array

It has already been pointed out that the components of the strain array show a close correspondence in form to those of the stress array. It will now be shown that a similar correspondence exists between their properties.

(a) *Symmetry*

As shown in Article 2.5 the shear strain γ_{xy} in the xy plane is associated half with each of the x and y axes. Therefore since γ_{xy} has one sign the two components $\tfrac{1}{2}\gamma_{xy}$ appearing in the x and y rows of the array (A2.1) are equal quantities. In the same way the components $\tfrac{1}{2}\gamma_{yz}$ in the y and z rows are equal to each other, as are those $\tfrac{1}{2}\gamma_{xz}$ in the x and z rows. Therefore the strain array (A2.1) is symmetrical about the diagonal line of *normal* (linear) strains (cf. Article 1.4).

(b) *Superposition*

In the discussion of infinitesimal strain in Article 2.4 it was shown that successive increments of deformation could all be expressed as strains relative to the undeformed or initial condition of the body. It follows that states of infinitesimal strain may be superimposed at a point in a material. The components of the strain arrays representing successive states of strain

$$R_1, \quad R_2, \quad \ldots, \quad R_i, \quad \ldots$$

may therefore be added provided that they have been measured with respect to the same coordinate axes (cf. Article 1.5).

(c) *Transformation of Axes*

The state of strain is, in general, as given by array (A2.1) for axes x, y, z. If one of these axes, say z, does not participate in a particular deformation process then the strain array take the simplified form

$$\begin{pmatrix} \epsilon_x & \tfrac{1}{2}\gamma_{xy} & 0 \\ \tfrac{1}{2}\gamma_{xy} & \epsilon_y & 0 \\ 0 & 0 & 0 \end{pmatrix} \qquad (A2.2)$$

which is termed plane strain and is analogous to the state of plane stress described in Article 1.6. The strain array (A2.2) may be transformed to axes x', y', z by a procedure analogous to that for stresses, leading to

$$\begin{pmatrix} \epsilon_{x'} & \tfrac{1}{2}\gamma_{x'y'} & 0 \\ \tfrac{1}{2}\gamma_{x'y'} & \epsilon_{y'} & 0 \\ 0 & 0 & 0 \end{pmatrix} \qquad (A2.3)$$

To establish the transformation consider a rectangular element in the xy plane with sides of length dx and dy as in Figure 2.5a. The diagonal OP is of

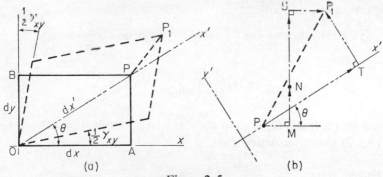

Figure 2.5

length dx' and lies along the new axis x' which makes θ to the x axis. The y' axis is normal to the x' axis. Then

$$dx = dx' \cos\theta; \quad dy = dx' \sin\theta \qquad (2.3)$$

After deformation the element assumes the dotted shape of Figure 2.5a, P being displaced to P_1. The problem is now to express the components of the two arrays in terms of each other and the angle θ. The displacement PP_1 is resolved in Figure 2.5b into components PM, MN, NS, SP_1 parallel to the x, y axes, and alternative components PT, TP_1 parallel to the x', y' axes. From these figures we see that PM $= \epsilon_x \, dx$, i.e. the product of strain in the x direction with the length experiencing that strain.

Similarly NS $= \epsilon_y \, dy$ and PT $= \epsilon_{x'} \, dx'$. Also MN $= \frac{1}{2}\gamma_{xy} \, dx$, i.e. the product of the angle of rotation of dx and its length.

Similarly $SP_1 = \frac{1}{2}\gamma_{xy} \, dy$ and $TP_1 = \frac{1}{2}\gamma_{x'y'} \, dx'$.

Each of the x, y, z displacements can be resolved into components in the x', y', z directions. Thus

$$PT = (PM + SP_1) \cos \theta + (MN + NS) \sin \theta$$

and, substituting

$$\epsilon_{x'} \, dx' = (\epsilon_x \, dx + \tfrac{1}{2}\gamma_{xy} \, dy) \cos \theta + (\tfrac{1}{2}\gamma_{xy} \, dx + \epsilon_y \, dy) \sin \theta$$

Applying equations (2.3) and simplifying

$$\epsilon_{x'} = \epsilon_x \cos^2 \theta + \epsilon_y \sin^2 \theta + 2(\tfrac{1}{2}\gamma_{xy}) \sin \theta \cos \theta \qquad (2.4a)$$

or,

$$\epsilon_{x'} = \frac{(\epsilon_x + \epsilon_y)}{2} + \frac{(\epsilon_x - \epsilon_y)}{2} \cos 2\theta + \tfrac{1}{2}\gamma_{xy} \sin 2\theta \qquad (2.4b)$$

By analogy with Article 1.6 $\epsilon_{y'}$ may be found by setting $2\theta = 2(\theta + \pi/2)$ in equations (2.4).

Similarly by resolving in the y' direction

$$\tfrac{1}{2}\gamma_{x'y'} \, dx' = -(\epsilon_x \, dx + \tfrac{1}{2}\gamma_{xy} \, dy) \sin \theta + (\tfrac{1}{2}\gamma_{xy} \, dx + \epsilon_y \, dy) \cos \theta$$

and so proceeding as above,

$$\tfrac{1}{2}\gamma_{x'y'} = -\epsilon_x \sin \theta \cos \theta + \epsilon_y \sin \theta \cos \theta + \tfrac{1}{2}\gamma_{xy} (\cos^2 \theta - \sin^2 \theta) \qquad (2.5a)$$

or,

$$\tfrac{1}{2}\gamma_{x'y'} = -\frac{(\epsilon_x - \epsilon_y)}{2} \sin 2\theta + \tfrac{1}{2}\gamma_{xy} \cos 2\theta \qquad (2.5b)$$

Equations (2.4) and (2.5) allow the array (A2.2) to be transformed to the form (A2.3) (compare Article 1.6).

(d) *Principal Axes*

If equation (2.4b) is differentiated with respect to θ and the stationary values determined, it is found that there is one set of axes for which the normal strains have extreme values and the shear strains are zero. These are the *principal axes of strain* and the corresponding normal strains are the *principal strains* denoted ϵ_1, ϵ_2, ϵ_3. As in the case of stress a complete analysis shows that these results are valid for the most general state of strain (cf. Article 1.7).

(e) *Maximum Shear Strain*

Similarly from equation (2.5b) the extreme values of shear strain are found to be associated with a set of axes at 45° to the principal axes; the maximum shear strain in the material (cf. Article 1.8) is the greatest of:

$$\tfrac{1}{2}|\epsilon_1 - \epsilon_2| \text{ or } \tfrac{1}{2}|\epsilon_2 - \epsilon_3| \text{ or } \tfrac{1}{2}|\epsilon_3 - \epsilon_1| \tag{2.6}$$

2.7 Mohr Strain Circle

Equations (2.4) and (2.5) have the same form as equations (1.3) and (1.4) for transformation of stress when corresponding variables are as in Table 2.1.

Table 2.1   Corresponding quantities for Mohr Circle Construction

Stress	σ_x	σ_y	τ_{xy}	θ	$\sigma_{x'}$	$\tau_{x'y'}$
Strain	ϵ_x	ϵ_y	$\tfrac{1}{2}\gamma_{xy}$	θ	$\epsilon_{x'}$	$\tfrac{1}{2}\gamma_{x'y'}$

It follows that a Mohr Strain Circle may be drawn to provide a graphical solution of the strain transformation equations, all procedures from Article 1.10 being translated according to Table 2.1. Thus for the state of strain

$$\begin{pmatrix} \epsilon_x & \tfrac{1}{2}\gamma_{xy} & 0 \\ \tfrac{1}{2}\gamma_{xy} & \epsilon_y & 0 \\ 0 & 0 & \epsilon_z \end{pmatrix} \tag{A2.4}$$

we set out to scale $OP \equiv \epsilon_x$, $OQ \equiv \epsilon_y$ on the ϵ axis as in Figure 2.6a.

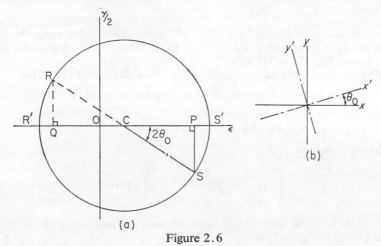

(a)

(b)

Figure 2.6

The centre of the Mohr Circle is at C the midpoint of PQ. PS is set out (using the same scale as above) to represent $\frac{1}{2}\gamma_{xy}$ directed down if $\frac{1}{2}\gamma_{xy}$ is positive, and directed up if $\frac{1}{2}\gamma_{xy}$ is negative. The circle with centre C and radius CS is the Mohr Strain Circle.

To find the strain array for alternative axes inclined at θ_0 to the x, y axes as in Figure 2.6b, a diameter S'CR' is drawn at an angle $2\theta_0$ to SCR

$$\begin{pmatrix} \epsilon_{x'} & \frac{1}{2}\gamma_{x'y'} & 0 \\ \frac{1}{2}\gamma_{x'y'} & \epsilon_{y'} & 0 \\ 0 & 0 & \epsilon_z \end{pmatrix} \qquad (A2.5)$$

Then $\epsilon_{x'} \equiv$ horizontal coordinate of S'

$\qquad \epsilon_{y'} \equiv$ horizontal coordinate of R'

$\qquad \frac{1}{2}\gamma_{x'y'} \equiv$ vertical coordinate of S' (positive down)

As with the Mohr Stress Circle, the principal strain array is obtained if S'CR' is horizontal as illustrated, and the maximum shear strain can be determined from the principal strain array using result (2.6).

Example 2.1 The state of strain at a point is measured with respect to axes x, y, z and found to be

$$\begin{pmatrix} 400 & 200 & 0 \\ 200 & -200 & 0 \\ 0 & 0 & 100 \end{pmatrix} \times 10^{-6}$$

Find the principal axes, the principal strain array, and the maximum shear strain.

Solution Referring to Figure 2.6a OP = 400 units, OQ = -200 units, PS is 200 units down, and C is at (100, 0) the midpoint of QP. With radius CS draw the Mohr Circle, from which CS = 361 units.

To find the principal axes consider the horizontal diameter S'CR'. Then

$\qquad \epsilon_{x'} \equiv$ horizontal coordinate of leading end S' = OS' = 461 units

$\qquad \epsilon_{y'} \equiv$ horizontal coordinate of end R' = OR' = -261 units

$\qquad \frac{1}{2}\gamma_{x'y'} \equiv$ vertical coordinate of S' = 0

Therefore the principal strain array is

$$\begin{pmatrix} 461 & 0 & 0 \\ 0 & -261 & 0 \\ 0 & 0 & 100 \end{pmatrix} \times 10^{-6}$$

$S\hat{C}S' = 2\theta_0 = 33°40'$ and so the principal axes are inclined at $16°50'$ to the original axes as in Figure 2.6b.

From result (2.6) the maximum shear strain is the greatest of

$$\tfrac{1}{2}|461-(-261)| \text{ or } \tfrac{1}{2}|-261-100| \text{ or } \tfrac{1}{2}|100-461| \text{ (all } \times 10^{-6})$$

i.e. 361×10^{-6}.

Problem for Solution 2.1

2.8 Volume Strain

A rectangular parallelepiped (Figure 1.4) of infinitesimal dimensions a, b, c is subjected to a state of strain expressed by the array (A2.1).

Then the side of length a increases in length by $\epsilon_x a$ so that its length changes to $a(1+\epsilon_x)$.

Similar changes occur in the other sides and so the new volume is

$$a(1+\epsilon_x) \times b(1+\epsilon_y) \times c(1+\epsilon_z)$$

Expanding and neglecting products of strains (infinitesimals) the new volume is $(abc(1+(\epsilon_x+\epsilon_y+\epsilon_z)))$.

The original volume was abc

and so the volume strain is $\dfrac{\text{change in volume}}{\text{original volume}}$

i.e.
$$\Delta = \epsilon_x + \epsilon_y + \epsilon_z \tag{2.7}$$

Since the axes were chosen at random and the volume strain cannot be affected by the directions of the axes, it follows that the sum of the three normal strains as expressed by equation (2.7) is an *invariant property* of the state of strain.

This result can be checked in the case of plane strain using equation (2.4b). By setting $\theta = (\theta+\pi/2)$ in that equation, one for $\epsilon_{y'}$ is obtained and when this is added to (2.4b) we find $\epsilon_{x'}+\epsilon_{y'} = \epsilon_x+\epsilon_y$ (compare Article 1.11). The general theory for transformation of all three axes results in a similar verification.[1]

Example 2.2 Referring to Example 2.1 the volume strain from the measured array is
$$(400+(-200)+100) \times 10^{-6} = 300 \times 10^{-6}$$

and from the principal strain array is
$$(461+(-261)+100) \times 10^{-6} = 300 \times 10^{-6} \text{ as before}$$

Problem for Solution 2.2

2.9 Strain Rosettes

The methods of strain measurement are now too sophisticated to be adequately discussed in the few pages which could be spared in a text of this type.

Introductory reading will be found in the list of references at the end of the chapter. However there is one aspect arising out of strain measurement techniques which it is appropriate to discuss here since it provides one application for the theory of strain array transformation dealt with in Articles 2.6 and 2.7.

The components of strain in any one surface comprise two normal strains and one shear strain, e.g. in an xy surface the strain components are ϵ_x, ϵ_y, $\frac{1}{2}\gamma_{xy}$. However, the methods generally available for strain measurement are confined to free surfaces and do not detect shear strains directly. The latter difficulty is overcome by measuring normal strain in three separate directions in the surface being considered. Referring to Figure 2.7a, A, B, C are each

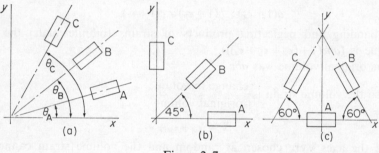

Figure 2.7

normal strain detectors, measuring strains ϵ_A, ϵ_B, ϵ_C in the directions defined by the angles θ_A, θ_B, θ_C. Such an arrangement is known as a strain rosette. These normal strains ϵ_A, ϵ_B, ϵ_C can each be related to the Cartesian components of the state of strain by substituting θ_A, θ_B, θ_C in the transformation equation (2.4a), i.e.

$$\epsilon_A = \epsilon_x \cos^2 \theta_A + \epsilon_y \sin^2 \theta_A + \gamma_{xy} \sin \theta_A \cos \theta_A$$

$$\epsilon_B = \epsilon_x \cos^2 \theta_B + \epsilon_y \sin^2 \theta_B + \gamma_{xy} \sin \theta_B \cos \theta_B \qquad (2.8)$$

$$\epsilon_C = \epsilon_x \cos^2 \theta_C + \epsilon_y \sin^2 \theta_C + \gamma_{xy} \sin \theta_C \cos \theta_C$$

Therefore if values of the θ_A, θ_B, θ_C together with the measured strains ϵ_A, ϵ_B, ϵ_C are inserted in these equations then they become three simultaneous equations in ϵ_x, ϵ_y, $\frac{1}{2}\gamma_{xy}$. Since these strain components are in a free surface, the z axis is a principal axis, and the strain array is as (A2.4).

The most general rosette is rarely used. In practice one or more of the gauges are made to coincide with the coordinate axes, so reducing the mathematical complexity. The two most common arrangements are shown in Figures 2.7b and c.

(a) *45/90° or Rectangular Rosette* (Figure 2.7b).

Here $\theta_A = 0°$, $\theta_B = 45°$, $\theta_C = 90°$ and so equations (2.8) simplify to

$$\epsilon_A = \epsilon_x$$

$$\epsilon_B = \tfrac{1}{2}\epsilon_x + \tfrac{1}{2}\epsilon_y + \tfrac{1}{2}\gamma_{xy}$$

$$\epsilon_C = \epsilon_y$$

Hence the Cartesian components of strain ϵ_x, ϵ_y, $\tfrac{1}{2}\gamma_{xy}$ are obtained from the measured normal strains ϵ_A, ϵ_B, ϵ_C by application of the equations

$$\epsilon_x = \epsilon_A; \quad \epsilon_y = \epsilon_C$$

$$\gamma_{xy} = 2\epsilon_B - \epsilon_A - \epsilon_C \tag{2.9}$$

(b) *Delta Rosette* (Figure 2.7c)

Here $\theta_A = 0$, $\theta_B = 120°$, $\theta_C = 240°$ and equations (2.8) become

$$\epsilon_A = \epsilon_x$$

$$\epsilon_B = \frac{\epsilon_x}{4} + \frac{3\epsilon_y}{4} - \gamma_{xy}\frac{\sqrt{3}}{4}$$

$$\epsilon_C = \frac{\epsilon_x}{4} + \frac{3\epsilon_y}{4} + \gamma_{xy}\frac{\sqrt{3}}{4}$$

from which expressions for the Cartesian components may be deduced.

Example 2.3 The following strains were measured using a rectangular rosette

$$\epsilon_A = -100 \times 10^{-6}; \quad \epsilon_B = 100 \times 10^{-6}; \quad \epsilon_C = 180 \times 10^{-6}.$$

Hence determine the principal axes and the principal strain array.

Solution Applying equations (2.9)

$$\epsilon_x = \epsilon_A = -100 \times 10^{-6}; \quad \epsilon_y = \epsilon_C = 180 \times 10^{-6}$$

$$\gamma_{xy} = 2\epsilon_B - \epsilon_A - \epsilon_C = 120 \times 10^{-6}$$

Therefore the state of strain for the x, y, z axes is

$$\begin{pmatrix} -100 & 60 & 0 \\ 60 & 180 & 0 \\ 0 & 0 & \epsilon_z \end{pmatrix} \times 10^{-6}$$

A Mohr Strain Circle is plotted with OP = -100, OQ = 180, PS = 60 units down, as in Figure 2.8a. The radius is found to be 152·5 and $\theta_0 = 11°36'$. Then the principal axes are as in Figure 2.8b and the principal strain array is

$$\begin{pmatrix} -112·5 & 0 & 0 \\ 0 & 192·5 & 0 \\ 0 & 0 & \epsilon_z \end{pmatrix} \times 10^{-6}$$

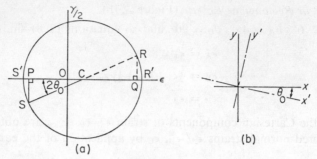

Figure 2.8

Problem for Solution 2.3

2.10   References

(a) *Cited*

1. Timoshenko, S. and Goodier, J. N., loc. cit. Article 1.15.

(b) *General*

1. Dove, R. C. and Adams, P. H., *Experimental Stress Analysis and Motion Measurement*, Merill, Columbus, Ohio, 1964.
2. Dally, J. W. and Riley, W. F., loc. cit. Article 1.15.
3. Perry, C. C. and Lissner, H. R., *The Strain Gage Primer*, McGraw-Hill, New York, 1962.

2.11   Problems for Solution

2.1.   Find the principal axes, principal strain array and maximum shear strain for the following states of strain.

$$\text{(a)} \begin{pmatrix} 0 & -200 & 0 \\ -200 & 400 & 0 \\ 0 & 0 & -400 \end{pmatrix} 10^{-6} \qquad \text{(b)} \begin{pmatrix} -100 & 100 & 0 \\ 100 & -300 & 0 \\ 0 & 0 & 0 \end{pmatrix} 10^{-6}$$

$$\text{(c)} \begin{pmatrix} 100 & 0 & 0 \\ 0 & 100 & 0 \\ 0 & 0 & 0 \end{pmatrix} 10^{-6}$$

2.2.   Determine the volume strain for the states of strain in Problem 2.1.

2.3.   In each of the following find the principal axes and the principal strain array: (a) A delta rosette (Figure 2.7c) gives $\epsilon_A = 150 \times 10^{-6}$; $\epsilon_B = -600 \times 10^{-6}$; $\epsilon_C = 400 \times 10^{-6}$. (b) A rectangular rosette (Figure 2.7b) gives $\epsilon_A = 0$; $\epsilon_B = 200 \times 10^{-6}$; $\epsilon_C = 0$.

Stress/Strain Relationships

3.1 Introduction

A great variety of materials is used in engineering, including metals, plastics (polymers) and concrete, and these have enormous differences between their atomic and molecular constitutions.[1,2] Despite this complexity it is found from experiment that the stress/strain relationships for most materials can be represented by one or more of a limited number of idealized types of material behaviour. In this chapter we first consider the more common of these, and then formulate the particular relationship which is most widely used as the basis of stress and structural analysis.

The classification of types of stress/strain behaviour to be given in Article 3.2 is based on the fundamental characteristics of time dependence and elasticity. Both of these can be described by considering the effect of uniform uniaxial tension in a rod, as described in Article 1.14 and illustrated in Figure 3.1a, for which the state of stress is

$$\begin{pmatrix} \sigma_x & 0 & 0 \\ 0 & 0 & 0 \\ 0 & 0 & 0 \end{pmatrix} \tag{A3.1}$$

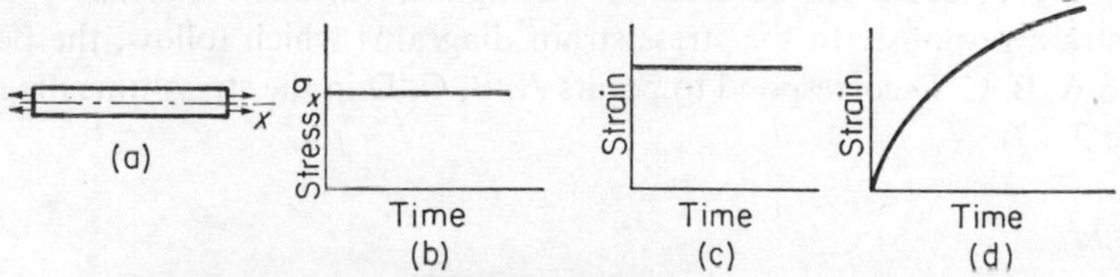

Figure 3.1

Suppose that end tensions are imposed instantaneously and then maintained at constant magnitude (as in Figure 3.1b) while some measure of deformation such as the strain component ϵ_x is measured continuously. If the strain is constant as in Figure 3.1c the material is said to be time independent;

such behaviour is characteristic of the common structural metals such as
steel and aluminium at ordinary temperatures. If the strain increases with
time (as in Figure 3.1d) the material is classed as time dependent; such
materials include metals at elevated temperatures and plastics at ordinary
temperatures; the behaviour described by Figures 3.1b and d is usually
termed creep.

Now let us consider the response of the rod when the tensions are imposed
as before but completely (and instantaneously) removed after the elapse of
a finite time interval $t_S = t_1 - t_0$ as shown in Figure 3.2a. If the rod is able
completely to recover its original state on removal of loads then the material
is said to be *elastic*; such behaviour is illustrated in Figure 3.2b. If part or all

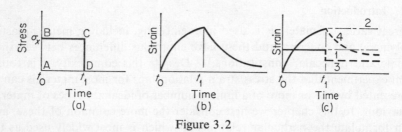

Figure 3.2

of the deformation is never recovered after removal of loads the material is
said to be *inelastic*; the irrecoverable deformation is termed plastic deformation. Curves 1 and 2 of Figure 3.2c refer to a completely inelastic material
while curves 3 and 4 are for one which is only partly so. Almost all the
common structural materials have a significant range of stresses for which the
material is elastic.

3.2   Types of Stress/Strain Relationship

The behaviour of a material when subjected to one or more loads of the
type of Figure 3.2a can be used to distinguish various idealized types of
stress/strain response. In the stress/strain diagrams which follow, the points
marked A, B, C, D correspond to points A, B, C, D in the stress/time diagram
(Figure 3.2a).

(a)   *Rigid*

This material does not deform, as in rigid body mechanics, and its response
is given by line 1 in Figure 3.3a. It is time independent.

(b)   *Linear Elastic*

Stress is proportional to strain (line 2 of Figure 3.3a), is independent of time
(B and C coincident) and the strain is completely recovered on load removal

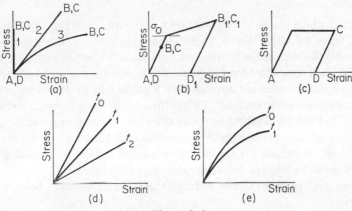

Figure 3.3

(A and D coincident). This model of material behaviour was first proposed by Robert Hooke in the 17th century and is still the foundation of most methods of stress and structural analysis. Most materials have some range of stresses for which this model is at least a reasonable approximation, see for example Figure 2.2.

(c) *Non-linear Elastic*

Illustrated by line 3 of Figure 3.3a this differs from case (b) only in that stress is not proportional to strain.

(d) *Elastic/Strain-hardening*

In this case (Figure 3.3b) there is linear elastic behaviour if the load (A, B, C, D) is less than some critical value denoted σ_0. If the load exceeds this value (A, B_1, C_1, D_1) stress increases at a lesser rate, and plastic deformation AD_1 occurs. Since the material is now linear elastic along the path D_1B_1 the effect of plastic deformation has been to raise the value of the critical stress (hence the term strain-hardening). All ductile materials show this effect to some degree but the straight line is an idealization.

(e) *Ideal Elastic/Plastic*

If BB_1 in Figure 3.3b is parallel to the strain axis we have as a limiting case the ideal elastic/plastic (Figure 3.3c). This material is linear elastic for stress less than the critical value but will undergo indefinite plastic deformation at the critical stress. This material corresponds quite closely to part of the stress/strain diagram of mild steel (Figure 2.2). It is often used as an approximation to a slightly strain-hardening material.

(f) *Linear Viscoelastic*

A material which creeps in response to a step load, as in Figure 3.2b, has no unique value of strain for a given stress. However if the deformation is recoverable the material is termed viscoelastic and there is a specific value of strain for each time after load application. Therefore if a series of step loads of different magnitudes are applied then a graph can be drawn of strain at a chosen time, against stress level. This is the so-called isochronous stress/strain diagram. In Figure 3.3d these plots are shown for increasing times $t_0, t_1, t_2 \ldots$; in this case stress is proportional to strain at each time and so the material is called linear viscoelastic. It should be noted that sufficient time for complete recovery must be allowed between successive step loads.

Perspex and rigid PVC are examples of materials which behave in this manner.

(g) *Non-linear Viscoelastic*

This is a viscoelastic material for which the isochronous stress/strain plots are non-linear (Figure 3.3e). Polyethylene and other crystalline thermoplastics behave in this manner.

For detailed discussion and bibliography the reader is referred to such works as references 3–5.

3.3 Linear Elastic Material

3.3.1 *Stress/Strain Equations*

By far the most useful model of material behaviour is the linear elastic material, which as we have seen (case (b) of Article 3.2) is time independent, elastic, and has stress proportional to strain. In the following discussion we shall also consider the material to be homogeneous (i.e. properties independent of position) and isotropic (i.e. properties at a point are independent of the direction in which they are measured).

(a) *Normal Stresses* Applying a state of stress

$$\mathbf{T}_1 = \begin{pmatrix} \sigma_x & 0 & 0 \\ 0 & 0 & 0 \\ 0 & 0 & 0 \end{pmatrix} \qquad (A3.2)$$

it is found qualitatively that the strain response is of the form

$$\mathbf{R}_1 = \begin{pmatrix} \epsilon_x & 0 & 0 \\ 0 & \epsilon_y & 0 \\ 0 & 0 & \epsilon_z \end{pmatrix} \qquad (A3.3)$$

If quantitative measurement is performed it is found that

$$R_1 = \begin{pmatrix} \dfrac{\sigma_x}{E} & 0 & 0 \\[2ex] 0 & \dfrac{-\nu\sigma_x}{E} & 0 \\[2ex] 0 & 0 & \dfrac{-\nu\sigma_x}{E} \end{pmatrix} \qquad (A3.4)$$

where E is a material constant called Young's Modulus of Elasticity, ν is a material constant called Poisson's Ratio.

Comparing arrays (A3.3) and (A3.4) we have

$$\epsilon_x = \frac{\sigma_x}{E}; \quad \epsilon_y = \epsilon_z = \frac{-\nu\sigma_x}{E} = -\nu\epsilon_x$$

as the stress/strain relations for the simple state of stress (A3.2) which corresponds to a tensile test on a rod (Figure 3.1a). While it may be noted that in this case a lateral contraction accompanies the extension in the x direction, it cannot be overemphasized that this is a *particular* result.

Since the material is isotropic, all directions are equivalent and so the stress state

$$T_2 = \begin{pmatrix} 0 & 0 & 0 \\ 0 & \sigma_y & 0 \\ 0 & 0 & 0 \end{pmatrix} \text{ causes } R_2 = \begin{pmatrix} \dfrac{-\nu\sigma_y}{E} & 0 & 0 \\[2ex] 0 & \dfrac{\sigma_y}{E} & 0 \\[2ex] 0 & 0 & \dfrac{-\nu\sigma_y}{E} \end{pmatrix}$$

and the stress state

$$T_3 = \begin{pmatrix} 0 & 0 & 0 \\ 0 & 0 & 0 \\ 0 & 0 & \sigma_z \end{pmatrix} \text{ causes } R_3 = \begin{pmatrix} \dfrac{-\nu\sigma_z}{E} & 0 & 0 \\[2ex] 0 & \dfrac{-\nu\sigma_z}{E} & 0 \\[2ex] 0 & 0 & \dfrac{\sigma_z}{E} \end{pmatrix}$$

Since the axes are the same and conditions at the same point are being considered, the principle of superposition may be applied (Articles 1.5 and

2.6). Thus, imposing a combined state of stress

$$\mathbf{T}_1 + \mathbf{T}_2 + \mathbf{T}_3 = \begin{pmatrix} \sigma_x & 0 & 0 \\ 0 & \sigma_y & 0 \\ 0 & 0 & \sigma_z \end{pmatrix} \tag{A3.5}$$

the measured state of strain will be

$$\mathbf{R}_1 + \mathbf{R}_2 + \mathbf{R}_3 = \begin{pmatrix} \dfrac{\sigma_x}{E} - \dfrac{\nu\sigma_y}{E} - \dfrac{\nu\sigma_z}{E} & 0 & 0 \\ 0 & -\dfrac{\nu\sigma_x}{E} + \dfrac{\sigma_y}{E} - \dfrac{\nu\sigma_z}{E} & 0 \\ 0 & 0 & -\dfrac{\nu\sigma_x}{E} - \dfrac{\nu\sigma_y}{E} + \dfrac{\sigma_z}{E} \end{pmatrix} \tag{A3.6}$$

From array (A3.6) the normal strain component ϵ_x is therefore

$$\epsilon_x = \frac{1}{E}(\sigma_x - \nu(\sigma_y + \sigma_z)) \tag{3.1a}$$

Similarly

$$\epsilon_y = \frac{1}{E}(\sigma_y - \nu(\sigma_z + \sigma_x)) \tag{3.1b}$$

and

$$\epsilon_z = \frac{1}{E}(\sigma_z - \nu(\sigma_x + \sigma_y)) \tag{3.1c}$$

It should be noted that normal stresses cause only normal strains.

(b) *Shear Stresses*   If the stress state

$$\mathbf{T}_4 = \begin{pmatrix} 0 & \tau_{xy} & 0 \\ \tau_{yx} & 0 & 0 \\ 0 & 0 & 0 \end{pmatrix} \tag{A3.7}$$

is imposed the strain state is found to be of the form

$$\mathbf{R}_4 = \begin{pmatrix} 0 & \tfrac{1}{2}\gamma_{xy} & 0 \\ \tfrac{1}{2}\gamma_{xy} & 0 & 0 \\ 0 & 0 & 0 \end{pmatrix} \tag{A3.8}$$

Measurement shows

$$\mathbf{R}_4 = \begin{pmatrix} 0 & \dfrac{\tau_{xy}}{2G} & 0 \\ \dfrac{\tau_{yx}}{2G} & 0 & 0 \\ 0 & 0 & 0 \end{pmatrix} \tag{A3.9}$$

where G is a material constant known as the Shear or Rigidity Modulus of elasticity.

Comparing (A3.8) and (A3.9) we have

$$\gamma_{xy} = \frac{\tau_{xy}}{G} \tag{3.2a}$$

so that this shear stress depends only on the corresponding shear strain. Then since the material is isotropic we can write similar relations for the other two shear strains.
i.e.

$$\gamma_{yz} = \frac{\tau_{yz}}{G} \tag{3.2b}$$

and

$$\gamma_{xz} = \frac{\tau_{xz}}{G} \tag{3.2c}$$

It should be noted that shear stresses cause only shear strains.

(c) *General State of Stress* Combining the results obtained in (a) and (b) above for normal and shear stresses we can say that when a general state of stress

$$\mathbf{T} = \begin{pmatrix} \sigma_x & \tau_{xy} & \tau_{xz} \\ \tau_{yx} & \sigma_y & \tau_{yz} \\ \tau_{zx} & \tau_{zy} & \sigma_z \end{pmatrix} \tag{A3.10}$$

acts at a point then the corresponding strain array for the same axes

$$\mathbf{R} = \begin{pmatrix} \epsilon_x & \frac{1}{2}\gamma_{xy} & \frac{1}{2}\gamma_{xz} \\ \frac{1}{2}\gamma_{xy} & \epsilon_y & \frac{1}{2}\gamma_{yz} \\ \frac{1}{2}\gamma_{xz} & \frac{1}{2}\gamma_{yz} & \epsilon_z \end{pmatrix} \tag{A3.11}$$

can be derived from the stress array using

$$\epsilon_x = \frac{1}{E}(\sigma_x - \nu(\sigma_y + \sigma_z)) \tag{3.3a}$$

$$\epsilon_y = \frac{1}{E}(\sigma_y - \nu(\sigma_z + \sigma_x)) \tag{3.3b}$$

$$\epsilon_z = \frac{1}{E}(\sigma_z - \nu(\sigma_x + \sigma_y)) \tag{3.3c}$$

$$\gamma_{xy} = \frac{\tau_{xy}}{G} \tag{3.3d}$$

$$\gamma_{yz} = \frac{\tau_{yz}}{G} \tag{3.3e}$$

3

$$\gamma_{xz} = \frac{\tau_{xz}}{G} \tag{3.3f}$$

These six equations express the six independent strain components in terms of the six independent stress components. It is possible to solve these equations to express the stress components in terms of the strain components as

$$\sigma_x = 2G\epsilon_x + \lambda(\epsilon_x + \epsilon_y + \epsilon_z) \tag{3.4a}$$

$$\sigma_y = 2G\epsilon_y + \lambda(\epsilon_x + \epsilon_y + \epsilon_z) \tag{3.4b}$$

$$\sigma_z = 2G\epsilon_z + \lambda(\epsilon_x + \epsilon_y + \epsilon_z) \tag{3.4c}$$

$$\tau_{xy} = G\gamma_{xy} \tag{3.4d}$$

$$\tau_{yz} = G\gamma_{yz} \tag{3.4e}$$

$$\tau_{xz} = G\gamma_{xz} \tag{3.4f}$$

where

$$\lambda = \frac{\nu E}{(1+\nu)(1-2\nu)} \tag{3.5}$$

is called Lamé's elastic constant. As will be proved in Section 3.3.3 the shear modulus can be expressed in terms of E and ν.

It will be seen from equations (3.4) that each stress component depends on the corresponding strain component with the shear modulus as operator but the normal stresses are also a function of the strain invariant $(\epsilon_x + \epsilon_y + \epsilon_z)$ described in Article 2.8 and called the volume strain. Thus it would appear that each stress component is at least partly associated with a shearing action but the normal stresses are also associated with volume changes.

Example 3.1 The principal stresses at a point in a plate are $-60, 20, 0$ (units MN/m²). Find the state of strain and the volume strain. Take $\nu = 0.3$ and $E = 210$ GN/m².

Solution The state of stress is

$$\begin{pmatrix} -60 & 0 & 0 \\ 0 & 20 & 0 \\ 0 & 0 & 0 \end{pmatrix} \text{MN/m}^2$$

Applying equations (3.3)

$$\epsilon_x = \frac{10^6}{E}(-60-0.3(20+0)) = -315 \times 10^{-6}$$

$$\epsilon_y = \frac{10^6}{E}(20-0.3(-60)) = 181 \times 10^{-6}$$

$$\epsilon_z = \frac{10^6}{E}(0-0.3(-60+20)) = 57 \times 10^{-6}$$

$$\gamma_{xy} = \gamma_{yz} = \gamma_{xz} = 0$$

Therefore the state of strain is

$$\begin{pmatrix} -315 & 0 & 0 \\ 0 & 181 & 0 \\ 0 & 0 & 57 \end{pmatrix} \times 10^{-6}$$

The volume strain is $\epsilon_x + \epsilon_y + \epsilon_z = 77 \times 10^{-6}$.

Example 3.2 An alloy steel bar of 10 mm × 10 mm cross-section and 60 mm length is subjected to compressive forces of 24 kN uniformly distributed over its end surfaces. The lateral surfaces are subjected to an unknown fluid pressure p. The lateral strains ϵ_y and ϵ_z are measured and are each found to be 100×10^{-6}. Deduce the value of p, and the volume strain. Take $\nu = 0.25$ and $E = 200$ GN/m².

Solution On the end (i.e. x) faces there is a compression intensity of 24,000/ (0.01×0.01), i.e. 240 MN/m², and on the y and z surfaces there is compression of intensity p.

Therefore the state of stress is

$$\begin{pmatrix} -240 & 0 & 0 \\ 0 & -p & 0 \\ 0 & 0 & -p \end{pmatrix} \text{MN/m}^2$$

Applying equations (3.3)

$$\epsilon_x = 10^6 \, (-240 - 0.25 \, (-p-p))/E = (-240 + 0.5p) \times 10^6 /E$$
$$\epsilon_y = 10^6 \, (-p - 0.25 \, (-240-p))/E = (60 - 0.75p) \times 10^6 /E = 100 \times 10^{-6}$$

From the latter $p = 53.3$ MN/m² and so $\epsilon_x = -1067 \times 10^{-6}$ and the state of strain is

$$\begin{pmatrix} -1067 & 0 & 0 \\ 0 & 100 & 0 \\ 0 & 0 & 100 \end{pmatrix} \times 10^{-6}$$

The volume strain is $(\epsilon_x + \epsilon_y + \epsilon_z) = -867 \times 10^{-6}$.

Problems for Solution 3.1–3.8

3.3.2 Bulk Modulus

Consider a state of uniform normal tension in three dimensions as described in Article 1.14. Then $\sigma_x = \sigma_y = \sigma_z = p$ where p is the value of the uniform tension.

The state of stress is given by the array

$$\begin{pmatrix} p & 0 & 0 \\ 0 & p & 0 \\ 0 & 0 & p \end{pmatrix} \qquad \text{(A3.12)}$$

Hence applying equations (3.3)

$$\epsilon_x = \epsilon_y = \epsilon_z = \frac{p(1-2\nu)}{E}$$

and

$$\gamma_{xy} = \gamma_{yz} = \gamma_{xz} = 0$$

Then the volume strain is

$$\Delta = (\epsilon_x + \epsilon_y + \epsilon_z) = \frac{3p(1-2\nu)}{E}$$

i.e.

$$\frac{p}{\Delta} = \frac{E}{3(1-2\nu)} \tag{3.6}$$

Noting that $p = (\sigma_x + \sigma_y + \sigma_z)/3$ equation (3.6) can also be interpreted as a relation between the stress and strain invariants (see Articles 1.11 and 2.8). The expression on the right-hand side is called the Bulk Modulus of elasticity, K
i.e.

$$K = \frac{E}{3(1-2\nu)} \tag{3.7}$$

When $\nu \to 0.5$, $K \to \infty$ and the material is said to be incompressible, which is closely approximated by rubber.

3.3.3 *Relations Between the Elastic Constants*

Five elastic constants have been introduced, viz.

$$E, \nu, G, \lambda, K$$

in order of appearance. λ, K were already expressed in terms of E and ν by equations (3.5) and (3.7). It will now be proved that G can also be expressed in terms of E and ν.

A state of stress for axes x, y, z is applied as follows

$$\mathbf{T} = \begin{pmatrix} 0 & B & 0 \\ B & 0 & 0 \\ 0 & 0 & 0 \end{pmatrix} \tag{A3.13}$$

The state of strain obtained by application of the stress/strain equations (3.3) is

$$\mathbf{R} = \begin{pmatrix} 0 & \dfrac{B}{2G} & 0 \\ \dfrac{B}{2G} & 0 & 0 \\ 0 & 0 & 0 \end{pmatrix} \tag{A3.14}$$

The principal stress array can be found by applying the stress transformation equations (1.3), (1.4) or by drawing a Mohr Circle (cf. Figure 1.13b). Hence we find

$$T = \begin{pmatrix} B & 0 & 0 \\ 0 & -B & 0 \\ 0 & 0 & 0 \end{pmatrix} \tag{A3.15}$$

Similarly using the corresponding procedures for strain we find the principal strain array from (A3.14) as

$$R = \begin{pmatrix} \dfrac{B}{2G} & 0 & 0 \\ 0 & \dfrac{-B}{2G} & 0 \\ 0 & 0 & 0 \end{pmatrix} \tag{A3.16}$$

However the principal strains can also be determined from the principal stresses in (A3.15) by using the stress/strain equations (3.3). Then

$$\epsilon_1 = B(1+\nu)/E; \quad \epsilon_2 = -B(1+\nu)/E; \quad \epsilon_3 = 0$$

Comparing these with the corresponding expressions in array (A3.16) we find

$$G = \frac{E}{2(1+\nu)} \tag{3.8}$$

The elastic constants λ, K, G have now all been expressed in terms of E, ν only by equations (3.5), (3.7) and (3.8). Any other two could of course be chosen as the basic quantities and all the others expressed in terms of them. Thus the elastic behaviour of the isotropic linear elastic material can be fully described by only two material properties. Table 3.1 lists typical values of the elastic constants for some representative materials.

Table 3.1 Values of elastic constants; units of E, G, K are GN/m²

Material	E	ν	G	K
Steel	210	0·29	81	167
Aluminium alloy	70	0·35	26	78
Titanium	114	0·32	43	105

3.3.4 *Plane Stress and Strain*

It is now appropriate to state the stress/strain equations for the special states of stress and strain termed *plane stress* and *plane strain*. Many real problems correspond closely to these simplified states of stress and strain which have previously been mentioned in Articles 1.6 and 2.6 respectively.

(a) *Plane Stress*   Here all stress components associated with one axis are assumed to be zero. The stress components then all have directions lying within the plane perpendicular to this axis—hence 'plane' stress and the state of stress has the form

$$\begin{pmatrix} \sigma_x & \tau_{xy} & 0 \\ \tau_{yx} & \sigma_y & 0 \\ 0 & 0 & 0 \end{pmatrix} \qquad\qquad (A3.17)$$

where z is the direction perpendicular to the 'plane'.

By introducing this state of stress in the stress/strain equations (3.3) the following are obtained:

$$\epsilon_x = \frac{1}{E}(\sigma_x - \nu\sigma_y) \qquad\qquad (3.9a)$$

$$\epsilon_y = \frac{1}{E}(\sigma_y - \nu\sigma_x) \qquad\qquad (3.9b)$$

$$\epsilon_z = -\frac{\nu}{E}(\sigma_x + \sigma_y) \qquad\qquad (3.9c)$$

$$\gamma_{xy} = \frac{\tau_{xy}}{G} = \frac{2(1+\nu)\tau_{xy}}{E} \qquad\qquad (3.9d)$$

Equations for the stresses in terms of the strains can be obtained by inversion of equations (3.9) or by substitution of the state of stress in equations (3.4). Thus:

$$\sigma_x = \frac{E}{(1-\nu^2)}(\epsilon_x + \nu\epsilon_y) \qquad\qquad (3.10a)$$

$$\sigma_y = \frac{E}{(1-\nu^2)}(\epsilon_y + \nu\epsilon_x) \qquad\qquad (3.10b)$$

$$\tau_{xy} = \gamma_{xy}G = \frac{E\gamma_{xy}}{2(1+\nu)} \qquad\qquad (3.10c)$$

Structures which have one dimension very much smaller than the other two are usually in a state of stress approximating to plane stress.

(b) *Plane Strain*   In this case all the strain components associated with one coordinate direction are zero, i.e. the state of strain is given by

$$\begin{pmatrix} \epsilon_x & \frac{1}{2}\gamma_{xy} & 0 \\ \frac{1}{2}\gamma_{xy} & \epsilon_y & 0 \\ 0 & 0 & 0 \end{pmatrix} \qquad\qquad (A3.18)$$

If these strains are introduced in the stress/strain equations (3.4) the following are obtained

$$\sigma_x = \lambda(\epsilon_x + \epsilon_y) + 2G\epsilon_x \tag{3.11a}$$

$$\sigma_y = \lambda(\epsilon_x + \epsilon_y) + 2G\epsilon_y \tag{3.11b}$$

$$\sigma_z = \lambda(\epsilon_x + \epsilon_y) \tag{3.11c}$$

$$\tau_{xy} = \gamma_{xy}G \tag{3.11d}$$

The inverse forms of these equations are

$$\epsilon_x = \frac{1+\nu}{E}\left((1-\nu)\sigma_x - \nu\sigma_y\right) \tag{3.12a}$$

$$\epsilon_y = \frac{1+\nu}{E}\left((1-\nu)\sigma_y - \nu\sigma_x\right) \tag{3.12b}$$

$$\gamma_{xy} = \frac{\tau_{xy}}{G} = \frac{2(1+\nu)\tau_{xy}}{E} \tag{3.12c}$$

The plane state of strain is approximated by cross-sectional planes near the centre of bodies having a large axial extent.

3.3.5 *Matrix Form of the Stress/Strain Equations*
The elements of matrix algebra are outlined in Appendix II.

(a) *Plane Stress Equations* The governing equations for the special state of plane stress discussed in Section 3.3.4 take the form

$$\begin{bmatrix} \epsilon_x \\ \epsilon_y \\ \gamma_{xy} \end{bmatrix} = \frac{1}{E}\begin{bmatrix} 1 & -\nu & 0 \\ -\nu & 1 & 0 \\ 0 & 0 & 2(1+\nu) \end{bmatrix}\begin{bmatrix} \sigma_x \\ \sigma_y \\ \tau_{xy} \end{bmatrix} \tag{3.13}$$

$$\begin{bmatrix} \sigma_x \\ \sigma_y \\ \tau_{xy} \end{bmatrix} = \frac{E}{(1-\nu^2)}\begin{bmatrix} 1 & \nu & 0 \\ \nu & 1 & 0 \\ 0 & 0 & \frac{(1-\nu)}{2} \end{bmatrix}\begin{bmatrix} \epsilon_x \\ \epsilon_y \\ \gamma_{xy} \end{bmatrix} \tag{3.14}$$

(b) *Plane Strain Equations*

$$\begin{bmatrix} \epsilon_x \\ \epsilon_y \\ \gamma_{xy} \end{bmatrix} = \frac{1+\nu}{E}\begin{bmatrix} (1-\nu) & -\nu & 0 \\ -\nu & (1-\nu) & 0 \\ 0 & 0 & 2 \end{bmatrix}\begin{bmatrix} \sigma_x \\ \sigma_y \\ \tau_{xy} \end{bmatrix} \tag{3.15}$$

$$
\begin{bmatrix} \sigma_x \\ \sigma_y \\ \tau_{xy} \end{bmatrix} = \frac{E}{(1+\nu)(1-2\nu)} \begin{bmatrix} (1-\nu) & \nu & 0 \\ \nu & (1-\nu) & 0 \\ 0 & 0 & \dfrac{(1-2\nu)}{2} \end{bmatrix} \begin{bmatrix} \epsilon_x \\ \epsilon_y \\ \gamma_{xy} \end{bmatrix} \quad (3.16)
$$

The general stress/strain equations (3.3) or (3.4) may similarly be written in matrix form if required.

Problem for Solution 3.9

3.4 References

1. Van Vlack, L. H., *Elements of Materials Science*, 2nd ed., Addison-Wesley, Reading, Mass, 1964.
2. Cottrell, A. H., *The Mechanical Properties of Matter*, Wiley, New York, 1964.
3. Freudenthal, A. M., *The Inelastic Behaviour of Engineering Materials and Structures*, Wiley, New York, 1964.
4. Ferry, J. D., *Viscoelastic Properties of Polymers*, Wiley, New York, 1961.
5. Eirich, F. R. (Ed.), *Rheology Theory and Application*, Vols. I–V, Academic, New York, 1956–1969.

3.5 Problems for Solution

For steel take $E = 210 \text{ GN/m}^2$, $\nu = 0.3$ if not otherwise indicated.

3.1. An alloy steel bar of 10 mm × 10 mm square cross-section, and 400 mm length is subjected to axial tensile forces of 20 kN. Find the state of strain and the volume strain and the actual volume change. Take $E = 200 \text{ GN/m}^2$, $\nu = 0.25$.

3.2. A piece of steel is subject to principal stresses σ_1, σ_2, σ_3 of 100, −60, 0 (units MN/m^2) respectively. (a) State the greatest shear stress in the material and find the state of strain and the volume strain. (b) If σ_2 is changed to −100 MN/m^2 find the new volume strain.

3.3. The bar of Problem 3.1 is subjected to axial compressive forces of 12 kN and a lateral fluid pressure p. (a) Find the change in length of the bar if p is made great enough to prevent any lateral strain. Find this value of p. (b) Find the change in length when $p = 0$.

3.4. Calculate the state of strain corresponding to the state of stress

$$
\begin{pmatrix} 200 & 100 & 0 \\ 100 & 100 & 0 \\ 0 & 0 & 0 \end{pmatrix} \text{MN/m}^2
$$

Find the principal stress array and the principal strain array. State the maximum shear stress. Take $E = 200 \text{ GN/m}^2$, $\nu = 0.25$.

3.5. Repeat Problem 3.4 but with $\sigma_z = -100 \text{ MN/m}^2$.

3.6. A sample of the material of Problem 3.2 yielded at 210 MN/m^2 in a tensile test. State the safety factor in parts (a) and (b) of that problem. If the loads are increased until yielding occurs calculate the volume strain at yield in each of cases (a) and (b).

3.7. Referring to Problem 1.13 (Chapter 1) determine the states of strain in the rod and the tube if they are steel.

3.8. Referring to Problem 2.3(a) (Chapter 2) find the state of stress if the material is steel, and $\sigma_z = 0$.

3.9. Rework Problem 3.4 using matrix notation.

4

Strain Energy

4.1 Strain Energy

4.1.1 *Introduction*

Consider the bar of length L and cross-sectional area A shown in Figure 4.1a when subject to uniaxial tension (see Figure 1.14a) with state of stress

$$\mathbf{T} = \begin{pmatrix} \sigma_x & 0 & 0 \\ 0 & 0 & 0 \\ 0 & 0 & 0 \end{pmatrix} \qquad (A4.1)$$

The ends of the bar are therefore subject to opposed forces X of magnitude $A\sigma_x$ acting in the x direction. Under stress the bar strains and its change of length u in the x direction is the product $L\epsilon_x$ of the normal strain and the original length. Now for a linear elastic material equation (3.3a) shows that

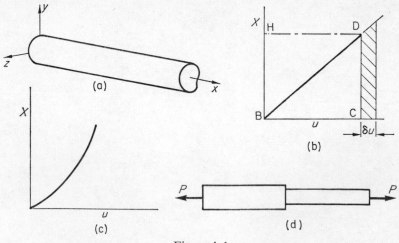

Figure 4.1

$\epsilon_x = \sigma_x/E$ since σ_y and σ_z are zero. Therefore we can write

$$X = A\sigma_x = AE\epsilon_x = \frac{AE}{L}u = ku \qquad (4.1)$$

where $k = AE/L$ may be termed a stiffness coefficient. From equation (4.1) we see that for the linear elastic material the applied forces X are proportional to their relative displacement u, as illustrated in Figure 4.1b. Now let the displacement u corresponding to forces X be increased an infinitesimal amount δu. Then the work done by the forces during this displacement is the hatched area in Figure 4.1b. To a first approximation this increment δW_e of work is given by $X\delta u$, and in the limit as δu tends to zero we have

$$dW_e = Xdu \qquad (4.2)$$

The total work done by the applied forces X in producing the final relative displacement u is therefore

$$W_e = \int_0^u Xdu \qquad (4.3)$$

For the linear elastic case, this quantity is the triangular area BCD in Figure 4.1b, of magnitude $\frac{1}{2}Xu$, and so, substituting for X and u, we can write

$$W_e = \tfrac{1}{2}Xu = \tfrac{1}{2}\sigma_x\epsilon_x AL \qquad (4.4)$$

Although the bar also has strains ϵ_y and ϵ_z in the y and z directions given by equations (3.3b and c) no work is done since there are no forces in these directions. Therefore the quantity in equation (4.4) is the total work done on the bar and by the Law of Conservation of Energy this work is either stored in the body or dissipated. Now in the case of an elastic material the original dimensions can be recovered by removing the loads thereby following the path DB in Figure 4.1b during which the body does work $-\frac{1}{2}Xu$ on the loads. Thus the work $\frac{1}{2}Xu$ done in attaining the relative displacement u must be stored in the body as a form of potential energy. This is called *strain energy* and can be interpreted physically as the internal energy gained when the distortion of the material causes the atoms to be moved against the interatomic force fields. Denoting the strain energy by the general symbol U, we can now write

$$U = W_e \qquad (4.5)$$

where W_e is given by (4.3) for any elastic material (e.g. Figure 4.1c), and by (4.4) for a linear elastic material. It should be noted that in the above it is assumed that the loads are applied in such a way that the system is always in equilibrium (see also Article 4.2).

Combining equations (4.4) and (4.5) and dividing by the volume AL of the bar we obtain

$$U_0 = \tfrac{1}{2}\sigma_x\epsilon_x \qquad (4.6)$$

where U_0 is the strain energy per unit volume, or *strain energy density*. In the particular case discussed above U_0 was uniform, i.e. constant throughout the body.

By substituting $\epsilon_x = \sigma_x/E$ in equation (4.6) the strain energy density in this case can be expressed in terms of stress as

$$U_0 = \frac{\sigma_x^2}{2E} \tag{4.7}$$

Example 4.1 A steel rod is 1 m long, and of uniform cross-sectional area 10^{-3} m^2. The rod is in simple tension as in Figure 1.14a. Calculate the strain energy when (a) $\sigma_x = 100$ MN/m^2. (b) $\sigma_x = 200$ MN/m^2. Take $E = 210$ GN/m^2.

Solution (a) Since $\sigma_y = \sigma_z = 0$ we have from equation (3.3a)

$$\epsilon_x = \sigma_x/E = 100 \times 10^6/(210 \times 10^9) = 0{\cdot}476 \times 10^{-3}$$

Then from (4.6) we obtain the strain energy density as

$$U_0 = \tfrac{1}{2} \times 100 \times 10^6 \times 0{\cdot}476 \times 10^{-3} = 23{\cdot}8 \times 10^3 \text{ Nm/m}^3$$

This is the same throughout the rod, which has volume 10^{-3} m^3, and so the strain energy is 23·8 Nm.

(b) Here $\epsilon_x = 200 \times 10^6/(210 \times 10^9) = 0{\cdot}952 \times 10^{-3}$ and $U = 95{\cdot}2$ Nm.

Note that doubling the stress quadruples the strain energy since the strain is also doubled. Appreciation of the amount of energy stored in the rod may be helped by noting that if it were raised 1·23 m it would gain the same amount of gravitational potential energy.

The strain energy of a bar in tension can be written directly in terms of the tensile forces X by combining (4.4) and (4.5) to give

$$U = \tfrac{1}{2}Xu$$

Using equation (4.1) we obtain the equivalent forms

$$U = \tfrac{1}{2}ku^2 \text{ and } U = \frac{\tfrac{1}{2}X^2}{k} \tag{4.8a, b}$$

where

$$k = \frac{AE}{L} \tag{4.9}$$

Example 4.2 Figure 4.1d shows an assembly consisting of two bars of equal length and cross-sectional areas $3A$ and A. Find an expression for the strain energy and compare with that stored in a uniform bar of the same total length and volume.

Solution The strain energy of the assembly is the sum of the strain energies of its parts. Thus, since both bars are subject to tensile forces P, we have, using (4.8) and (4.9),

$$U_{(i)} = \frac{P^2(0\cdot 5L)}{2(3A)E} + \frac{P^2(0\cdot 5L)}{2(A)E} = \frac{P^2L}{3AE}$$

A uniform bar of length L and the same volume has area $2A$ and so

$$U_{(ii)} = \frac{P^2L}{4AE}$$

Then

$$U_{(i)} : U_{(ii)} = 4 : 3$$

This result is another example of the effect of the dependence of strain energy on the square of the stress (equation 4.7).

Problems for Solution   4.1–4.4

4.1.2 *General Expression (Linear Elastic Material)*

We begin by determining the strain energy of an infinitesimal element of dimensions dx, dy, dz (Figure 4.2a) when subject to shearing stress in the xy plane only, as in Figure 1.6d. These stresses cause shear strain γ_{xy} in the xy plane as shown in Figure 4.2b. Shear stress and strain are proportional for a

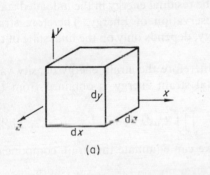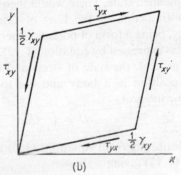

Figure 4.2

linear elastic material (equation 3.3d) and therefore as in Section 4.1.1, the work done on the element is half the product of the forces and their relative displacements, i.e.

$$\tfrac{1}{2}(\tau_{xy}\, dy\, dz)\,(\tfrac{1}{2}\gamma_{xy}\, dx) + \tfrac{1}{2}(\tau_{yx}\, dx\, dz)\,(\tfrac{1}{2}\gamma_{xy}\, dy)$$

Since $dx \times dy \times dz$ is the volume of the element the strain energy density is

$$U_0 = \tfrac{1}{2}\tau_{xy}\gamma_{xy} \qquad\qquad (4.10)$$

We can now generalize expressions (4.6) and (4.10) to obtain the strain energy density under general loading conditions for which the stress array has the form

$$\begin{pmatrix} \sigma_x & \tau_{xy} & \tau_{xz} \\ \tau_{yx} & \sigma_y & \tau_{yz} \\ \tau_{zx} & \tau_{zy} & \sigma_z \end{pmatrix} \qquad\qquad (A4.2)$$

The corresponding strain array is (A2.1) (Article 2.5).

Let us for the moment assume that the load system is applied in such a way that all the stress components increase together in the same proportion until the final values are attained. The strains then vary in the same proportions according to the stress/strain equations (3.3). A load displacement graph of the form of Figure 4.1b could therefore be drawn for each normal stress or set of shear stresses, and the total strain energy of the element is then

$$dU = U_0 \, dx \, dy \, dz \tag{4.11}$$

where

$$U_0 = \tfrac{1}{2}(\sigma_x \epsilon_x + \sigma_y \epsilon_y + \sigma_z \epsilon_z + \tau_{xy} \gamma_{xy} + \tau_{yz} \gamma_{yz} + \tau_{xz} \gamma_{xz}) \tag{4.12}$$

While it is convenient for the stresses to be applied in the manner described above, this is not necessary. For instance if we first apply σ_x and then all the other stresses simultaneously, the strain energy density is still given by equation (4.12). If this were not so the final strain energy could depend on the sequence of stressing and it would be possible to store energy on a 'high-energy' path to the loaded state and then use a 'low-energy' path to return to the unloaded state. There would then be residual energy in the unloaded state, in contradiction of the Law of Conservation of Energy. Therefore strain energy, being a form of potential energy, depends only on the final state of the body as expressed by equation (4.12).

In general, the state of stress and therefore the strain energy density vary with position in a body and the total strain energy is obtained from the volume integral

$$U = \int_V U_0 \, dV = \iiint U_0 \, dx \, dy \, dz \tag{4.13}$$

Substituting from equations (3.3) we can eliminate the strain components from (4.12) giving

$$U_0 = \frac{1}{2E}((\sigma_x^2 + \sigma_y^2 + \sigma_z^2) - 2\nu(\sigma_x \sigma_y + \sigma_y \sigma_z + \sigma_z \sigma_x)) + \frac{1}{2G}(\tau_{xy}^2 + \tau_{yz}^2 + \tau_{xz}^2) \tag{4.14}$$

The dependence of the strain energy density on products of stress components should be noted and comparison made with the solution of Example 4.1.

If stresses are eliminated using equations (3.4) we obtain

$$U_0 = \tfrac{1}{2}\lambda\Delta^2 + G(\epsilon_x^2 + \epsilon_y^2 + \epsilon_z^2) + \tfrac{1}{2}G(\gamma_{xy}^2 + \gamma_{yz}^2 + \gamma_{xz}^2) \tag{4.15}$$

where $\Delta = (\epsilon_x + \epsilon_y + \epsilon_z)$ and λ is Lamé's elastic constant (equation 3.5). In this form we see that the strain energy density and hence the strain energy is always a positive quantity. If for example the signs of the stress components in a stress array were all reversed the strain energy would be unchanged.

Equations (4.14) and (4.15) are intended only for reference. The basic expression (4.12) taken in conjunction with the strain/stress equations is sufficient for most purposes.

Example 4.3 A thin-walled tube 2 m long and of internal and external diameters 100 mm and 104 mm respectively is in the uniform state of stress expressed by the array

$$\begin{pmatrix} 100 & 100 & 0 \\ 100 & 0 & 0 \\ 0 & 0 & 0 \end{pmatrix} \text{MN/m}^2$$

Calculate the strain energy (a) directly (b) by first finding the principal stress array. $E = 210 \text{ GN/m}^2$; $\nu = 0.3$.

Solution (a) From equation (3.8) $G = 80.8 \times 10^9 \text{ N/m}^2$.
Using the stress/strain equations

$$\epsilon_x = 0.476 \times 10^{-3}; \quad \epsilon_y = \epsilon_z = -0.3\,\epsilon_x;$$
$$\gamma_{xy} = 1.237 \times 10^{-3}; \quad \gamma_{yz} = \gamma_{xz} = 0$$

Hence using (4.12) $U_0 = 85.7 \times 10^3 \text{ Nm/m}^3$.
Since the state of stress is the same throughout the tube

$$U = U_0 \times (\text{volume of tube}) = 110 \text{ Nm}$$

(b) By drawing a Mohr Circle as described in Article 1.10 we find the principal stress array as

$$\begin{pmatrix} 161.7 & 0 & 0 \\ 0 & -61.7 & 0 \\ 0 & 0 & 0 \end{pmatrix} \text{MN/m}^2$$

Determining strains and substituting in (4.12) we find $U_0 = 85.7 \times 10^3 \text{ Nm/m}^3$ as in part (a) as is to be expected since the strain energy for a given system cannot depend on the particular choice of coordinate axes.

Example 4.4 A body is in the uniform state of stress expressed by the array

$$\begin{pmatrix} 100 & 0 & 0 \\ 0 & \sigma_y & 0 \\ 0 & 0 & 0 \end{pmatrix} \text{MN/m}^2$$

Find the limiting values of strain energy density at the onset of failure using the maximum shear stress theory of failure. $E = 100 \text{ GN/m}^2$; $\nu = 0.25$; $\tau_0 = 80 \text{ MN/m}^2$.

Solution Since all shear stresses are zero the given array is for the principal axes. Therefore by result (1.12)

$$\tau_{max} \text{ is the greatest of } \left|\frac{100 - \sigma_y}{2}\right| \text{ or } \left|\frac{\sigma_y - 0}{2}\right| \text{ or } \left|\frac{0 - 100}{2}\right|$$

Then since $\tau_{max} \leqslant \tau_0 = 80 \text{ MN/m}^2$ the critical conditions are

$$80 = \tfrac{1}{2}|100 - \sigma_y| \quad \text{or} \quad 80 = \tfrac{1}{2}|\sigma_y| \tag{i}$$

Case (a) When $\sigma_y > 0$ the second condition is critical and $\sigma_y \leqslant 160 \text{ MN/m}^2$. Using

$\sigma_y = 160$ MN/m² and proceeding as in the previous example we find

$$U_0 = 138 \times 10^3 \text{ Nm/m}^3$$

Case (b) When $\sigma_y < 0$ the first of conditions (i) is critical and $\sigma_y \geqslant -60$ MN/m². Then

$$U_0 = 83 \times 10^3 \text{ Nm/m}^3$$

We note that at failure according to the maximum shear stress theory the strain energy density does not have a unique value.

Problems for Solution 4.5, 4.6

4.1.3 *Matrix Forms for Strain Energy*

The strain energy density expressed by equation (4.12) can be written in matrix form as follows

$$U_0 = \tfrac{1}{2}[\sigma_x \quad \sigma_y \quad \sigma_z \quad \tau_{xy} \quad \tau_{yz} \quad \tau_{xz}] \begin{bmatrix} \epsilon_x \\ \epsilon_y \\ \epsilon_z \\ \gamma_{xy} \\ \gamma_{yz} \\ \gamma_{xz} \end{bmatrix} \qquad (4.16)$$

i.e.

$$U_0 = \tfrac{1}{2}\boldsymbol{\sigma}^T \boldsymbol{\epsilon} \qquad (4.17)$$

where

$$\boldsymbol{\epsilon} = \{\epsilon_x \quad \epsilon_y \quad \epsilon_z \quad \gamma_{xy} \quad \gamma_{yz} \quad \gamma_{xz}\} \qquad (4.18)$$

$$\boldsymbol{\sigma} = \{\sigma_x \quad \sigma_y \quad \sigma_z \quad \tau_{xy} \quad \tau_{yz} \quad \tau_{xz}\} \qquad (4.19)$$

and row matrix $\boldsymbol{\sigma}^T$ is the transpose of column matrix $\boldsymbol{\sigma}$. Note that braces { } are used for a column matrix written as a row matrix to save space, e.g. (4.18) (see Appendix II).

Equation (4.16) contains the product of a (1×6) and a (6×1) matrix, the result of which is a (1×1) matrix, i.e. a scalar quantity, as is expected for energy.

Stress or strain could be eliminated from equations (4.17) using equations (3.3) or (3.4) written in matrix form. For instance we could write (3.3) in the form $\boldsymbol{\epsilon} = \mathbf{C}\boldsymbol{\sigma}$ where $\boldsymbol{\epsilon}$ and $\boldsymbol{\sigma}$ are as in (4.18, 4.19) and $\mathbf{C}$ is a square matrix of material properties. The particular form of $\mathbf{C}$ for plane stress is shown in equations (3.13). Then (4.17) becomes

$$U_0 = \tfrac{1}{2}\boldsymbol{\sigma}^T \mathbf{C}\boldsymbol{\sigma} \qquad (4.20)$$

Example 4.5 As an illustration we rework part (a) of Example 4.3.

This is a plane stress situation and

$$\boldsymbol{\sigma} = \{\sigma_x \quad \sigma_y \quad \tau_{xy}\} = \{10^8 \quad 0 \quad 10^8\}$$

Taking **C** from equations (3.13) and substituting we have

$$U_0 = \frac{1}{2E} [10^8 \quad 0 \quad 10^8] \begin{bmatrix} 1 & -0.3 & 0 \\ -0.3 & 1 & 0 \\ 0 & 0 & 2.6 \end{bmatrix} \begin{bmatrix} 10^8 \\ 0 \\ 10^8 \end{bmatrix} = 85.7 \text{ kNm/m}^3.$$

Problem for Solution 4.7

4.2 Non-equilibrium Loading

As was pointed out in Section 4.1.1 the expressions for strain energy in terms of work done were derived on the assumption that the system is in equilibrium at all stages of the loading process. As is shown by Figure 4.1b this means that for a linear elastic material, load must be proportional to deformation. Now let us examine what happens if the load is applied in some other way; suppose, for example that a tensile load equal to the final value X_1 is applied instantaneously to the bar of Figure 4.1a, and then held constant. The bar begins to deform and during the first increment δu of elongation the load does work $X_1 \delta u$ (Figure 4.3a). However, as we have seen, the strain

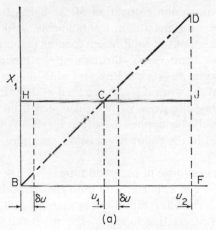
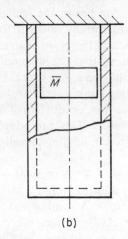

Figure 4.3

energy of the material increases only in proportion to the elongation as represented by the chain dashed line in Figure 4.3a. The strain energy stored during the increment δu can be obtained by substituting in equation (4.8a). Clearly the work done by the load during this increment is not all absorbed as strain energy. Therefore if energy is to be conserved, the rest of the work done must be absorbed as kinetic energy of the bar.

Thus we see that when the load is applied in a non-equilibrium manner acceleration of the bar takes place. During subsequent increments of deformation the work done continues to exceed the strain energy stored in the bar until point C in Figure 4.3a is attained at some elongation u_1. In the next increment δu we see that the work done by the load is less than the energy stored as strain energy. The difference is therefore abstracted from the kinetic energy, i.e. deceleration begins. This process continues until point D is attained at which the total work done is equal to the strain energy stored and so the bar is at rest. Equating the areas BDF and BHJF we see that $u_2 = 2u_1$. However, by comparison with Figure 4.1b, we see that u_1 is the *equilibrium* elongation of the bar due to load X_1. Thus when the load is applied in the manner of Figure 4.3a the maximum elongation is twice that attained when the load is gradually increased from zero as in Figure 4.1b.

Although the bar has come to rest at D of Figure 4.3a, it is not in equilibrium; since $u_2 > u_1$ the bar has excess potential energy and will start to accelerate into contraction. If conditions are frictionless the bar will in fact execute a vibration between limits of $u = 0$ and $u = 2u_1$. However, under real conditions the vibration will decay more or less quickly until the bar adopts the equilibrium deformation u_1, the surplus work represented by the area BHC being dissipated.

From the point of view of stress analysis the significance of the above discussion lies in the fact that the maximum elongation of the bar is twice the equilibrium value. It follows that the maximum instantaneous value of strain, and therefore of stress, is twice that for equilibrium loading (assuming of course that the material remains linear elastic throughout the process). It should be noted that this conclusion was reached by first obtaining u_2 by equating strain energy of the bar and the work done by the load at the instantaneous rest point (D in Figure 4.3a); the stress is then calculated from the strain u_2/L. This procedure can easily be applied to other forms of non-equilibrium loading as will be shown in the following classical example.

Example 4.6   A mass $\bar{M} = 100$ kg is a loose fit in a steel tube (Figure 4.3b) and is allowed to fall a height $H = 0 \cdot 1$ m under gravity onto the sealed base of the tube. Assuming that the base is rigid and that the tube material remains linear elastic, assess the maximum stress developed in the tube if it has length 1 m and cross-sectional area 1000 mm². $E = 210$ GN/m².

Solution   Since the base is rigid all the energy of the falling mass is absorbed in elongating the tube. Let u be the elongation of the tube when the mass is first brought to rest. Then the work done on the tube is the loss of potential energy of the mass, i.e. $\bar{M}(H+u)$g, and this has been wholly absorbed as strain energy of the bar. Therefore using (4.8a) for the strain energy we have

$$\bar{M}(H+u)\text{g} = \tfrac{1}{2}ku^2$$

Substituting numerical quantities and solving we obtain $u = 0 \cdot 976 \times 10^{-3}$ m.

Hence the strain is

$$\epsilon_x = \frac{u}{L}$$

and the stress is

$$\sigma_x = E\epsilon_x = 205 \text{ MN/m}^2$$

When, after the impact energy has been dissipated, the system attains equilibrium the stress is

$$(\sigma_x)_{eq} = \frac{\bar{M}g}{A} = 0.98 \text{ MN/m}^2$$

Thus we see that under impact loading conditions very large stresses may be developed (in the above example about 200 times the equilibrium stress). However it must be emphasized that the argument given is an oversimplification; in particular we have of course neglected the effect of the mass of the elastic material. Nevertheless the procedure described does permit an assessment of the order of the effect, and can also be adapted to a situation in which a mass $\bar{M}$ moving with velocity v is brought to rest by impact with an elastic bar such as the tube of Example 4.6. In this case the kinetic energy $\frac{1}{2}\bar{M}v^2$ is assumed converted to strain energy $\frac{1}{2}ku^2$. Equating these expressions we can find the axial deformation u and hence the stress in the bar.

Problems for Solution 4.8–4.10

4.3 Complementary Energy

Consider the bar of Figure 4.1a subject again to tensile forces X giving a uniform state of stress as in array (A4.1) with $\sigma_x = X/A$. The relative displacement of the ends, i.e. the elongation of the bar, is $u = \epsilon_x L$ as before. Let us suppose, to begin with, that the material of the bar is non-linear elastic (see line 3 of Figure 3.3a) so that σ_x is not proportional to ϵ_x. We might then obtain a graph of tensile force against elongation as in Figure 4.4a. Now if X is increased by a small amount δX the area above the graph increases by the amount shown hatched, which, to a first approximation, is

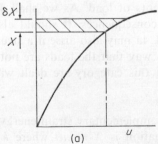

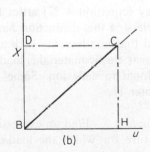

(a)   (b)

Figure 4.4

$u\delta X$. Denoting this by $\delta \overline{W}_e$, then in the limit, we obtain

$$d\overline{W}_e = u\,dX \qquad (4.21)$$

The total area above the graph is then

$$\overline{W}_e = \int_0^X u\,dX \qquad (4.22)$$

This quantity $\overline{W}_e$ is called the complementary work done by the loads on the bar. We now introduce a quantity $\overline{U}$ called the complementary energy of the body and define it as being equal to the complementary work done when the loads are applied in an equilibrium manner, i.e.

$$\overline{U} = \overline{W}_e \qquad (4.23)$$

This statement is therefore tantamount to the adoption of a 'Law of Conservation of Complementary Energy'. For the moment complementary quantities must appear fictitious, but in Section 12.5.1 we shall be able to give some physical interpretation.

It is clear from Figure 4.4a that the complementary energy due to a given set of loads is not, in general, equal to the strain energy. However, in the particular case of a linear elastic material (Figure 4.4b) the complementary energy (area BCD) is equal to the strain energy (area BCH). Therefore in this particular case we shall be able to *evaluate* the complementary energy by using the corresponding expressions for strain energy. In particular the complementary energy per unit volume (i.e. the complementary energy density $\overline{U}_0$) in a *linear elastic* material is obtained using equation (4.12) as

$$\overline{U}_0 = \tfrac{1}{2}(\sigma_x\epsilon_x + \sigma_y\epsilon_y + \sigma_z\epsilon_z + \tau_{xy}\gamma_{xy} + \tau_{yz}\gamma_{yz} + \tau_{xz}\gamma_{xz}) \qquad (4.24)$$

The total complementary energy of an extensive body is the summation over all volume elements i.e.

$$\overline{U} = \int_V \overline{U}_0\,dV \qquad (4.25)$$

Despite the numerical equality of U and $\overline{U}$ for a linear elastic material it is not trivial to maintain the distinction because strain energy is defined through (4.3) in terms of increments of displacement whereas complementary energy (equation 4.22) arises from increments of load. As we shall see in Article 4.4 this distinction has important consequences. It should also be noted that non-linearity such as in Figure 4.4a may also arise if a structure of linear elastic material is loaded in such a way that the loads are not proportional to deflexion. Some situations in this category are dealt with in Chapter 11.

Example 4.7 Find an expression for the complementary strain energy of a structure for which the load/deflexion equation is $X = ku^3$ where k is a constant.

Solution   Applying equations (4.22, 4.23) we have

$$\bar{U} = \int_0^X u\,\mathrm{d}X = \int_0^X (X/k)^{1/3}\,\mathrm{d}X = 0{\cdot}75 X^{4/3}/k^{1/3}$$

Problem for Solution   4.11

4.4   Deflexion Theorem (Castigliano's Theorem)

Consider a body subject to a set of applied loads F_1, F_2, ... F_i, ... F_n (see Figure 4.5a) and maintained in equilibrium by supports which also prevent displacement as a rigid body. The applied loads may be either forces or

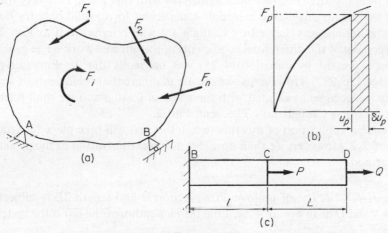

Figure 4.5

moments as shown. Suppose that one of the loads, say F_i, is increased a small amount δF_i. Then from Figure 4.4a we see that the complementary work done by the load F_i is, to a first approximation, given by

$$u_i \delta F_i \tag{i}$$

where u_i is the displacement of the point of application of load F_i. This displacement is a translation or rotation according to whether F_i is a force or a moment.

The load increment δF_i will also cause increments of displacement at the points of application of the other loads. For example the displacement of load F_p increases by δu_p as shown in Figure 4.5b. During this displacement *real* work is done, equal to the hatched area in the figure, but there is no complementary work since F_p does not change. (Refer again to the definition of an increment of complementary work in equation (4.21).) Thus the quantity (i) above is the total complementary work due to the increment δF_i

of load F_i. By equation (4.23) this is equal to the increment $\delta \bar{U}$ of complementary energy

i.e. $$\delta \bar{U} = u_i \delta F_i$$

Then in the limit as δF_i tends to zero we have

$$\frac{\partial \bar{U}}{\partial F_i} = u_i \qquad (4.26)$$

in which the partial derivative is written because $\bar{U}$ is a function of all the loads. Equation (4.26) states that the partial derivative of the complementary energy (expressed in terms of the applied loads) with respect to any load, is equal to the relative displacement associated with that load. This very useful result is valid for *any* elastic material. A particular form, valid only for *linear* elastic materials, was derived by Castigliano[1] who referred to it as Part 2 of his 'Theorem of the differential coefficients of the internal work'. The generalization expressed by equation (4.26) was originally due to Engesser (see reference 2, p. 292). However, since the use of differential coefficients of strain energy has become associated with the name of Castigliano we shall refer to result (4.26) as Castigliano's Theorem: Part 2.

The main application of this theorem in this text will take place in Articles 7.2 and 7.3. However, we shall now give a simple illustration of the power of this method.

Example 4.8   A bar of uniform cross-section A and length $2L$ is subject to forces P and Q as in Figure 4.5c. Find the elongation of the bar if the material is linear elastic.

Solution   For equilibrium the support at B must exert a force $(P + Q)$ to the left. Then it is apparent that part BC is in tension of magnitude $(P + Q)$ and part CD in tension of value Q. Using (4.8) and (4.9) we have

$$U = ((P + Q)^2 L + Q^2 L)/(2AE)$$

For a linear elastic material $U = \bar{U}$ and so applying (4.26) the displacement of Q, which is the same as the elongation of the bar, is

$$\frac{\partial \bar{U}}{\partial Q} = \frac{(P + 2Q)L}{(AE)}$$

Problems for Solution   4.12, 4.13

4.5   Strain Energy of Distortion

By the method of superposition (Article 1.5) a general state of stress can be expressed as follows (all arrays being measured at the same point with respect to the same coordinate axes).

$$\begin{pmatrix} \sigma_x & \tau_{xy} & \tau_{xz} \\ \tau_{yx} & \sigma_y & \tau_{yz} \\ \tau_{zx} & \tau_{zy} & \sigma_z \end{pmatrix} = \begin{pmatrix} \bar{\sigma} & 0 & 0 \\ 0 & \bar{\sigma} & 0 \\ 0 & 0 & \bar{\sigma} \end{pmatrix} + \begin{pmatrix} s_x & \tau_{xy} & \tau_{xz} \\ \tau_{yx} & s_y & \tau_{yz} \\ \tau_{zx} & \tau_{zy} & s_z \end{pmatrix}$$

i.e.
$$\mathbf{T} = \mathbf{T_1} + \mathbf{T_2}$$
where
$$\sigma_x = \bar{\sigma} + s_x; \quad \sigma_y = \bar{\sigma} + s_y; \quad \sigma_z = \bar{\sigma} + s_z \tag{4.27}$$
Here
$$\bar{\sigma} = (\sigma_x + \sigma_y + \sigma_z)/3 \tag{4.28}$$

is the mean normal stress at the point, and s_x, s_y, s_z are the deviations of the normal stresses from the mean value. Referring to equation (3.6) we see that the state of stress $\mathbf{T_1}$ causes volume strain but $\mathbf{T_2}$ does not since the sum of its normal stresses $(s_x + s_y + s_z)$ is zero, as is easily confirmed by adding equations (4.27). On the other hand because of result (1.12) no shearing stresses are associated with $\mathbf{T_1}$ which cannot therefore cause any distortion. We therefore conclude that $\mathbf{T_1}$ causes only volume change and $\mathbf{T_2}$ only distortion.

The total strain energy can now be considered as the sum of the energy stored in volume change, i.e. dilatation, and that stored in distortion. The dilatational strain energy density is found by substituting $\mathbf{T_1}$ in equation (4.14)

i.e.
$$(U_0)_{\text{dil}} = \frac{3\bar{\sigma}^2(1-2\nu)}{2E}$$

The distortional strain energy density can also be found by substituting $\mathbf{T_2}$ in (4.14). The algebra is more easily performed if the state of stress is expressed relative to the principal axes, when we obtain finally,

$$(U_0)_{\text{dist}} = \frac{1+\nu}{6E} \left((\sigma_1 - \sigma_2)^2 + (\sigma_2 - \sigma_3)^2 + (\sigma_3 - \sigma_1)^2 \right) \tag{4.29}$$

This quantity has been made the basis of a hypothesis of failure (see Section 5.6.1).

4.6 St. Venant's Principle

This principle, which was first expressed by St. Venant, is concerned with the effects of loadings which are statically equivalent to zero. To illustrate what is meant by such a loading we consider the case of a bar in simple tension due to forces of magnitude P. In Figure 4.6a this tension is applied by concentrated axial forces, and in Figure 4.6b by forces distributed uniformly over the ends. Clearly, the stress distributions in the bar will not be the same in these two cases although from the point of view of simple statics the load systems

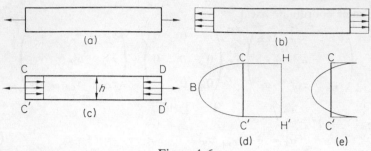

Figure 4.6

are equivalent. Now the actual difference between the load systems can be represented by the combination of axial tensile forces P and distributed compressive forces P shown in Figure 4.6c. (This is easily proved by adding Figure 4.6c to Figure 4.6b to obtain Figure 4.6a.) Furthermore, the loads on each end of the bar in Figure 4.6c have zero resultant. We therefore conclude that the difference between the actual load systems (Figures 4.6a and b) is represented by a load system (Figure 4.6c) which is 'statically equivalent to zero force and moment'.

Situations similar to that described above occur frequently in practice and St. Venant asserted that the effect of loadings statically equivalent to zero force and moment is negligible beyond a distance from the loaded area comparable to its own dimensions. In the case of the bar in tension this means that the stresses due to the load system of Figure 4.6c can be considered negligible at distances from the ends greater than h. Consequently there is no significant difference between the stress distributions due to concentrated or distributed tensile forces except close to the ends.

St. Venant held that the principle was justified by experience. With modern techniques its validity can be demonstrated experimentally and also by numerical methods (see for example Figure 13.9). In addition Goodier[3] has pointed out how this result can be predicted from assessment of the strain energy of the bar.

For example consider the system of Figure 4.6c and let σ be the order of magnitude of the stresses in the bar due to this load system which is statically equivalent to zero force and moment. Then the forces applied to the ends of the bar are of the order σh^2. To find the work done by these forces we must have an estimate of the distances through which they move. To obtain this we hold DD' so that rigid body displacement of the bar is prevented. Then we would expect that the displacements of end CC' due to the distributed compressive forces and to the concentrated tensile force, would be of the form of CHH'C' and CBC', respectively, in Figure 4.6d. The net displacement would then be of the form of Figure 4.6e, and its inclination to CC' at the points of zero displacement would be of the order of the local shear strain.

The shear stresses will be of the same order as σ and G is of the same order as E. Therefore the required rotation is of the order σ/E and so the displacement of the forces is of the order $h(\sigma/E)$. Finally the work done on the bar is of the order

$$\sigma h^2(\sigma h/E) \tag{i}$$

From equation (4.14) the strain energy density in the bar is of the order σ^2/E and so the strain energy is $((\sigma^2/E) \times \text{volume})$. Equating this to (i) we see that the volume involved in the stress distribution can only be of the order of h^3. Thus the stresses are negligible beyond a distance h from the end as required by St. Venant's Principle.

It should be noted that the principle does not apply to thin-walled tubular forms and statically determinate frames (see for example reference 4).

4.7 References

1. Castigliano, C. A. P., *The Theory of Equilibrium of Elastic Systems, and its Applications*, 1879, transl. by E. S. Andrews, Dover, New York, 1966.
2. Timoshenko, S. P., *History of Strength of Materials*, McGraw-Hill, New York, 1953.
3. Goodier, J. N., 'A general proof of St. Venant's Principle', *Phil. Mag.*, Ser. 7, **23**, 607; **24**, 325 (1937).
4. Hoff, N. J., 'The applicability of St. Venant's Principle to aircraft structures', *J. Aero. Sciences*, **12**, 455 (1945).

4.8 Problems for Solution

Steel: E, ν, ρ (210 GN/m², 0·3, 7900 kg/m³). Aluminium: E, ν, ρ (70 GN/m², 0·35, 2700 kg/m³).

4.1. Calculate the strain energy of the system of Problem 1.13 when at the permissible load, if the material is steel and the tube and rod have lengths of 1 m and 0·8 m respectively.

4.2. Find the strain energy stored in an aluminium rod having the same diameter and mass as the steel rod in Example 4.1 when the stress is 100 MN/m².

4.3. Find the stress σ_x for an aluminium rod to have the same strain energy as the steel rod of Example 4.1 part (a), both rods having the same dimensions.

4.4. A uniform rod has length L, cross-sectional area A, and is subject only to normal stresses $\sigma_x = bx$ where b is a constant. Obtain an expression for the strain energy.

4.5. Find the strain energy densities corresponding to the states of stress given in Problems 1.5b, and 1.8b and c if the material is steel.

4.6. Find the strain energy in the steel plate of Problem 1.15 when t has the smallest safe value.

4.7. Rework Problem 4.5 using matrix notation.

4.8. A weight carrier has a 6 mm diameter steel rod of length 250 mm and its flange can be assumed rigid. Find the height from which a mass of 10 kg can be dropped onto the flange without exceeding the yield stress of 240 MN/m².

4.9. A mass of 400 kg with velocity 1 m/sec is suddenly brought to rest by a 10 m long cable of 25 mm diameter. Estimate the greatest stress in the cable if $E = 100$ GN/m².

4.10. Repeat Problem 4.9 when the body is a ship of mass 10^6 kg and the cable is 0·1 m in diameter and 100 m long.

4.11. The load/deflexion relation for a structure is $P = ku^{0·8}$ where k is constant. Find $\bar{U}/U$.

4.12. A bar has the general arrangement of Figure 4.5c but CD is a tube with cross-section A, one-third that of BC. Find the elongation of the assembly.

4.13. Find the displacement at C for the system of Problem 4.12 when Q is such that the displacement at D is zero.

5

Bars: Direct Loading

5.1 Structures Formed of Bars

The simplest type of structure able to resist both moments and forces is a straight bar of uniform cross-section as in Figure 5.1. The dimensions of the bar in one direction are large compared with the others; an axis parallel to this direction and acting through the centre of area of the cross-section is called the longitudinal axis, denoted as x in the figure.

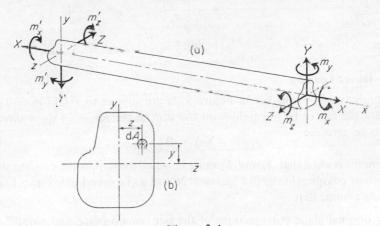

Figure 5.1

Forces acting on a bar can always be resolved into components X, Y, Z parallel to the x, y, z axes (Figure 5.1a); similarly, moments can be resolved into components m_x, m_y, m_z acting in planes normal to the x, y, z axes. The conditions for static equilibrium of a bar are listed in Appendix 1; for this purpose force components are considered positive when parallel to the positive direction of the corresponding axis; moments are positive if clockwise when viewed in the positive direction of the axis named in the subscript (i.e. the right-hand screw rule).

Systems of loads in static equilibrium may cause tension/compression, bending, or twisting of a bar; these effects are considered separately in Chapters 5–8, and in combination in Chapter 9. The behaviour of a great many engineering structures and components (including beams, shafts, frames, springs) can be investigated by considering them to consist essentially of bars loaded as discussed, or of combinations of bars. Some of the procedures developed in the analysis of systems of bars will be found in Chapter 13 to be adaptable to the treatment of bodies of more general shape. We begin our study by considering in this chapter the effect which we call direct loading and which is caused by forces acting along the longitudinal axis.

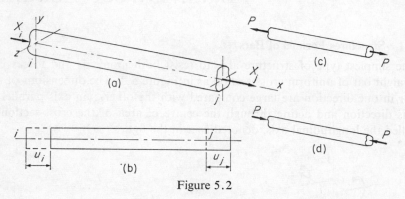

Figure 5.2

5.2 Direct Loading

The ends i and j of the bar in Figure 5.2a are subject to axial forces X_i and X_j. For the bar to be in equilibrium the sum of the forces in the x direction has to be zero, i.e.

$$\sum X = 0 \tag{5.1}$$

Hence it is clear that X_i and X_j must be equal and opposite, causing simple tension or compression as in Figures 5.2c and d. To investigate direct loading we now assume that

(a) original plane cross-sections of the bar remain plane and parallel after application of the loads
(b) σ_x is the only non-zero stress component.

End displacement is only allowed in the x direction and is represented by u. The displacements of the left- and right-hand ends are then denoted u_i and u_j (Figure 5.2b). Then the change of length (deformation) of the rod is (Article 2.2)

$$\Delta L = u_j - u_i$$

and the strain of the material in the x direction is

$$\epsilon_x = \frac{\Delta L}{L} = \frac{1}{L}(u_j - u_i) \tag{5.2}$$

Since the end cross-sections are assumed to remain parallel under load the strain ϵ_x is the same throughout the bar. By assumption (a) the state of stress is

$$\begin{pmatrix} \sigma_x & 0 & 0 \\ 0 & 0 & 0 \\ 0 & 0 & 0 \end{pmatrix} \tag{A5.1}$$

Using this information in the strain/stress equations (3.3) we find

$$\epsilon_x = \frac{\sigma_x}{E}; \quad \epsilon_y = \epsilon_z = \frac{-\nu\sigma_x}{E} \tag{5.3}$$

Also

$$\gamma_{xy} = \gamma_{yz} = \gamma_{xz} = 0$$

and so the state of strain is

$$\begin{pmatrix} \epsilon_x & 0 & 0 \\ 0 & \epsilon_y & 0 \\ 0 & 0 & \epsilon_z \end{pmatrix} \tag{A5.2}$$

Combining equations (5.2) and (5.3)

$$\sigma_x = \frac{E}{L}(u_j - u_i) \tag{5.4}$$

The states of stress and strain in the bar are uniform and so the total force on an end face (area A) is found by considering an element dA as in Figure 5.1b and is

$$\int_A \sigma_x dA \quad \text{i.e.} \quad A\sigma_x$$

The moment of the force on the element dA about the z axis is $((\sigma_x dA)y)$ and for the whole cross-section is

$$\sigma_x \int_A y dA$$

However

$$\int_A y dA = 0$$

since the z axis passes through the centre of area of the cross-section. A similar result holds for moment about the y axis. Thus the stress distribution on the ends due to the assumed displacements is statically equivalent to an axial force of magnitude $(A\sigma_x)$, as actually shown in Figure 5.2a. Although this concentrated force does not correspond with the distribution of force required for the assumed strain/stress system, it follows from St. Venant's Principle (see Article 4.6) that the effect of this difference is negligible at axial distances from the ends greater than a typical dimension of the cross-section. Therefore for slender bars the stress distribution may be assumed uniform throughout the bar regardless of the precise distribution of end force.

When the bar is in tension (compare Figure 5.2c) the normal stress at end j and the statically equivalent axial force both act in the same direction and so

$$X_j = A\sigma_x \qquad (5.5)$$

Example 5.1 A 2 m long steel tube has outer and inner diameters 0·2 m and 0·1 m. Find the direct loading to cause an axial contraction of 1 mm if $E = 210 \text{ GN/m}^2$.

Solution Here $u_j - u_i = -10^{-3}$ m and from (5.4) $\sigma_x = -105 \text{ MN/m}^2$. The cross-sectional area A is $\pi(0\cdot1^2 - 0\cdot05^2) \text{ m}^2$ and so from (5.5)

$$X_j = -2\cdot47 \text{ MN}$$

The force at end i is equal and opposite giving a system as in Figure 5.2d.

Failure Condition

Array (A5.1) is clearly a principal stress array and so utilizing equation (1.21) the Maximum Shear Stress Theory predicts failure of the bar when

$$|\sigma_x| = \sigma_0$$

where σ_0 is the stress at failure in a simple tensile test.

For instance if the material of the tube in Example 5.1 has $\sigma_0 = 210 \text{ MN/m}^2$ then the greatest possible compressive stress is -210 MN/m^2, and using (5.5), the greatest compressive load is $-4\cdot94$ MN.

Collinear Assembly

A set of bars joined together in series (e.g. Figure 5.3) will be called a collinear assembly.

Referring to Figure 5.3a the clamp exerts a force X_1 which for equilibrium (equation 5.1) must have magnitude P and act to the left; each bar clearly experiences tensile direct loading of magnitude P; stress and elongation can be obtained by application of equations (5.5) and (5.4). Now suppose that the same set of bars is clamped and loaded as in Figure 5.3b. The clamps exert forces X_1 and X_3 and for equilibrium

$$X_1 + P + X_3 = 0$$

It is therefore impossible to find the values of X_1 and X_3 from equilibrium considerations only; a system such as this is said to be statically indeterminate. Forces such as X_1 and X_3 exerted by clamps are known as reactive loads. The number of reactive loads in excess of one is called the degree of indeterminacy of the structure. The additional equations needed to determine the extra

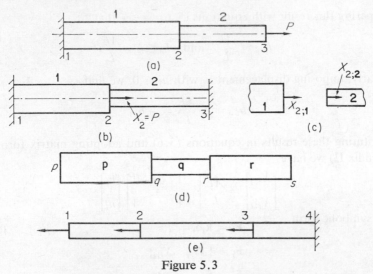

Figure 5.3

reactive loads are obtained by considering the deformation of the system. This is most conveniently carried out by a stiffness procedure which will be described in the rest of this chapter.

Problems for Solution 5.1, 5.2

5.3 Stiffness Equations for Direct Loading

5.3.1 *Stiffness Equations for One Bar*

Here we seek to determine equations which express the end forces on a bar in terms of the displacements of the ends. The end forces will now be denoted $X_{i;j}$ and $X_{j;j}$ where the second subscript j (printed in bold type) signifies that the force acts on the jth bar. Thus $X_{i;j}$ is the axial force on end i of the jth bar. It is clear from equation (5.1) that $X_{i;j} = -X_{j;j}$.

Then the required equations can be written in the form

$$X_{i;j} = b_{11}u_i + b_{12}u_j$$
$$X_{j;j} = b_{21}u_i + b_{22}u_j$$

(5.6)

The coefficients b_{11} etc. are stiffnesses, i.e. forces per unit displacement, and equations (5.6) are called the stiffness equations of the bar. Now if we impose end displacement u_j with $u_i = 0$, we have, using equations (5.5), (5.4) and (5.1)

$$X_{j;j} = \sigma_x A = \frac{AE}{L} u_j = -X_{i;j}$$

Comparing this result with equations (5.6) we see that

$$b_{22} = \frac{AE}{L} \quad \text{and} \quad b_{12} = -\frac{AE}{L}$$

Similarly, imposing displacement u_i with $u_j = 0$, we find

$$b_{21} = -\frac{AE}{L} = -b_{11}$$

Substituting these results in equations (5.6) and adopting matrix form (see Appendix II) we have

$$\begin{bmatrix} X_{i;\mathbf{j}} \\ X_{j;\mathbf{j}} \end{bmatrix} = \frac{AE}{L} \begin{bmatrix} 1 & -1 \\ -1 & 1 \end{bmatrix} \begin{bmatrix} u_i \\ u_j \end{bmatrix} \tag{5.7}$$

or in symbolic form

$$\mathbf{F_j} = \mathbf{k_j} \boldsymbol{\delta_j} \tag{5.7a}$$

In (5.7a)

$$\mathbf{F_j} = \{X_{i;\mathbf{j}} \quad X_{j;\mathbf{j}}\} \tag{5.8}$$

$$\boldsymbol{\delta_j} = \{u_i \quad u_j\} \tag{5.9}$$

are the column matrices of end loads and displacements respectively for the jth bar. The square matrix

$$\mathbf{k_j} = \frac{AE}{L} \begin{bmatrix} 1 & -1 \\ -1 & 1 \end{bmatrix} = \begin{bmatrix} k_{ii;\mathbf{j}} & k_{ij;\mathbf{j}} \\ k_{ji;\mathbf{j}} & k_{jj;\mathbf{j}} \end{bmatrix} \tag{5.10}$$

is symmetric and is called the element stiffness matrix. The quantity AE/L may be termed a stiffness parameter. Its value for the bar considered in Example 5.1 is $2 \cdot 47 \times 10^9$ N/m. Then if $u_i = 0$ and $u_j = -10^{-3}$ m we obtain from equations (5.7) $X_{j;\mathbf{j}} = -X_{i;\mathbf{j}} = -2 \cdot 47$ MN. If the bar were for example numbered **1** and its ends numbered 1 and 2 then these forces would be described as $X_{2;\mathbf{1}} = -X_{1;\mathbf{1}} = -2 \cdot 47$ MN.

In (5.10) the elements of the stiffness matrix have also been given formal designations $k_{ii;\mathbf{j}}$ etc. Referring to the stiffness equations (5.7) we see that for example $k_{ij;\mathbf{j}} = -AE/L$ is the stiffness coefficient which relates force at the i end of the bar to displacement at the j end. The bold-type subscript represents the number of the bar itself. If the bar of Example 5.1 is numbered as in the previous paragraph then

$$k_{ij;\mathbf{j}} = k_{12;\mathbf{1}} = -\frac{AE}{L} = -2 \cdot 47 \times 10^9 \text{ N/m}.$$

Problem for Solution   5.3

5.3.2 *Stiffness Equations for a Collinear Assembly*

A collinear assembly of bars has two or more joined together in series as in Figure 5.3; the bars are numbered in bold type. The restriction is imposed that external forces are only to be exerted at the ends of a bar. These positions are called nodes and are numbered in ordinary type. The forces exerted at the n nodes are denoted $X_1, X_2, \ldots, X_i, \ldots X_n$ and may be written as a column matrix or vector

$$\mathbf{F} = \{ X_1 \quad X_2 \ldots X_i \ldots X_n \} \tag{5.11}$$

where $\mathbf{F}$ represents the assembly nodal load vector. Similarly we can write an assembly vector of nodal displacements

$$\boldsymbol{\delta} = \{ u_1 \quad u_2 \ldots u_i \ldots u_n \} \tag{5.12}$$

Adopting a similar approach to that followed in Section 5.3.1 we now seek to represent the structural behaviour of the assembly by a set of linear equations expressing nodal loads in terms of nodal displacements.

$$X_1 = a_{11}u_1 + a_{12}u_2 + \ldots \quad \ldots + a_{1n}u_n$$

$$X_2 = a_{21}u_1 + a_{22}u_2 + \ldots \quad \ldots + a_{2n}u_n$$

$$\tag{5.13}$$

$$X_n = a_{n1}u_1 \quad a_{n2}u_2 \quad \ldots \quad \ldots + a_{nn}u_n$$

where a_{11} etc. are stiffness coefficients. These equations can be written in matrix form as

$$
\begin{bmatrix} X_1 \\ X_2 \\ \cdot \\ \cdot \\ \cdot \\ X_n \end{bmatrix}
=
\begin{bmatrix} a_{11} & a_{12} & \cdots & a_{1n} \\ a_{21} & a_{22} & \cdots & a_{2n} \\ \cdot & \cdot & & \cdot \\ \cdot & \cdot & & \cdot \\ \cdot & \cdot & & \cdot \\ a_{n1} & a_{n2} & \cdots & a_{nn} \end{bmatrix}
\begin{bmatrix} u_1 \\ u_2 \\ \cdot \\ \cdot \\ \cdot \\ u_n \end{bmatrix}
\tag{5.14}
$$

Using equations (5.11) and (5.12) these may be written symbolically as

$$\mathbf{F} = \mathbf{K}\boldsymbol{\delta} \tag{5.14a}$$

where $\mathbf{K}$ is a square matrix of stiffness coefficients, size $(n \times n)$, and is called the stiffness matrix of the assembly.

The assembly stiffness equations are built up by combining the stiffness equations of the individual bars in accordance with twin conditions of compatibility of displacement, and of equilibrium of forces, at the nodes. The former is a formal statement of the fact that the displacement of a node is the same as the displacement of the end(s) of the bar(s) forming the node. For example in Figure 5.3a the nodal displacement u_2 is the displacement of the right-hand end of bar **1** and of the left-hand end of bar **2**.

4

The condition of equilibrium is best introduced through an example. Consider node 2 of Figure 5.3b on which there is some nodal force X_2. The bar **1** has its ends at nodes 1 and 2 and from the discussion in Section 5.3.1 there is a force $X_{j;j}$ on its right-hand end. Here $j = 2$ and $\mathbf{j} = 1$ so that this force is $X_{2;1}$. Similarly the left end of bar **2** is subject to a force $X_{i;j}$ where $i = 2$ and $\mathbf{j} = 2$. These forces are shown in Figure 5.3c. If the system is to be in equilibrium the external force at the node must be the resultant of the forces exerted on the two bars meeting at the node, i.e.

$$X_2 = X_{2;1} + X_{2;2}$$

If we consider the general system in Figure 5.3d then at any node q

$$X_q = X_{q;\mathbf{p}} + X_{q;\mathbf{q}} \tag{5.15}$$

An equation of this type can be written for each node. It should be noted that if the external load X_q is zero then the end forces on the bars must be equal and opposite.

The end forces $X_{q;\mathbf{p}}$ and $X_{q;\mathbf{q}}$ appearing in equation (5.15) can now be eliminated by substituting in terms of displacements from the stiffness equations of the form (5.7) for the bars **p** and **q**. To illustrate this process consider the system of Figure 5.3b for which the nodes are 1, 2, 3 and the bars are **1** and **2**. Only bar **1** is present at node 1 and the nodal equilibrium equation is

$$X_1 = X_{1;1} \tag{5.16}$$

The stiffness equations for bar **1** can be written out formally using equations (5.7) and (5.10) with $i = 1, j = 2, \mathbf{j} = 1$. Then

$$\begin{bmatrix} X_{1;1} \\ X_{2;1} \end{bmatrix} = \begin{bmatrix} k_{11;1} & k_{12;1} \\ k_{21;1} & k_{22;1} \end{bmatrix} \begin{bmatrix} u_1 \\ u_2 \end{bmatrix} \tag{5.17}$$

and so (5.16) becomes

$$X_1 = k_{11;1}u_1 + k_{12;1}u_2 \tag{5.18}$$

At node 2 bars **1** and **2** are present and (5.15) becomes

$$X_2 = X_{2;1} + X_{2;2} \tag{5.19}$$

Using $i = 2, j = 3, \mathbf{j} = 2$ in (5.7) and (5.10) the stiffness equations of bar **2** can be written as

$$\begin{bmatrix} X_{2;2} \\ X_{3;2} \end{bmatrix} = \begin{bmatrix} k_{22;2} & k_{23;2} \\ k_{32;2} & k_{33;2} \end{bmatrix} \begin{bmatrix} u_2 \\ u_3 \end{bmatrix} \tag{5.20}$$

Substituting from this and from (5.17) in (5.19)

$$X_2 = k_{21;1}u_1 + (k_{22;1} + k_{22;2})u_2 + k_{23;2}u_3 \tag{5.21}$$

Similarly we can show that at node 3

$$X_3 = k_{32;2}u_2 + k_{33;2}u_3 \qquad (5.22)$$

Comparing equations (5.18), (5.21) and (5.22) with either (5.13) or (5.14) it is clear that for example

$$a_{11} = k_{11;1}; \qquad a_{22} = (k_{22;1} + k_{22;2}); \qquad a_{32} = k_{32;2}$$

Thus each assembly stiffness coefficient is the sum of all the coefficients having the same nodal subscripts in the stiffness equations of the bars. Writing (5.18), (5.21) and (5.22) in matrix form corresponding to equation (5.14a) we have

$$\begin{bmatrix} X_1 \\ X_2 \\ X_3 \end{bmatrix} = \begin{bmatrix} k_{11;1} & k_{12;1} & 0 \\ k_{21;1} & (k_{22;1} + k_{22;2}) & k_{23;2} \\ 0 & k_{32;2} & k_{33;2} \end{bmatrix} \begin{bmatrix} u_1 \\ u_2 \\ u_3 \end{bmatrix}$$

These are the assembly stiffness equations the square matrix being the assembly stiffness matrix **K**. It will be noted the assembly stiffness coefficients a_{13} and a_{31} in this example are both zero. This means that force and displacement at nodes 1 and 3 are not directly interrelated.

The above procedure of substitution in nodal equilibrium equations (5.15) is the fundamental method of forming the assembly stiffness equations. However the actual numerical procedure can be speeded as follows. The nodal load and displacement vectors are written directly as column matrices of size $(n \times 1)$ where n is the number of nodes. A square null matrix (see Appendix II) is formed of size $(n \times n)$. The stiffness matrices of the bars are computed using the first of identities (5.10). Then using the second of (5.10) the components of all the element stiffness matrices in turn are added to the components of **K** which have the same nodal subscripts.

Example 5.2 A steel and an aluminium bar each of 400 mm² cross-section, and length 1 m are joined to form a 2 m bar. If E is 210 GN/m² for steel and 70 GN/m² for aluminium write down the assembly stiffness equations.

Solution Let the steel bar be element **1** joining nodes 1 and 2, and the aluminium bar be element **2**, joining nodes 2 and 3. Since there are 3 nodes $F = \{X_1 \quad X_2 \quad X_3\}$ and $\delta = \{u_1 \quad u_2 \quad u_3\}$. The **K** matrix is formed as a 3×3 null matrix. For bar **1** AE/L is $400 \times 10^{-6} \times 210 \times 10^9$, i.e. 84×10^6 N/m and from (5.10)

$$k_1 = 10^6 \begin{bmatrix} 84 & -84 \\ -84 & 84 \end{bmatrix} = \begin{bmatrix} k_{11;1} & k_{12;1} \\ k_{21;1} & k_{22;1} \end{bmatrix} \qquad (i)$$

Entering the components of k_1 in the locations of **K** having the same nodal subscripts, e.g. $k_{21;1}$, in row 2 column 1 we obtain

$$10^6 \begin{bmatrix} 84 & -84 & 0 \\ -84 & 84 & 0 \\ 0 & 0 & 0 \end{bmatrix}$$

Now

$$\mathbf{k}_2 = 10^6 \begin{bmatrix} 28 & -28 \\ -28 & 28 \end{bmatrix} = \begin{bmatrix} k_{22;2} & k_{23;2} \\ k_{32;2} & k_{33;2} \end{bmatrix} \tag{ii}$$

Entering $\mathbf{k}_2$ in $\mathbf{K}$, we place for example $k_{23;2}$ in row 2 column 3 and finally the assembly stiffness equations are

$$\begin{bmatrix} X_1 \\ X_2 \\ X_3 \end{bmatrix} = 10^6 \begin{bmatrix} 84 & -84 & 0 \\ -84 & 112 & -28 \\ 0 & -28 & 28 \end{bmatrix} \begin{bmatrix} u_1 \\ u_2 \\ u_3 \end{bmatrix} \tag{iii}$$

The component 112×10^6 in $\mathbf{K}$ represents the coupling of the 2 elements at node 2.

The process of substitution in equation (5.15) illustrated above by numerical examples can now be stated formally. However this is not essential to the understanding of the rest of the chapter; it is presented here only for logical completeness and may be omitted until referred to in Chapter 13.

Let two adjacent bars $\mathbf{p}$ and $\mathbf{q}$ of a collinear assembly have ends at nodes p, q and q, r respectively. Then applying (5.10) and (5.7) to bar $\mathbf{p}$ with $i = p$, $j = q, \mathbf{j} = \mathbf{p}$ we obtain

$$X_{q;\mathbf{p}} = k_{qp;\mathbf{p}} u_p + k_{qq;\mathbf{p}} u_q$$

Similarly for bar $\mathbf{q}$ with $i = q, j = r, \mathbf{j} = \mathbf{q}$

$$X_{q;\mathbf{q}} = k_{qq;\mathbf{q}} u_q + k_{qr;\mathbf{q}} u_r$$

and so (5.15) becomes

$$X_q = k_{qp;\mathbf{p}} u_p + (k_{qq;\mathbf{p}} + k_{qq;\mathbf{q}}) u_q + k_{qr;\mathbf{q}} u_r$$

The coefficient of each nodal displacement can be written as a summation over the $\mathbf{n}$ bars of the assembly

$$\sum_{\mathbf{p}=1}^{\mathbf{n}} k_{qp;\mathbf{p}}$$

and then the nodal force X_q can be written by a second summation over the n nodes, as

$$X_q = \sum_{p=1}^{n} \left(\sum_{\mathbf{p}=1}^{\mathbf{n}} k_{qp;\mathbf{p}} \right) u_p \tag{5.23}$$

Then the assembly stiffness equations (5.14) could be expressed by (5.23) when p and q take all values from 1 to n, and $\mathbf{p}$ takes all values from $\mathbf{1}$ to $\mathbf{n}$, assuming of course that nodes and bars have been numbered in sequence.

Problem for Solution   5.4

5.3.3 *Applications*

The nodal loads can be determined from the assembly stiffness equations (5.14) if all the nodal displacements are known. To determine stress in any bar the end forces on it are first determined using the stiffness equations (5.7) for the bar, and then equation (5.5) is used.

Example 5.3 The assembly of Example 5.2 is subjected to nodal displacements $u_1 = u_3 = 0$; $u_2 = 10^{-3}$ m. Calculate the nodal loads, and the stress in element **2**.

Solution Here $\delta = \{0 \quad 10^{-3} \quad 0\}$ and substituting in equation (iii) of Example 5.2

$$\begin{bmatrix} X_1 \\ X_2 \\ X_3 \end{bmatrix} = 10^6 \begin{bmatrix} 84 & -84 & 0 \\ -84 & 112 & -28 \\ 0 & -28 & 28 \end{bmatrix} \begin{bmatrix} 0 \\ 10^{-3} \\ 0 \end{bmatrix} = 10^3 \begin{bmatrix} -84 \\ 112 \\ -28 \end{bmatrix} \text{N}$$

Using k_2 from (ii) of Example 5.2, in equations (5.7)

$$\begin{bmatrix} X_{2;2} \\ X_{3;2} \end{bmatrix} = 10^6 \begin{bmatrix} 28 & -28 \\ -28 & 28 \end{bmatrix} \begin{bmatrix} 10^{-3} \\ 0 \end{bmatrix} = 10^3 \begin{bmatrix} 28 \\ -28 \end{bmatrix} \text{N}$$

Now using equations (5.5) $(\sigma_x)_2 = -28 \times 10^3/(400 \times 10^{-6}) = -70 \times 10^6$ N/m².

However if we knew all the nodal forces and tried to use the complete equations (5.14) to find all the nodal displacements we should find that this could not be done; the matrix **K** would prove to be singular and so could not be inverted (Appendix II). The reason for this can be found by looking back to the development of the stiffness equations for one bar. There end force and displacement were related by using equations (5.4) and (5.5) which, taken together, express end force in terms of *difference* of end displacement. Therefore if both ends of the bar discussed in Section 5.3.1 were moved the same distance along the x axis the end forces would be unchanged. In other words a rigid body translation in the x direction does not alter the forces (compare the discussion in Article 2.1). Equally then we would expect the same to be true for an assembly of bars. It is left as an exercise for the reader to show that if all the nodal displacements in Example 5.3 are changed by +1 mm the nodal forces are unchanged. We can express this in another way by saying that there is not a unique set of displacements for a given set of forces, only a unique deformation, i.e. differences of displacements.

Since in direct loading the only displacement of the mass centre of a bar is along the x axis we say the system has one degree of freedom of rigid body displacement. Then if we know the position of one node the positions of the others depend only on deformation $(u_j - u_i)$ and can be determined if the nodal forces are known. We say that we must *prescribe* at least one nodal displacement before we can determine the others. Referring to equations (5.13) let say u_2 be prescribed. Then the terms $a_{i2}u_2$ $(i = 1, 2, \ldots n)$ become known values and may be transferred to the left-hand sides

$$(X_1 - a_{12}u_2) = a_{11}u_1 + a_{13}u_3 + \ldots + a_{1n}u_n$$

$$(X_2 - a_{22}u_2) = a_{21}u_1 + a_{23}u_3 + \ldots + a_{2n}u_n$$

$$\quad\vdots\qquad\qquad\vdots\qquad\vdots\qquad\vdots\qquad\vdots \tag{5.24}$$

$$(X_n - a_{n2}u_2) = a_{n1}u_1 + a_{n3}u_3 + \ldots + a_{nn}u_n$$

We note that since there are n equations but $(n-1)$ unknowns $(u_1, u_3, \ldots u_n)$, one of the nodal forces need not be known. This is taken to be at the node with known displacement and is called a reactive load. We have then n equations in the n unknowns $(X_2; u_1, u_3, \ldots u_n)$. The equation containing the unknown nodal load is now excluded leaving

$$\begin{bmatrix} X_1 \\ X_3 \\ \cdot \\ \cdot \\ \cdot \\ X_n \end{bmatrix} - u_2 \begin{bmatrix} a_{12} \\ a_{32} \\ \cdot \\ \cdot \\ \cdot \\ a_{n2} \end{bmatrix} = \begin{bmatrix} a_{11} & a_{13} & \ldots & a_{1n} \\ a_{31} & a_{33} & \ldots & a_{3n} \\ \cdot & \cdot & \cdot & \cdot \\ \cdot & \cdot & \cdot & \cdot \\ \cdot & \cdot & \cdot & \cdot \\ a_{n1} & a_{n3} & \ldots & a_{nn} \end{bmatrix} \begin{bmatrix} u_1 \\ u_3 \\ \cdot \\ \cdot \\ \cdot \\ u_n \end{bmatrix} \tag{5.25}$$

or symbolically,

$$\mathbf{F}_R' = \mathbf{K}_{RR}\boldsymbol{\delta}_R \tag{5.26}$$

These reduced equations can now be solved for the unknown displacements i.e.

$$\boldsymbol{\delta}_R = \mathbf{K}_{RR}^{-1}\mathbf{F}_R' \tag{5.27}$$

Reactive loads are then found by introducing the nodal displacements in the complete stiffness equations (5.14) and the stresses in the bars are found as previously described. Equations (5.25) can be obtained directly from (5.14) by:

(a) eliminating the row (2 in the above example) corresponding to the prescribed displacement.

(b) removing from the stiffness matrix the column (2 in the above example) corresponding to the prescribed displacement.

(c) subtracting from the left side the products of the prescribed displacement (u_2 in the example) with the elements of the column displaced from K in (b) (i.e. $a_{i2}u_2$ ($i = 1, 3, \ldots n$) in the above example).

In the particular case when the prescribed displacement is zero ($u_2 = 0$ in the example) the products in (c) are all zero. If N_p nodal displacements are prescribed, the procedure described above is applied for each and the reduced stiffness matrix has size $(n - N_p) \times (n - N_p)$.

Example 5.4 The assembly of Example 5.2 is subjected to a force of 40 kN at node 3 (Figure 5.3a). Determine the stress in each bar.

Solution Referring to Figure 5.3a the nodal loads and displacements are

$$\mathbf{F} = \{X_1 \quad X_2 \quad X_3\} = \{X_1 \quad 0 \quad 4 \times 10^4\}$$

$$\boldsymbol{\delta} = \{u_1 \quad u_2 \quad u_3\} = \{0 \quad u_2 \quad u_3\}$$

Equations (iii) of Example 5.2 are reduced by eliminating the first row. The first column of $\mathbf{K}$ is then removed and since $u_1 = 0$ it makes no contribution to the left side of the equations. Then

$$\begin{bmatrix} X_2 \\ X_3 \end{bmatrix} = \begin{bmatrix} 0 \\ 4 \times 10^4 \end{bmatrix} = 10^6 \begin{bmatrix} 112 & -28 \\ -28 & 28 \end{bmatrix} \begin{bmatrix} u_2 \\ u_3 \end{bmatrix}$$

Solving, $u_2 = 0.476 \times 10^{-3}$ m; $u_3 = 4 \times 0.476 \times 10^{-3}$ m.
For bar 1

$$\begin{bmatrix} X_{1;1} \\ X_{2;1} \end{bmatrix} = 10^6 \begin{bmatrix} 84 & -84 \\ -84 & 84 \end{bmatrix} \begin{bmatrix} 0 \\ 0.476 \times 10^{-3} \end{bmatrix} = 10^3 \begin{bmatrix} -40 \\ 40 \end{bmatrix} \text{ N.}$$

Using (5.5) $(\sigma_x)_1 = 40 \times 10^3 / (400 \times 10^{-6}) = 100 \times 10^6 \text{ N/m}^2.$

Similarly $(\sigma_x)_2 = 100 \text{ MN/m}^2.$

Example 5.5 The assembly of Example 5.2 is subject to a force of $+40$ kN at node 2 when arranged as in Figure 5.3b. Find reactive loads and element stresses when (a) $u_1 = 0$, (b) $u_1 = 0.5 \times 10^{-3}$ m.

Solution (a) $u_1 = u_3 = 0$ and $X_2 = 40 \times 10^3$ N. Therefore equations (iii) of Example 5.2 become

$$\begin{bmatrix} X_1 \\ 40 \times 10^3 \\ X_3 \end{bmatrix} = 10^6 \begin{bmatrix} 84 & -84 & 0 \\ -84 & 112 & -28 \\ 0 & -28 & 28 \end{bmatrix} \begin{bmatrix} 0 \\ u_2 \\ 0 \end{bmatrix} \tag{i}$$

In reduced form these become $[40 \times 10^3] = 10^6 [112] [u_2]$
i.e. $u_2 = 0.357 \times 10^{-3}$ m
Substituting in (i) $X_1 = -30 \times 10^3$ N; $X_3 = -10^4$ N.

$$\begin{bmatrix} X_{1;1} \\ X_{2;1} \end{bmatrix} = 10^6 \begin{bmatrix} 84 & -84 \\ -84 & 84 \end{bmatrix} \begin{bmatrix} 0 \\ 0.357 \times 10^{-3} \end{bmatrix} = 10^3 \begin{bmatrix} -30 \\ 30 \end{bmatrix} \text{ N.}$$

Then $(\sigma_x)_1 = 75 \text{ MN/m}^2$ and similarly $(\sigma_x)_2 = -25 \text{ MN/m}^2.$
(b) With $u_1 = 0.5 \times 10^{-6}$ m equations (i) reduce to

$$[40 \times 10^3] - 10^6 [-84 \, u_1] = 10^6 [112] [u_2]$$

i.e. $u_2 = 0.732 \times 10^{-3}$ m. Proceeding as before,

$$X_1 = -19.5 \times 10^3 \text{ N}; \quad X_3 = -20.5 \times 10^3 \text{ N};$$
$$(\sigma_x)_1 = 48.8 \text{ MN/m}^2; \quad (\sigma_x)_2 = -51.2 \text{ MN/m}^2.$$

In a more formal treatment of the reduction process we partition (see Appendix II) the assembly stiffness equations as

$$\begin{bmatrix} \mathbf{F_R} \\ \mathbf{F_P} \end{bmatrix} = \begin{bmatrix} \mathbf{K_{RR}} & \mathbf{K_{RP}} \\ \mathbf{K_{PR}} & \mathbf{K_{PP}} \end{bmatrix} \begin{bmatrix} \boldsymbol{\delta_R} \\ \boldsymbol{\delta_P} \end{bmatrix} \tag{5.28}$$

where $\boldsymbol{\delta_R}$ and $\boldsymbol{\delta_P}$ are vectors of unknown and prescribed nodal displacements respectively. Then $\mathbf{F_R}$ is a vector of known nodal forces, and $\mathbf{F_P}$ one of reactive forces. The reduced equations are then

$$\mathbf{F_R} = \mathbf{K_{RR}}\boldsymbol{\delta_R} + \mathbf{K_{RP}}\boldsymbol{\delta_P} \tag{5.29}$$

Inverting these we have

$$\boldsymbol{\delta_R} = \mathbf{K_{RR}^{-1}}(\mathbf{F_R} - \mathbf{K_{RP}}\boldsymbol{\delta_P}) \tag{5.30}$$

in which a formal expression has been obtained for the matrix $\mathbf{F_R'}$ appearing in equation (5.27).

For example when the system of Figure 5.3a (Examples 5.2, 5.4) is subject to a force of 40 kN at node 3 the stiffness equations can be partitioned as

$$\begin{bmatrix} X_1 \\ \hline 0 \\ 4 \times 10^4 \end{bmatrix} = 10^6 \begin{bmatrix} 84 & -84 & 0 \\ \hline -84 & 112 & -28 \\ 0 & -28 & 28 \end{bmatrix} \begin{bmatrix} 0 \\ u_2 \\ u_3 \end{bmatrix} \tag{5.31}$$

corresponding to

$$\begin{bmatrix} \mathbf{F_P} \\ \mathbf{F_R} \end{bmatrix} = \begin{bmatrix} \mathbf{K_{PP}} & \mathbf{K_{PR}} \\ \mathbf{K_{RP}} & \mathbf{K_{RR}} \end{bmatrix} \begin{bmatrix} \boldsymbol{\delta_P} \\ \boldsymbol{\delta_R} \end{bmatrix} \tag{5.32}$$

In this case the reduced stiffness equations (5.29) are the second row of the partitioned equations (5.32) and comparison may be made with Example 5.4.

Problems for Solution   5.5–5.8

5.3.4 *Parallel and Coaxial Assemblies*

The methods described above can also be applied to certain other systems of bars which have parallel or common longitudinal axes (Figures 5.4a and b), and some examples will now be given. It should be noted that systems of bars with longitudinal axes inclined to each other are known as trusses and are dealt with separately in Chapter 14.

Example 5.6   A steel rod and an aluminium tube mounted coaxially (Figure 5.4a) are both 1 m long and are rigidly joined together at the ends. The cross-sectional areas are 1000 mm² and 500 mm² respectively and the Youngs

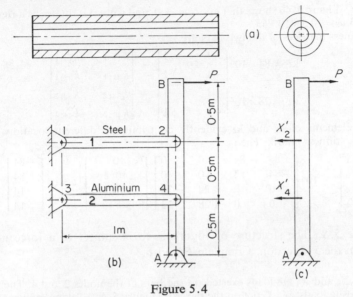

Figure 5.4

Moduli 210 GN/m² and 70 GN/m². If one end is clamped find the force P at the other to cause a stress of 100 MN/m² in the tube.

Solution If we clamp node 1 and apply P at node 2 then

$$\mathbf{F} = \{X_1 \quad P\} \qquad \delta = \{0 \quad u_2\}$$

The stiffness matrices of the two bars are from (5.10)

$$\mathbf{k_1} = 210 \times 10^6 \begin{bmatrix} 1 & -1 \\ -1 & 1 \end{bmatrix} = \begin{bmatrix} k_{11;1} & k_{12;1} \\ k_{21;1} & k_{22;1} \end{bmatrix}$$

$$\mathbf{k_2} = 35 \times 10^6 \begin{bmatrix} 1 & -1 \\ -1 & 1 \end{bmatrix} = \begin{bmatrix} k_{11;2} & k_{12;2} \\ k_{21;2} & k_{22;2} \end{bmatrix}$$

and the assembly stiffness equations are

$$\begin{bmatrix} X_1 \\ P \end{bmatrix} = 10^6 \begin{bmatrix} (210+35) & -(210+35) \\ -(210+35) & (210+35) \end{bmatrix} \begin{bmatrix} 0 \\ u_2 \end{bmatrix}$$

Hence $P = 245 \times 10^6 u_2$. Substituting from this in equation (5.4) with $\sigma_x = 100$ MN/m² and $E = 70$ GN/m² we find $P = 350$ kN.

Parallel assemblies can also be treated using the methods of Section 5.3.2 provided that connexions such as AB in Figure 5.4b are not deformable and exert only axial forces on the bars.

Example 5.7 Steel and aluminium bars **1** and **2** each of 400 mm² cross-section and 1 m length are loaded as in Figure 5.4b. Node 2 is displaced 1 mm to the right. Find the nodal loads assuming that AB is rigid and that the pinned joints at the ends of the bars exert only axial forces on them.

Solution   Figure 5.4b shows that $u_1 = u_3 = 0$, and since AB rotates as a rigid body $u_4 = 0.5\, u_2 = 0.5 \times 10^{-3}$ m.

The stiffness matrices of the two bars are

$$\mathbf{k_1} = 84 \times 10^6 \begin{bmatrix} 1 & -1 \\ -1 & 1 \end{bmatrix} = \begin{bmatrix} k_{11;1} & k_{12;1} \\ k_{21;1} & k_{22;1} \end{bmatrix}$$

$$\mathbf{k_2} = 28 \times 10^6 \begin{bmatrix} 1 & -1 \\ -1 & 1 \end{bmatrix} = \begin{bmatrix} k_{33;2} & k_{34;2} \\ k_{43;2} & k_{44;2} \end{bmatrix}$$

All the elements of $\mathbf{k_1}$ and $\mathbf{k_2}$ evidently contribute to different locations in the assembly stiffness matrix. Then

$$\mathbf{F} = \begin{bmatrix} X_1 \\ X_2 \\ X_3 \\ X_4 \end{bmatrix} = 10^6 \begin{bmatrix} 84 & -84 & 0 & 0 \\ -84 & 84 & 0 & 0 \\ 0 & 0 & 28 & -28 \\ 0 & 0 & -28 & 28 \end{bmatrix} \begin{bmatrix} 0 \\ 10^{-3} \\ 0 \\ 0.5 \times 10^{-3} \end{bmatrix} = 10^3 \begin{bmatrix} -84 \\ 84 \\ -14 \\ 14 \end{bmatrix} \text{N} \quad \text{(i)}$$

Example 5.8   The structure of Figure 5.4b is subject to a force at B of 40 kN in the x direction. Complete $\mathbf{F}$ and $\boldsymbol{\delta}$.

Solution X_2 and X_4 are loads exerted on the bars at the nodes 2 and 4. Then equal and opposite loads X_2', X_4' act on the bar AB (Figure 5.4c). Taking moments about A for AB

$$X_4' \times 0.5 + X_2' \times 1 - 40 \times 10^3 \times 1.5 = 0$$

Hence we can write

$$X_4 + 2\,X_2 - 120 \times 10^3 = 0 \tag{i}$$

As in the previous example

$$u_1 = u_3 = 0; \qquad u_4 = 0.5\, u_2$$

Hence using $\boldsymbol{\delta} = \{0 \quad u_2 \quad 0 \quad 0.5\, u_2\}$ in (i) of Example 5.7 we obtain

$$X_2 = 84 \times 10^6\, u_2; \qquad X_4 = 14 \times 10^6\, u_2$$

Substituting for X_2 and X_4 in (i) $u_2 = 2u_4 = 0.658 \times 10^{-3}$ m.

Hence $\mathbf{F} = 10^3 \{-55.4 \quad 55.4 \quad -9.2 \quad 9.2\}$ N.

Problems for Solution   5.9, 5.10

5.4   Thermal Stresses

5.4.1 *Introduction*

The linear expansion ϵ_T per unit length of an isotropic solid is taken to be proportional to temperature change T (units °C)

i.e. $\epsilon_T = \alpha T$ (5.33)

where α is the coefficient of linear expansion of the material (units 1/°C). For most materials used for structural purposes this relation is valid over a sufficiently wide range of temperatures.

Provided that the dimensional changes required by equation (5.33) are free to occur no stresses are developed in a body due to change of temperature. However, if the body is prevented from adopting the dimensions appropriate to its temperature it is effectively in a state of mechanical strain. The stresses required to produce these strains are called thermal stresses.

For example the rod AB in Figure 5.5a originally at temperature T_0 is free to contract when cooled T. However, if the support at B is held as in Figure 5.5b, this contraction is prevented and the rod has an effective tensile strain equal to αT. The corresponding tensile stress can be determined by application of the stress/strain equations for the material.

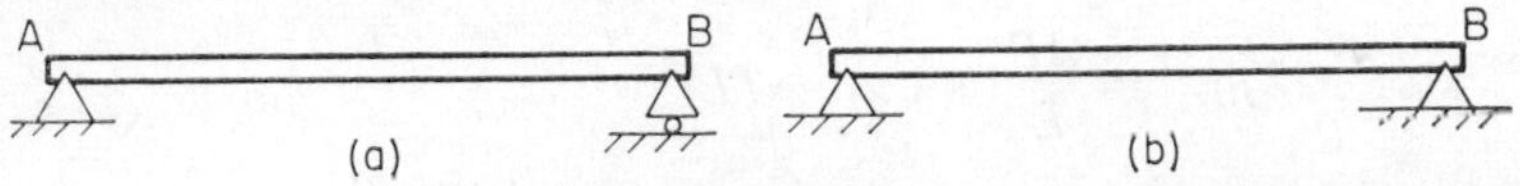

Figure 5.5

Thermal stresses can also occur due to non-uniform heating. Suppose that in Figure 5.4a the rod and tube are made from the same material and rigidly connected at the ends while both are at temperature T_0. If the rod is raised to a temperature T_1 and the tube to temperature T_2, their free expansions are different. However the end connexions impose a common actual expansion so that both rods cannot adopt their free dimensions. The hotter material is prevented from fully attaining its free dimensions and is thus in compression, while the cooler material is extended beyond its free dimensions and is therefore in tension. An important example is that of a tube containing hot fluid. The hot parts of the tube wall are restrained by the colder outer parts, which are themselves in tension due to the underlying warmer material.

If the rod and tube of Figure 5.4a are now made of materials having different values of α and both are brought to a new temperature T_1, then the bar with the greater expansion coefficient increases its free dimensions more than the other. Since each has the same actual change of length the rods are mechanically strained and in this way thermal stresses are developed under uniform temperature conditions due to *inhomogeneity* of the structure.

5.4.2 *Stiffness Equations for One Bar*

Let the bar of Figure 5.2a experience a temperature change T relative to its initial state. Then extending the method of Section 5.3.1 we seek relations between end loads, end displacements and temperature change in the form

$$X_{i;j} = b_{11}u_i + b_{12}u_j + b_{13}T$$
$$X_{j;j} = b_{21}u_i + b_{22}u_j + b_{23}T \tag{5.34}$$

If $T = 0$ these equations are equivalent to (5.7) and so

$$b_{11} = b_{22} = -b_{12} = -b_{21} = \frac{AE}{L}$$

We now hold end i ($u_i = 0$) and permit free thermal expansion (or contraction) due to the temperature change T, i.e. we let $X_{j;i} = 0$. Then the displacement at the end j is equal to the thermal expansion

i.e. $u_j = \epsilon_T L = \alpha T L$

Now putting $u_i = X_{j;i} = 0$ in (5.34) $b_{22}u_j = -b_{23}T$ and so $b_{23} = -AE\alpha$
Similarly we find $b_{13} = AE\alpha$ and equations (5.34) becomes

$$X_{i;i} = \frac{AE}{L}(u_i - u_j + \alpha T L)$$

$$X_{j;i} = \frac{AE}{L}(u_i + u_j - \alpha T L)$$

i.e. $\begin{bmatrix} X_{i;i} \\ X_{j;i} \end{bmatrix} = \frac{AE}{L}\begin{bmatrix} 1 & -1 \\ -1 & 1 \end{bmatrix}\begin{bmatrix} u_i \\ u_j \end{bmatrix} + AE\alpha T \begin{bmatrix} 1 \\ -1 \end{bmatrix}$

$$= \begin{bmatrix} k_{ii;j} & k_{ij;j} \\ k_{ji;j} & k_{jj;j} \end{bmatrix}\begin{bmatrix} u_i \\ u_j \end{bmatrix} + \begin{bmatrix} Q_{i;j} \\ Q_{j;j} \end{bmatrix} \qquad (5.35)$$

Symbolically $\mathbf{F_j} = \mathbf{k_j}\boldsymbol{\delta_j} + \mathbf{Q_j}$ (5.35a)

where $\mathbf{F_j}$, $\boldsymbol{\delta_j}$, $\mathbf{k_j}$ are end load, end displacement and stiffness matrices for the bar, and

$$\mathbf{Q_j} = AE\alpha T \{1 \qquad -1\} = \{Q_{i;j} \qquad Q_{j;j}\} \qquad (5.36)$$

is a matrix of 'thermal forces'.

The stress in the rod is obtained using equation (5.5) for the jth bar

i.e. $X_{j;i} = A\sigma_x$.

Substituting from (5.35) we note the following particular form

$$\sigma_x = \frac{E}{L}(u_j - u_i) - E\alpha T = E(\epsilon_x - \epsilon_T) \qquad (5.37)$$

Example 5.9 A steel rod (nodes 1, 2) of 1 m length and 400 mm² cross-sectional area is clamped at node 2 and subject to a force of -20 kN at node 1. Find the elongation and the stress in the rod when its temperature changes $-40°C$.
Take $E = 210$ GN/m²; $\alpha = 11 \times 10^{-6}/°C$.

Solution Here $AE/L = 400 \times 10^{-6} \times 210 \times 10^9 = 84 \times 10^6$ N/m
and $A\alpha E = 400 \times 10^{-6} \times 11 \times 10^{-6} \times 210 \times 10^9 = 924$ N/°C.
Substituting in equations (5.35) and calling the rod 1

$$\begin{bmatrix} -2 \times 10^4 \\ X_{2;1} \end{bmatrix} = 84 \times 10^6 \begin{bmatrix} 1 & -1 \\ -1 & 1 \end{bmatrix}\begin{bmatrix} u_1 \\ 0 \end{bmatrix} + (924 \times -40)\begin{bmatrix} 1 \\ -1 \end{bmatrix}$$

Then $u_1 = 0{\cdot}202 \times 10^{-3}$ m and $X_{2;1} = 20$ kN.
Applying (5.5) $\sigma_x = 50$ MN/m^2.

Problems for Solution 5.11, 5.12

5.4.3 *Collinear Assembly*

For an assembly of **n** bars with n nodes we extend equations (5.13) by using Q_i for the thermal force at the ith node, obtaining typically

$$X_i = a_{i1}u_1 + a_{i2}u_2 + \ldots + a_{ii}u_i + \ldots + a_{in}u_n + Q_i \qquad (5.38)$$

and for the assembly

$$\mathbf{F} = \mathbf{K}\boldsymbol{\delta} + \mathbf{Q} \qquad (5.39)$$

Another equation for X_i can be obtained by applying conditions of compatibility and equilibrium at the ith node, in conjunction with stiffness equations for the elements meeting at the ith node. The coefficients a_{ij} were determined by this procedure in Section 5.3.2. Now proceeding in a similar manner by reference to a two-component assembly as in Figure 5.3a the equilibrium equation for node 2 is as before (Figure 5.3c).

$$X_2 = X_{2;1} + X_{2;2} \qquad (5.40)$$

Using equation (5.35) the stiffness equations of the bars are

$$\begin{bmatrix} X_{1;1} \\ X_{2;1} \end{bmatrix} = \begin{bmatrix} k_{11;1} & k_{12;1} \\ k_{21;1} & k_{22,1} \end{bmatrix} \begin{bmatrix} u_1 \\ u_2 \end{bmatrix} + \begin{bmatrix} Q_{1;1} \\ Q_{2;1} \end{bmatrix} \qquad (5.41)$$

$$\begin{bmatrix} X_{2;2} \\ X_{3;2} \end{bmatrix} = \begin{bmatrix} k_{22;2} & k_{23;2} \\ k_{32;2} & k_{33;2} \end{bmatrix} \begin{bmatrix} u_2 \\ u_3 \end{bmatrix} + \begin{bmatrix} Q_{2;2} \\ Q_{3;2} \end{bmatrix} \qquad (5.42)$$

and so substituting in (5.40)

$$X_2 = k_{21;1}u_1 + (k_{22;1} + k_{22;2})u_2 + k_{23;2}u_3 + (Q_{2;1} + Q_{2;2})$$

and in this example $Q_2 = Q_{2;1} + Q_{2;2}$.

In practice then, $\mathbf{K}$ is generated as in Section 5.3.2 and $\mathbf{Q}$ is obtained in a similar way by forming a column matrix

$$\mathbf{Q} = \{Q_1 \quad Q_2 \quad \ldots \quad Q_n\} \qquad (5.43)$$

The thermal force components for all the bars are accumulated in the rows of $\mathbf{Q}$ given by their nodal subscript.

As discussed in Section 5.3.3 at least one nodal displacement must be prescribed in order to eliminate rigid body displacement. Corresponding to (5.27) we have

$$\boldsymbol{\delta}_\mathrm{R} = \mathbf{K}_{\mathrm{RR}}^{-1}(\mathbf{F}_\mathrm{R}' - \mathbf{Q}_\mathrm{R}) \qquad (5.44)$$

Example 5.10   A collinear assembly consists of a steel rod **1** with $A = 400$ mm², $E = 210$ GN/m², $L = 1$ m, and an aluminium rod **2**, $A = 800$ mm², $E = 70$ GN/m², $L = 2$ m. The materials have linear expansion coefficients of $11 \times 10^{-6}/°C$ and $23 \times 10^{-6}/°C$. Form the structure stiffness equations and calculate the stresses in the two materials when the temperature is raised $40°C$ with no displacement permitted at nodes 1 and 3.

Solution   The stiffness and thermal force matrices of the bars are obtained using (5.35)

$$\mathbf{k_1} = 84 \times 10^6 \begin{bmatrix} 1 & -1 \\ -1 & 1 \end{bmatrix} ; \quad \mathbf{Q_1} = 36{,}960 \begin{bmatrix} +1 \\ -1 \end{bmatrix} = \begin{bmatrix} Q_{1;1} \\ Q_{2;1} \end{bmatrix}$$

$$\mathbf{k_2} = 28 \times 10^6 \begin{bmatrix} 1 & -1 \\ -1 & 1 \end{bmatrix} ; \quad \mathbf{Q_2} = 51{,}520 \begin{bmatrix} +1 \\ -1 \end{bmatrix} = \begin{bmatrix} Q_{2;2} \\ Q_{3;2} \end{bmatrix}$$

The nodal displacement matrix is $\delta = \{u_1 \quad u_2 \quad u_3\} = \{0 \quad u_2 \quad 0\}$
The nodal force matrix is $\mathbf{F} = \{X_1 \quad X_2 \quad X_3\} = \{X_1 \quad 0 \quad X_3\}$

The assembly stiffness equations are now generated by accumulating stiffness coefficients and thermal forces with the following result

$$\begin{bmatrix} X_1 \\ 0 \\ X_3 \end{bmatrix} = 10^6 \begin{bmatrix} 84 & -84 & 0 \\ -84 & (84+28) & -28 \\ 0 & -28 & 28 \end{bmatrix} \begin{bmatrix} 0 \\ u_2 \\ 0 \end{bmatrix} + \begin{bmatrix} 36{,}960 \\ (-36{,}960+51{,}520) \\ -51{,}520 \end{bmatrix} \quad \text{(i)}$$

For instance, referring to equations (5.41) and (5.42), $Q_{2;1}$, which is $-36{,}960$, goes in row 2 of $\mathbf{Q}$ as does $Q_{2;2}$ which is $+51{,}520$, so that $Q_2 = (-36{,}960+51{,}520)$ as shown in (i). Solving the latter, we obtain $u_2 = -0.13 \times 10^{-3}$ m. Then, using (5.37), we find stresses of -119.7 MN/m² and -59.85 MN/m² in the steel and aluminium respectively. Both rods are in compression since neither is free to expand.

Example 5.11   A steel tube of external diameter 24 mm and bore 16 mm encloses a copper rod of 14 mm diameter to which it is rigidly joined at each end. If at $10°C$ there is no stress, calculate the stresses when the temperature is raised to $200°C$. If an axial tension of 10 kN is then applied find the final stresses in each material. $E(\text{copper}) = 120$ GN/m²; $\alpha(\text{copper}) = 17 \times 10^{-6}/°C$.

Solution   This is a coaxial assembly of the type discussed in Section 5.3.4, both bars joining nodes 1 and 2. If we take $u_1 = 0$ as prescribed displacement the bar stiffness equations are, taking the tube as bar **1** and the rod as bar **2**, and unit length

$$\begin{bmatrix} X_{1;1} \\ X_{2;1} \end{bmatrix} = 52.8 \times 10^6 \begin{bmatrix} 1 & -1 \\ -1 & 1 \end{bmatrix} \begin{bmatrix} 0 \\ u_2 \end{bmatrix} + 110 \times 10^3 \begin{bmatrix} +1 \\ -1 \end{bmatrix} \quad \text{(i)}$$

$$\begin{bmatrix} X_{1;2} \\ X_{2;2} \end{bmatrix} = 18.5 \times 10^6 \begin{bmatrix} 1 & -1 \\ -1 & 1 \end{bmatrix} \begin{bmatrix} 0 \\ u_2 \end{bmatrix} + 59.6 \times 10^3 \begin{bmatrix} +1 \\ -1 \end{bmatrix} \quad \text{(ii)}$$

The assembly stiffness matrix is of size $n \times n$, i.e. 2×2. Also

$$\mathbf{F} = \{X_1 \quad X_2\} = \{X_1 \quad 0\} \quad \text{and} \quad \delta = \{u_1 \quad u_2\} = \{0 \quad u_2\}$$

Then the assembly stiffness equations are

$$\begin{bmatrix} X_1 \\ 0 \end{bmatrix} = 10^6 \begin{bmatrix} (52\cdot8+18\cdot5) & (-52\cdot8-18\cdot5) \\ (-52\cdot8-18\cdot5) & (52\cdot8+18\cdot5) \end{bmatrix} \begin{bmatrix} 0 \\ u_2 \end{bmatrix} + 10^3 \begin{bmatrix} (+110+59\cdot6) \\ -(110+59\cdot6) \end{bmatrix} \quad \text{(iii)}$$

whence $u_2 = 2\cdot38 \times 10^{-3}$ m. Using (5.37) we find stresses of $62\cdot1$ MN/m^2 and -101 MN/m^2 in the tube and rod.

Superposition of axial tension of 10 kN is equivalent to setting $X_2 = 10^4$ N in equations (iii) whereupon $u_2 = 2\cdot52 \times 10^{-3}$ m and the stresses are $91\cdot6$ MN/m^2 and $-84\cdot3$ MN/m^2.

Problems for Solution 5.13–5.15

5.5 Initial Stresses

When some preliminary strain is necessary to complete an assembly a state of initial stress is said to exist. This condition, which may arise for example due to misfit of components, can be formulated in a manner similar to that for the thermal effects. Let us suppose that the bar ij (Figure 5.2a) is found to be ΔL less than the specified length L, and let direct loads be applied to maintain it at length L. Then the initial strain is

$$\epsilon_I = \frac{\Delta L}{L} \tag{5.45}$$

Then as in Section 5.4.2 we write for the bar

$$X_{i;j} = b_{11}u_i + b_{12}u_j + b_{13}\epsilon_I$$
$$X_{j;j} = b_{21}u_i + b_{22}u_j + b_{23}\epsilon_I \tag{5.46}$$

where $b_{11}, b_{12}, b_{21}, b_{22}$ are as previously determined. Let the bar be load free and held at i. Then $u_i, X_{i;j}, X_{j;j}$ are zero. Since there is no load the rod has its actual length of $(L-\Delta L)$, and since $u_i = 0$, $u_j = -\Delta L = -\epsilon_I L$. But from (5.46) $0 = b_{12}u_j + b_{13}\epsilon_I$ where $b_{12} = -AE/L$, and so $b_{13} = -AE$. Similarly $b_{23} = AE$.

Then

$$\begin{bmatrix} X_{i;j} \\ X_{j;j} \end{bmatrix} = \frac{AE}{L} \begin{bmatrix} 1 & -1 \\ -1 & 1 \end{bmatrix} \begin{bmatrix} u_i \\ u_j \end{bmatrix} + AE\epsilon_I \begin{bmatrix} -1 \\ +1 \end{bmatrix}$$

$$= \begin{bmatrix} k_{ii;j} & k_{ij;j} \\ k_{ji;j} & k_{jj;j} \end{bmatrix} \begin{bmatrix} u_i \\ u_j \end{bmatrix} + \begin{bmatrix} c_{i;j} \\ c_{j;j} \end{bmatrix} \tag{5.47}$$

i.e.
$$\mathbf{F}_j = \mathbf{k}_j \boldsymbol{\delta}_j + \mathbf{c}_j \tag{5.47a}$$

The vector $\mathbf{c}_j$ represents the end forces needed to overcome misfit. In view of the similarity of equations (5.47) and (5.35) it is evident that assembly equations

$$\mathbf{F} = \mathbf{K}\boldsymbol{\delta} + \mathbf{c} \tag{5.48}$$

can be derived as discussed in Section 5.4.3 with **Q** replaced by **c**. The stress
is

$$\sigma_x = \frac{X_{j;j}}{A} = \frac{E}{L}(u_j - u_i) + \epsilon_I E \tag{5.49}$$

Example 5.12 If rod **1** of Example 5.10 was initially 1 mm too long find the
initial stresses in the assembly.

Solution k_1 and k_2 and **K** are as obtained in Example 5.10. Rod **1** has to be
shortened to complete the assembly, i.e. $\epsilon_I = -10^{-3}$. Therefore

$$\mathbf{c}_1 = AE\epsilon_I \{-1 \quad +1) = 84{,}000 \{+1 \quad -1\} \text{ N.}$$

Rod **2** requires no initial strain so that $\mathbf{c}_2 = \{0 \quad 0\}$.

In forming vector **c** the components of, for example, $\mathbf{c}_2$ are placed in rows 2, 3
since rod **2** joins nodes 2 and 3.

.e. $\mathbf{c} = \{+84{,}000 \quad (-84{,}000+0) \quad 0\}$ \hfill (i)

Replacing **Q** in (i) of Example 5.10 by **c** from (i) above and solving we find

$$u_2 = 0{\cdot}75 \times 10^{-3} \text{ m.}$$

Then applying (5.49) stresses are $-52{\cdot}5$ MN m^2 and $-26{\cdot}25$ MN/m^2.

Example 5.13 A steel bolt of cross-sectional area 100 mm^2 and thread pitch
1 mm is enclosed in a sleeve of area 200 mm^2. The nut is turned so that it
advances $\frac{1}{4}$ turn from the 'just tight' position. Find the stress in the bolt if
the sleeve has $E = 70$ GN/m^2 (see Figure 5.6a). $L = 200$ mm.

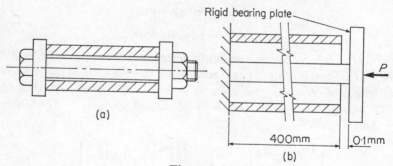

Figure 5.6

Solution Turning the nut along the thread an axial distance of $0{\cdot}25 \times 10^{-3}$ m can
be interpreted as an initial strain, the bolt as it were having been too short. Taking
the bolt as rod **1** we have $(\epsilon_I)_1 = 0{\cdot}25 \times 10^{-3}/0{\cdot}2 = 1{\cdot}25 \times 10^{-3}$.

There is no initial strain in the sleeve (rod **2**), i.e. $(\epsilon_I)_2 = 0$.

For rod **1** $(AE/L) = 105 \times 10^6$ N/m and $\mathbf{c}_1 = 26{\cdot}25 \times 10^3 \{-1 \quad +1)$.

For rod **2** $(AE/L) = 70 \times 10^6$ N/m and $\mathbf{c}_2 = \{0 \quad 0\}$.

Both rods have ends i, j at nodes 1, 2 giving assembly stiffness equations

$$\begin{bmatrix} X_1 \\ X_2 \end{bmatrix} = 10^6 \begin{bmatrix} (105+70) & (-105-70) \\ (-105-70) & (105+70) \end{bmatrix} \begin{bmatrix} u_1 \\ u_2 \end{bmatrix} + 26{\cdot}25 \times 10^3 \begin{bmatrix} -1 \\ +1 \end{bmatrix} \tag{i}$$

Let $u_1 = 0$ to eliminate rigid body displacement. Then, since there are no external forces at the nodes, $X_1 = X_2 = 0$, and $u_2 = -0 \cdot 15 \times 10^{-3}$ m. Applying equation (5.49) the stresses are 105×10^6 N/m² and $-52 \cdot 5 \times 10^6$ N/m² respectively.

Problems for Solution 5.16, 5.17

5.6 Elastic/Plastic Analysis

5.6.1 *Introduction*

Plasticity is the study of the mechanics of deformation when part or all of the deformation is not recoverable. The subject includes processes of metal working but these will not be considered here. Attention will be directed only to the effect of material plasticity in relation to the main structural and stress analysis problems treated. In particular we shall study the calculation of limit loads and the development of residual states of stress.

In Article 3.2 various models were proposed for material undergoing plastic (i.e. non-recoverable) deformation. Of these only the ideal elastic/plastic model will be considered (Figure 3.3c). The strain hardening model is too complex for this introductory account and is treated in specialized texts, e.g. references 1 and 2.

The onset of plastic behaviour is called yielding and, as was noted in Article 1.12, has been found by experiment to occur when some function of the state of stress attains a critical value which is a property of the material. Since the state of stress is most concisely expressed by the principal stress array, we define this yield function as $\Gamma(\sigma_1, \sigma_2, \sigma_3)$ where σ_1, σ_2, σ_3 are the principal stresses. Note that some authors choose σ_1, σ_2, σ_3 such that $\sigma_1 \geqslant \sigma_2 \geqslant \sigma_3$; when this meaning is required Roman numerals will be substituted, i.e. $\sigma_I \geqslant \sigma_{II} \geqslant \sigma_{III}$ as in Article 1.7.

Using C for the critical value of the yield function we have at yield the condition

$$\Gamma(\sigma_1, \sigma_2, \sigma_3) = C \tag{5.50}$$

If $\Gamma < C$ yielding does not occur. The ideal elastic/plastic material is defined as perfectly elastic when $\Gamma < C$, and as perfectly plastic when $\Gamma = C$.

We shall now discuss the two forms of the yield condition which are widely used. The maximum shear stress or Tresca theory predicts initiation of yielding at a critical value of shear stress. As shown in Article 1.12 this is expressed in terms of σ_0, the yield stress in simple tension, as follows

$$\sigma_I - \sigma_{III} = \sigma_0 \tag{5.51}$$

The other widely used yield condition is the distortional strain energy or von Mises criterion (see Article 4.5) which is

$$\frac{(1+\nu)}{6E} ((\sigma_1 - \sigma_2)^2 + (\sigma_2 - \sigma_3)^2 + (\sigma_3 - \sigma_1)^2) = C$$

The quantity on the left is that part of the strain energy which is due only to distortion as opposed to dilatation. Again C is evaluated by considering the case of simple tension where

$$\sigma_1 = \sigma_0, \quad \sigma_2 = \sigma_3 = 0$$

Then

$$C = (1 + \nu)\sigma_0^2/3E$$

and the yield condition becomes

$$(\sigma_1 - \sigma_2)^2 + (\sigma_2 - \sigma_3)^2 + (\sigma_3 - \sigma_1)^2 = 2\sigma_0^2 \tag{5.52}$$

The yield conditions (5.51) and (5.52) define surfaces in a space with coordinates σ_1, σ_2, σ_3. This 'stress space', called the Haigh–Westergaard stress space,[3] can be visualized in two dimensions (Figure 5.7) by considering plane stress, for which $\sigma_3 = 0$. The maximum shear stress condition plots in the stress space as follows. In the first quadrant both σ_1 and σ_2 are positive

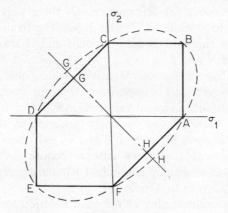

Figure 5.7

so that $\sigma_{III} = \sigma_3 = 0$ and $\sigma_I = \sigma_1$ or $\sigma_I = \sigma_2$. Therefore (5.51) is either $\sigma_1 = \sigma_0$, i.e. line AB, or $\sigma_2 = \sigma_0$ represented by line BC. In the second quadrant $\sigma_2 > 0$ and $\sigma_1 < 0$ so that $\sigma_I = \sigma_2$ and $\sigma_{III} = \sigma_1$. Thus equation (5.51) becomes $\sigma_2 = \sigma_1 + \sigma_0$, a line of slope 1 between C $(0, \sigma_0)$ and D $(-\sigma_0, 0)$ as shown. Similarly the remaining combinations are represented by lines DE, EF and FA. Then according to the maximum shear stress theory all combinations of σ_1 and σ_2 which lie on ABCDEFA initiate yielding, while those combinations which lie within this yield locus do not.

The von Mises yield condition (5.52) in the plane stress form ($\sigma_3 = 0$) reduces to

$$\sigma_0^2 = \sigma_1^2 - \sigma_1\sigma_2 + \sigma_2^2$$

the equation of the ellipse in Figure 5.7. When σ_1 or σ_2 is zero, or when

$\sigma_1 = \sigma_2$, i.e. at the six vertices of the hexagon ABCDEF the two yield conditions coincide. Otherwise the von Mises yield locus always lies outside that for the Tresca condition. The greatest difference between the loci is when $\sigma_1 = -\sigma_2$ (points G and H) corresponding to pure shear in the σ_1, σ_2 plane. Then the von Mises and Tresca predictions are respectively

$$\sigma_1 = -\sigma_2 = \sigma_0/\sqrt{3}$$

and

$$\sigma_1 = -\sigma_2 = \sigma_0/2$$

a difference of about 15%.

Many experiments, such as the classic work of Taylor and Quinney,[4] have shown closer agreement with the von Mises condition. However, since the Tresca condition is more exacting (Figure 5.7) it remains a satisfactory condition for limit design purposes. On the other hand in the study of plastic deformation the von Mises condition may be preferred since it does not require foreknowledge of the relative magnitudes of the principal stresses.

5.6.2 *Basic Plastic Response*

The state of stress in a bar subject to direct loading (Figure 5.2) was discussed in Article 5.2 and is considered to be uniform and of the form of array (A5.1). The principal stresses are $\sigma_1 = \sigma_x$, $\sigma_2 = \sigma_3 = 0$ and both yield conditions predict yielding at $|\sigma_x| = \sigma_0$. Then in accordance with the definition of the ideal elastic/plastic material (Article 3.2) there is perfect elasticity for $-\sigma_0 < \sigma_x < \sigma_0$ and unlimited yielding for $|\sigma_x| = \sigma_0$.

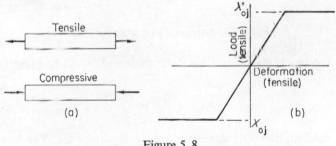

Figure 5.8

We now introduce the symbol X_j to represent states of balanced direct load in the jth bar (Figure 5.8a), with the tensile condition taken as positive, the compressive as negative. Then equation (5.5) can be rewritten as

$$\sigma_x = \frac{X_j}{A} \qquad (5.53)$$

The axial deformation $(u_j - u_i)_j$ we now represent by u_j. Then, since

$X_j = X_{j;j}$ we can rewrite the second equation of (5.7) as

$$X_j = \left(\frac{AE}{L}\right)u_j \qquad (5.54)$$

Thus the load is proportional to deformation (Figure 5.8b) until $|\sigma_x| = \sigma_0$. Denoting this yield load by X_{0j} we have from equation (5.53)

$$|X_{0j}| = A\sigma_0 \qquad (5.55)$$

While the load has this value the material is plastic and can yield indefinitely. Thus the yield load of a single bar is also its limit (or collapse) load

$$|\hat{X}_j| = |X_{0j}| = A\sigma_0 \qquad (5.56)$$

Consider now the response of a bar to the process illustrated in Figure 5.9a.

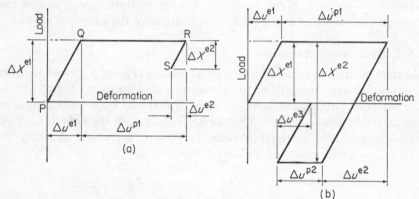

Figure 5.9. Subscript **j** omitted for clarity

Over PQ the bar is elastic and the deformation at the yield point Q is from (5.54)

$$u_{0j} = \frac{X_{0j}L}{AE} = \frac{\sigma_0 L}{E} \qquad (5.57)$$

Plastic deformation (QR in Figure 5.9) is now imposed. The load is then reduced, causing the stress given by equation (5.53) to be less than σ_0. The material response therefore becomes elastic again. However, equation (5.54) cannot be employed since at R the deformation includes the plastic contribution QR. Nevertheless the elastic response does make changes of load proportional to changes of deformation so that following (5.54) we write

$$\Delta u_j^{ei} = \left(\frac{L}{AE}\right)\Delta X_j^{ei} \qquad (5.58)$$

Δu_j^{ei} and ΔX_j^{ei} are the increments of deformation and load respectively during the elastic process. The superscript e denotes an elastic process and i the number of that process. For example in Figure 5.9a the increments of deformation during the processes PQ and RS are Δu_j^{e1} and Δu_j^{e2}. Introducing Δu_j^{pi} as the increment of deformation in the ith plastic process, we recognize that the total deformation of a bar is the sum of the increments of elastic and plastic deformation.

$$u_j = \sum \Delta u_j = \Delta u_j^{e1} + \Delta u_j^{e2} + \ldots + \Delta u_j^{p1} + \Delta u_j^{p2} + \ldots \qquad (5.59)$$

In the case of Figure 5.9a, the final deformation is

$$u_j = \Delta u_j^{e1} + \Delta u_j^{e2} + \Delta u_j^{p1}$$

If the final load is $0\cdot5\,X_{0j}$ we have, applying equation (5.58)

$$u_j = \frac{L}{AE}(\Delta X_j^{e1} + \Delta X_j^{e2}) + \Delta u_j^{p1} = \frac{L}{AE}(X_{0j} + 0\cdot5(-X_{0j})) + \Delta u_j^{p1}$$

Similarly the final deformation in Figure 5.9b is

$$u_j = \frac{L}{AE}(X_{0j} + (-2X_{0j}) + X_{0j}) + \Delta u_j^{p1} + \Delta u_j^{p2} = \Delta u_j^{p1} + \Delta u_j^{p2}$$

Similarly the final load is a sum of increments during both elastic and plastic processes. However for an ideal elastic/plastic material the load is constant during plastic deformation and so for example in the case of Figure 5.9b

$$X_j = \Delta X_j^{p1} + \Delta X_j^{e2} + \Delta X_j^{e0} + \Delta X_j^{p1} + \Delta X_j^{p2}$$
$$= X_{0j} + (-2X_{0j}) + X_{0j} + 0 + 0 = 0$$

In conclusion we repeat, what is obvious from the study of Figure 5.9, that there is no unique relation between X_j and u_j during an elastic/plastic process.

Example 5.14 A rod of 15 mm diameter, $0\cdot4$ m length, has $E = 210$ GN/m^2, $\sigma_0 = 240$ MN/m^2. The rod is given a plastic deformation of $+4$ mm and half of the load removed. Find the final deformation.

Solution This process corresponds to that in Figure 5.9a. The factor (L/AE) is equal to $1\cdot08 \times 10^{-8}$ m/N. Applying equation (5.58)

$$\Delta u_j^{e1} = 1\cdot08 \times 10^{-8} \times \Delta X_j^{e1} = 1\cdot08 \times 10^{-8} \times X_{0j}; \quad \Delta u_j^{e2} = 1\cdot08 \times 10^{-8} \times (-0\cdot5X_{0j})$$

From equation (5.55)

$$X_{0j} = A\sigma_0 = 42{,}600 \text{ N}.$$

The increment during the plastic process is given by $\Delta u_j^{p1} = 4 \times 10^{-3}$ m.
Then $u_j = \Delta u_j^{e1} + \Delta u_j^{e2} + \Delta u_j^{p1} = 10^{-3}(0\cdot46 - 0\cdot23 + 4) = 4\cdot23 \times 10^{-3}$ m.

5.6.3 *Collinear Assemblies* (*Limit Analysis*)

When the elastic behaviour of collinear assemblies was discussed in Article 5.3 the limit load was considered to be that required to initiate yielding anywhere.

In the elastic/plastic analysis the limit or collapse load is that at which yielding of the whole structure begins. As we have seen in equation (5.56) there is no distinction between these loads for a single bar in direct loading. However for assemblies of bars this is not always the case.

The assembly of Figure 5.3a is statically determinate, as noted in Article 5.2 and it can be seen that both bars are in tension with $X_1 = X_2 = P$. Let us suppose that the bars have different limit loads with $\hat{X}_1 < \hat{X}_2$, and let P be increased until it attains the value $\hat{X}_1$. Then $\hat{P} = \hat{X}_1$ is the collapse load of the assembly since gross deformation is possible (Figure 5.10a). The system is now analogous to a mechanism, with bar **1** acting as a slide. Note that in this example the maximum load for elastic behaviour, P_0, is the same as the collapse load $\hat{P}$.

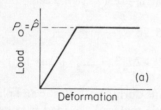 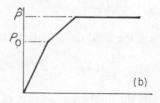

Figure 5.10

The assembly of Figure 5.3b is statically indeterminate since at node 2 there are three forces, i.e. $P = X_{2;1} + X_{2;2}$, only one of which can be independently specified. However it can be seen that bar **1** is in tension and **2** is in compression. Suppose that bar **1** yields first then at node 2 we have

$$P = P_0 = \hat{X}_1 + X_{2;2} = \text{Constant} + X_{2;2}$$

The system has thus become determinate. It is possible to increase P beyond P_0, the additional load being absorbed by bar **2**. When bar **2** also yields, the equilibrium equation for node 2 becomes $P = \hat{P} = \hat{X}_1 + \hat{X}_2 = $ constant.

Thus $\hat{P}$ is the collapse load and the system is again equivalent to a mechanism. In Figure 5.10b we see the change of slope when yielding begins in one bar, and the deformation at constant load when both bars have yielded. In this example $\hat{P}$ exceeds P_0. The plastic analysis therefore permits an increase in the design load for the same structure, or a reduction in the size of the bars for the same design load.

From the above examples we see that plastic collapse proceeds by the elimination of constraints until a mechanism is formed. In the system of Figure 5.3a only one constraint had to be removed, achieved by the yielding of one bar, whereas for that in Figure 5.3b two bars had to yield. In a more complex assembly there may be more than one way in which sufficient constraints can be removed. The collapse mechanism is that requiring the least load. For instance if the support at 3 of Figure 5.3b has a finite collapse load, then a mechanism can also be formed by yielding of bar **1** and collapse of the support.

Example 5.15 Two bars arranged as in Figure 5.3b have cross-sectional areas of 200 and 400 mm², lengths of 1 and 2 m, yield stresses of 300 and 200 MN/m². If $E = 200$ GN/m² determine the yield and limit loads.

Solution Referring to Figure 5.3b we see that the deformation of each bar has the same magnitude. Bar **1** yields at a deformation of $\pm L\sigma_0/E = \pm 1.5 \times 10^{-3}$ m; similarly bar **2** yields at $\pm 2 \times 10^{-3}$ m. Therefore P_0 is the force to produce a deformation of $\pm 1.5 \times 10^{-3}$ m. Using the procedures of Section 5.3.3 we find this to be ± 120 kN.

The assembly limit load $\hat{P}$ is, as discussed above, the sum of the limit loads of the bars. Using equation (5.56) we find

$$\hat{P} = \pm(\hat{X}_1 + \hat{X}_2) = 140 \text{ kN}$$

which is $16\frac{2}{3}\%$ greater than P_0.

Problem for Solution 5.18, 5.19

5.6.4 *Collinear Assemblies* (*Residual Stresses*)

The assembly of Figure 5.3b is elastic for $P < P_0$ (Figure 5.10b), but when $P_0 \leqslant P < \hat{P}$ one bar (say **1**) is in a plastic condition. However since both bars have the same magnitude of deformation the plastic deformation of bar **1** is equal in magnitude to the increment of elastic deformation of bar **2**. Let us suppose the load to be removed when P just attains $\hat{P}$. Then since bar **1** has undergone plastic (i.e. non-recoverable) deformation the total unloaded length of the free assembly would exceed the distance between the end points 1 and 3. In occupying this space both bars are therefore in compression, the resulting stresses being called residual stresses. This problem is equivalent to that of initial stresses due to misfit discussed in Article 5.5.

Example 5.16 The assembly of Example 5.15 is brought just to the limit load and then released. Determine the residual stresses.

Solution The yield deformation of each bar is the product of its length L with its yield strain σ_0/E.
i.e. $\qquad u_{01} = 1.5 \times 10^{-3}$ m; $\qquad u_{02} = 2 \times 10^{-3}$ m.

As discussed above, in this system the difference of these quantities (0.5×10^{-3} m) is the plastic deformation of bar **1**. Treating this as equivalent to an initial misfit and proceeding as in Example 5.12 we find

$$\begin{bmatrix} X_{1;1} \\ X_{2;1} \end{bmatrix} = 200 \times 10^{-6} \times 200 \times 10^9 \begin{bmatrix} 1 & -1 \\ -1 & 1 \end{bmatrix} \begin{bmatrix} u_1 \\ u_2 \end{bmatrix}$$

$$+ 200 \times 10^{-6} \times 200 \times 10^9 \times -0 \cdot 5 \times 10^{-3} \begin{bmatrix} -1 \\ +1 \end{bmatrix} \quad \text{(i)}$$

$$\begin{bmatrix} X_{2;2} \\ X_{3;2} \end{bmatrix} = \frac{400 \times 10^{-6} \times 200 \times 10^9}{2} \begin{bmatrix} 1 & -1 \\ -1 & 1 \end{bmatrix} \begin{bmatrix} u_2 \\ u_3 \end{bmatrix} \quad \text{(ii)}$$

$$\begin{bmatrix} X_1 \\ X_2 \\ X_3 \end{bmatrix} = 10^6 \begin{bmatrix} 40 & -40 & 0 \\ -40 & (40+40) & -40 \\ 0 & -40 & 40 \end{bmatrix} \begin{bmatrix} u_1 \\ u_2 \\ u_3 \end{bmatrix} + 20 \times 10^3 \begin{bmatrix} 1 \\ -1 \\ 0 \end{bmatrix} \quad \text{(iii)}$$

Here $u_1 = u_3 = 0$ and $X_2 = 0$ so that $u_2 = (1/4) \times 10^{-3}$.

Using this value, $X_{2;1}$ is found from (i) and then from (5.5) the stress in bar 1 is found to be -50 MN/m²; similarly the stress in bar 2 is -25 MN/m².

Problems for Solution 5.20, 5.21

5.7   References

(a) *Cited*

1. Mendelson, A., *Plasticity: Theory and Applications*, Macmillan, New York, 1968.
2. Hoffman, C. and Sachs, G., *Introduction to the Theory of Plasticity for Engineers*, McGraw-Hill, New York, 1953.
3. Westergaard, H. M., 'On the resistance of ductile materials ...', *J. Franklin Institute*, **189**, 627 (1920).
4. Taylor, G. I. and Quinney, H., 'The plastic distortion of metals', *Phil. Trans. Roy. Soc. (Lond.)*, A**230**, 323 (1931).

(b) *General*

Johns, D. J., *Thermal Stress Analyses*, Pergamon, Oxford, 1965.

5.8   Problems for Solution

E (steel) = 210 GN/m²; E (aluminium) = 70 GN/m²

α (steel) = $11 \times 10^{-6}/°C$; α (aluminium) = $23 \times 10^{-6}/°C$.

5.1. If the bars in Figure 5.3a are steel, have cross-sectional areas of 2000 mm² and 1000 mm², and each is 1 m long, find the stresses when P is 100 kN, and calculate the relative displacement of points 1 and 3. Find the safety factor when $\sigma_0 = 240$ MN/m².

5.2. Repeat Problem 5.1 if bar 1 is aluminium, for which $\sigma_0 = 80$ MN/m², instead of steel.

5.3. Write down the stiffness equations for each of the steel bars in Problem 5.1.

5.4. Obtain the assembly stiffness equations for the systems of Problems 5.1 and 5.2.

5.5. In Example 5.2 the cross-sectional area of the aluminium bar is doubled and the length of the steel bar is doubled. Construct the stiffness equations and calculate the nodal loads for nodal displacements $u_1 = u_3 = 0$; $u_2 = 10^{-3}$ m. Find the stresses in the bars.

5.6. An assembly consists of a steel tube of diameters 120 mm and 80 mm, length 2 m, and a steel rod of diameter 100 mm and length 1 m. If the allowable normal stress is 100 MN/m² find the greatest force which can be applied at the junction of the rod and tube when the ends are clamped.

5.7. A 3 m long steel bar of uniform cross-section 50 mm × 20 mm is subject to three forces of 50 kN, as shown in Figure 5.3e, and evenly spaced. Determine the elongation of the bar and the greatest stress.

5.8. If in Problem 5.7 the force of 50 kN at node 1 is replaced by a clamp find the nodal displacements and the greatest stress.

5.9. If the allowable stresses in the steel and aluminium are 240 MN/m² and 180 MN/m² respectively, find the greatest load for the assembly of Example 5.6.

5.10. If the steel and aluminium rods in Figure 5.4b are interchanged, find the nodal loads and displacements for (a) $u_2 = 10^{-3}$ m, (b) force of 40 kN at B.

5.11. A length of welded rail is laid when the temperature is 10°C. Find the stress in the rail when heated to 30°C, and when cooled to −5°C, assuming that the ends are unable to move.

5.12. An aluminium rod of 20 mm diameter and 2 m length is subject to tensile forces of 10 kN. Find the temperature change for deformation of ±2 mm.

5.13. Rework Example 5.10 for the case when one support yields 0·25 mm.

5.14. The steel tube and copper rod of Example 5.11 are held together by shear pins. Find the least diameter of pin needed to withstand a temperature change of 220°C if the allowable shear stress is 80 MN/m² for the pin.

5.15. Find the force which must act at node 2 of the system in Example 5.10 to prevent displacement of that node. What are then the stresses in the rods?

5.16. Three steel bars of cross-section 20 mm × 80 mm are to be bolted together to form a single bar. It is found that an error has been made in one bar so that the distance between centres is 1 mm more than the specified 2 m. A hydraulic jack is used to complete the assembly. Determine the stresses in the bars when the assembly is subject to tension of 400 kN.

5.17. A steel rod projects 0·1 mm from a hollow copper cylinder as shown in Figure 5.6b. Find the greatest load P which may be applied if the cross-sectional areas are each 1000 mm² and the permissible normal stresses are 200 and 150 MN/m² in the steel and copper respectively. Take E (copper) $= 120$ GN/m².

5.18. A rod and coaxial tube of length 0·2 m and cross-sectional areas 200 and 400 mm² respectively are rigidly connected together and compressed between platens. Sketch the load/deformation diagram and calculate the assembly collapse load. The yield stresses are 80 and 200 MN/m² respectively and $E - 200$ GN/m².

5.19. Calculate limit load for assembly of Example 5.2 (clamped at nodes 1 and 3) if the yield stresses are 240 MN/m² and 180 MN/m² for steel and aluminium respectively.

5.20. Repeat Example 5.16 with bar 2 of aluminium alloy having $\sigma_0 = 105$ MN/m².

5.21. If the load is removed just at collapse for system of Problem 5.18 determine the residual stresses.

Beams—I (Stresses)

6.1 Introduction

A bar subjected to loads (forces and moments) in a transverse plane as in Figure 6.1a undergoes bending; a bar primarily intended to resist bending action is known as a beam. In its most simple form an unloaded beam has its longitudinal axis straight, and has uniform cross-section with two perpendicular axes of symmetry (Figure 6.1c, d and e) which are taken as y and z axes; the longitudinal axis x passes through the centre of area of the cross-section.

Bending may be caused by loads in either of the transverse planes xy and xz. Loads in any other transverse plane can always be resolved into components in these two planes. We specifically refer to Chapters 9 and 11 those cases in which there are any forces with a component in the x direction. The loads acting may consist of concentrated forces, distributed forces and concentrated moments (A, B and C of Figure 6.1a). Concentrated forces and moments are considered to act at a mathematical point while distributed

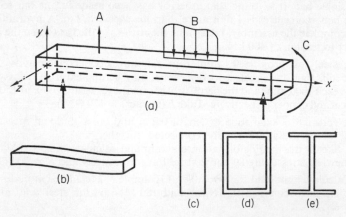

Figure 6.1 (a) Loads on beam. (b) Effect of loads on (a)—(half scale). (c) Rectangular section. (d) Hollow (box) section. (e) I-section

forces are exerted over a finite length of the beam. Concentrated forces parallel to the y, z axes will be denoted by Y, Z respectively. The intensities of distributed force (i.e. force per unit length) in the xy and xz planes will, where necessary for clarity, be denoted by w_y and w_z respectively; the subscript will be omitted if the distributed force in the problem acts only in one plane. Similarly, concentrated moments in the xy and xz planes are denoted m_z and m_y, respectively, in general, but where not necessary for clarity the subscript will be dropped. The sign convention for forces and moments follows the definitions in Article 5.1.

The conditions for equilibrium of a beam loaded in the xy plane are

$$\sum Y = 0; \qquad \sum m_z = 0 \qquad (6.1a)$$

and for loading in the xz plane are

$$\sum Z = 0; \qquad \sum m_y = 0 \qquad (6.1b)$$

A beam subject to a set of loads in a transverse plane can always be maintained in equilibrium by two concentrated forces or by a concentrated force and a moment in that plane. Since these loads are usually considered to be exerted by the supports of the beam they are known as reactive loads.

Referring to Figure 6.2a a roller support is considered to exert a concen-

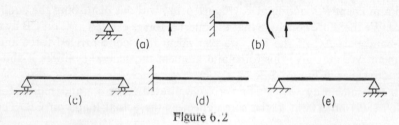

Figure 6.2

trated force on the beam, while the fixed support in Figure 6.2b exerts both a concentrated force and a moment. In order to maintain equilibrium we therefore require as a minimum, either two roller supports as in Figure 6.2c, or, one fixed support as in Figure 6.2d, the latter being known as a cantilever. More than this number of supports can, of course, be used but it should be clear that the number of reactive loads will then exceed the number of equilibrium equations (6.1).

However the supports have the additional property that they specify, i.e. prescribe displacement(s) at their points of action. For instance, for loading in the xy plane a roller prescribes the linear displacement (translation) in the y direction at the supported point; at a fixed support the values of both linear and angular displacement are prescribed. In the latter case, which corresponds to a cantilever (Figure 6.2d), the support is sufficient to prevent displacements of the beam as a rigid body, i.e. without change of size or shape. However, in the case of the beam in Figure 6.2c displacement as a rigid body in the x direction is possible. Therefore one roller is replaced by a hinge (Figure 6.2e) which prescribes displacement in the x direction at one point but does not introduce a reactive load since, as previously stated, there are no forces in the x direction; this system is called simply supported.

The two support systems in Figures 6.2d and e are therefore the minimum to satisfy the twin requirements of static equilibrium and prevention of rigid body displacements. This duality can be established from a single argument by using the Principle of Virtual Work (see Appendix I). Let either of the systems in Figures 6.2d and e be subject to loads and be considered to be rigid bodies. Then it is clear that since no virtual displacements are possible the virtual work is zero and the systems are in equilibrium.

6.2 Shearing Force and Bending Moment

6.2.1 *Shearing Force and Bending Moment Diagrams*

Figure 6.3a shows a bar in equilibrium under the action of a system of loads in a transverse plane. The material of the bar must be subject to internal forces since it must transmit the effects of the loads between their lines of action. Now the force P_1 acting to the left of C could be replaced by a force P_1 and a moment P_1x acting at C (Figure 6.3b) without disturbing the equilibrium. We therefore conclude that the internal forces exerted at C on CB by the left-hand part AC of the bar, are equivalent to a concentrated force and a moment. We now give this force and moment the general symbols V and M respectively as in Figure 6.3c.

In the same way we can find a concentrated force V' and moment M' which exert on AC at C the same effect as the actual loads on CB. These

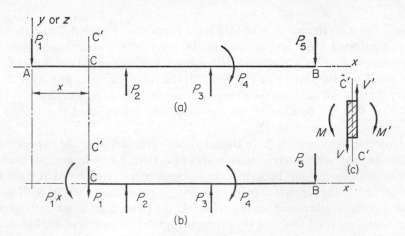

Figure 6.3

quantities are also shown in Figure 6.3c, where *for illustrative purposes only* a short length of bar at point C is shown. However the forces V and V' and moments M and M' must be thought of as actually acting at the geometric point C. Then since the bar is in equilibrium the forces V and V' are equal and opposite, as are the moments M and M'. The opposed forces V and V' exerted at C by the two parts of the bar tend to shear the material at C as in Figure 6.4a, and are called *shearing forces*. The opposed moments M and M' exerted at C tend to bend the material as in Figure 6.4c and are therefore called *bending moments*. As will be shown later in this chapter these internal actions can be related to the stresses acting on the cross-section of the bar. Formal definitions are now given.

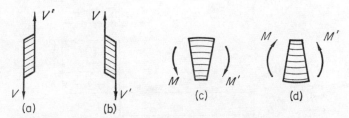

Figure 6.4   (a) Positive, (b) negative, shearing force. (c) Positive
('hogging'), (d) negative ('sagging'), bending moment

The *shearing force* at any position along a bar is the resultant of all the forces acting to one side of that position.

The *bending moment* at any position along a bar is the resultant moment about that position of all the loads acting to one side of it.

The sign convention which will be used for shearing force and bending moment is illustrated in Figure 6.4, and is the same for both xy and xz transverse planes.

Graphs showing the variation of these quantities over the whole length of a beam are called shearing force and bending moment diagrams and their construction will now be illustrated by some examples.

Example 6.1   A beam of length L is in equilibrium under the action of concentrated forces as shown in Figure 6.5a. Draw the shearing force and bending moment diagrams.

Solution   Extend down the lines of action of the concentrated forces as shown in the figure, draw base lines for shearing force and for bending moment, and choose suitable scales for these quantities.

Introduce a section line aa at x from the left-hand end. The procedure is now to move this section line from $x = 0$ to $x = L$ while plotting the values obtained.

From the definition given earlier the shearing force is the resultant force either to left or right of the section line aa. When aa is anywhere between A and B the resultant of the forces to the left is Pc/L acting up. According to the sign convention

(Figure 6.4b) this is a negative shearing force and is therefore plotted at constant value $-Pc/L$ over AB as shown in Figure 6.5b. When aa is anywhere between B and C the resultant force to the left of the section line is, using the sign convention, $-Pc/L+P$, i.e. $+Pb/L$, and is plotted as shown. It will be seen that there is an abrupt change in shearing force at the line of action of each concentrated force. The shearing force diagram could also be drawn by looking to the right of the section line aa. For instance when aa is between B and C the resultant force to the right is Pb/L acting up, which from Figure 6.4a is positive.

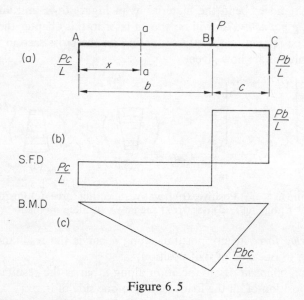

Figure 6.5

For any position of aa between A and B and distant x from A the resultant moment to the left of aa is Pcx/L acting clockwise. Referring to the sign convention in Figure 6.4d this is a negative bending moment, and, since it is linearly dependent on x, varies from zero at A to $-Pbc/L$ at B as shown in Figure 6.5c. When aa is between B and C the resultant moment to the left is

$$-\frac{Pcx}{L}+P(x-b)$$

where the force at B causes anticlockwise moment to the left (positive by the sign convention). The bending moment changes linearly from $-Pbc/L$ at B to zero at C ($x = L$) as shown. Moments could of course be taken to the right of aa with the same final result. In this example the bending moment is negative over the entire length of the beam corresponding to so-called 'sagging' curvature. If $b = c = L/2$ the maximum value of the bending moment is $-PL/4$.

Finally it should be noted that the forces at A and C could be reactive loads exerted by supports as in Figure 6.2e.

Example 6.2 Draw the shearing force and bending moment diagrams for the beam of Figure 6.6a. The moment of 20 kNm acts on the beam at A.

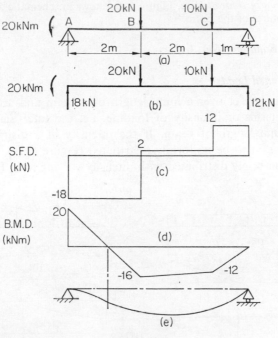

Figure 6.6

Solution Following Figure 6.2a the supports exert concentrated forces, say Y_A and Y_D, acting upwards. These reactive loads are determined from static equilibrium. Thus taking moments about D

$$20 - Y_A \times 5 + 20 \times 3 + 10 \times 1 = 0$$

and so $Y_A = 18$ kN. Similarly we find $Y_D = 12$ kN. The effective loads on the beam are therefore as in Figure 6.6b.

At any position between B and C the shearing force is $(-18 + 20)$ kN and between C and D it is $(-18 + 20 + 10)$ kN as shown in Figure 6.6c.

At any point between A and B the bending moment is given by

$$M = +20 - 18x$$

In this equation the quantity 20 kNm is the contribution of the clockwise moment at A which would tend to cause 'hogging' (Figure 6.4c); $-18x$ is the contribution of the force at A which would tend to cause 'sagging'. The bending moment is seen to vary linearly over AB from $+20$ kNm at A to -16 kNm at B, and is zero at $x = 10/9$ m. It will be noted that at A the bending moment changes abruptly to 20 kNm due to the concentrated moment. Similarly, by writing general expressions for bending moment applicable between B and C and between C and D, the bending moment diagram (Figure 6.6d) can be completed. As a particular example let us find the bending moment at C by looking to the right of C. Then the bending moment is (-12×1) kNm since the moment of the force at D would tend to cause sagging (Figure 6.4d). The position where the bending moment changes sign is known as a *point of contraflexure*. Figure 6.6e shows in schematic form the corresponding curvatures of the beam.

Problems for Solution   6.1, 6.2

6.2.2 *Distributed Loads*

Distributed forces act over a finite length of a beam and are conveniently expressed in terms of intensity of loading, i.e. the total amount of force exerted per unit length of beam. If the intensity of a distributed load is constant it is said to be uniformly distributed (Figure 6.7a); otherwise the load is non-uniformly distributed with intensity varying as in Figure 6.1a.

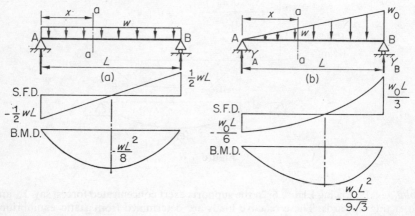

Figure 6.7

Example 6.3   Draw the shearing force and bending moment diagrams for the beam in Figure 6.7a.

Solution   From the equilibrium conditions (6.1) the reactive loads are each $\frac{1}{2}wL$. To the left of a section line aa at x from A there is the concentrated force at A of $\frac{1}{2}wL$ acting up, and a total downward force wx due to the distributed load. Then, applying the sign convention, the shearing force at aa is

$$V = -\frac{wL}{2} + wx \tag{i}$$

Thus the shearing force changes linearly between the values $-\frac{1}{2}wL$ at A and $\frac{1}{2}wL$ at B, as shown in Figure 6.7a.

The bending moment at aa is

$$M = -\left(\frac{wL}{2}\right)x + wx\left(\frac{x}{2}\right) \tag{ii}$$

where the first term is the moment of the reactive load and the second term consists of the product of the resultant of the distributed load wx and the distance from aa to its line of action. Equation (ii) is quadratic so that the bending moment diagram is a curve with zero values at A and B and a maximum value of $-wL^2/8$ at the centre of the beam.

The maximum bending moment can be written as $-(wL)(L/8)$ where wL is recognized as the total load on the beam. In Example 6.1 the maximum bending moment for a central load was noted to be $-PL/4$, thus if the same total force acts in each case, i.e. if $P = wL$, the maximum bending moment is twice as great for the concentrated force. While this result is particular to the support system used, it is often found that peak bending moments are reduced if loads are distributed rather than concentrated.

Example 6.4   Draw the shearing force and bending moment diagrams for the beam of Figure 6.7b.

Solution   The distributed load intensity at x from A is

$$w = w_0 \frac{x}{L} \qquad \text{(i)}$$

The total load is $\frac{1}{2}w_0L$, i.e. the average intensity times the length. It can readily be proved from statics that the resultant of the load acts through the centre of area of the load intensity diagram, i.e. at $\frac{2}{3}L$ fro them left end. Taking moments about B

$$-Y_AL + \tfrac{1}{2}w_0L(\tfrac{1}{3}L) = 0$$

i.e.
$$Y_A = \frac{w_0L}{6}$$

and so
$$Y_B = \tfrac{1}{2}w_0L - \frac{w_0L}{6} = \tfrac{1}{3}w_0L$$

The distributed load to the left of aa has resultant $\frac{1}{2}wx$ acting $x/3$ from aa. Therefore the shearing force is $(-Y_A + \frac{1}{2}wx)$ and substituting from (i) we find

$$V = -\frac{w_0L}{6} + \frac{w_0x^2}{2L} \qquad \text{(ii)}$$

At $x = 0$ $V_A = -w_0L/6$;   at $x = L$ $V_B = +\frac{1}{3}w_0L$; $V = 0$ at $x = L/\sqrt{3}$.

The bending moment at aa is

$$M = -Y_Ax + \tfrac{1}{2}wx(\tfrac{1}{3}x) = -\frac{w_0Lx}{6} + \frac{w_0x^3}{6L}$$

and its greatest magnitude is $-w_0L^2/9\sqrt{3}$ at $x = L/\sqrt{3}$.

In the following two examples the principles are as already described and only difficult points will be noted as a guide.

Example 6.5   Draw shearing force and bending moment diagrams for the beam in Figure 6.8a.

Solution   Following Figure 6.2b the reactive loads are a concentrated force Y_A and a moment m_A. From equilibrium of forces $Y_A = wb$, and by taking moments about A we find $m_A = (wb) \times (3b/2)$.

5

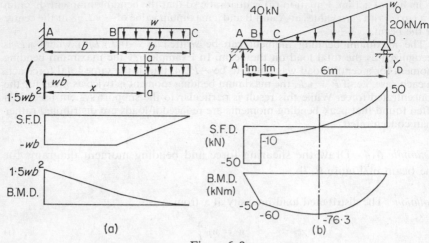

Figure 6.8

Between B and C, looking to the left of aa,
$$V = -wb + w(x-b); \quad M = 1\cdot5\,wb^2 - wbx + \tfrac{1}{2}w(x-b)^2$$

Example 6.6   Draw the shearing force and bending moment diagrams for the beam of Figure 6.8b.

Solution   Here $w = w_0 \left(\dfrac{x-2}{6}\right)$ for $x \geqslant 2$

Taking moments about D
$$- Y_A \times 8 + 40 \times 7 + \tfrac{1}{2}w_0 \times 6 \times 2 = 0$$
Hence $Y_A = 50$ kN and $Y_D = (40 + 60 - 50)$ kN.

Over CD
$$V = -Y_A + 40 + \tfrac{1}{2}w(x-2)$$
and
$$M = -Y_A x + 40(x-1) + \tfrac{1}{2}w(x-2) \times \tfrac{1}{3}(x-2)$$

It should be noted that shearing force and bending moment diagrams for a beam subject to several loads can be obtained by algebraic summation of diagrams drawn for each load acting separately.

Example 6.7   A uniformly distributed load of 4 kN/m intensity is added to the beam of Figure 6.6a. Find the shearing forces and bending moments at B and C.

Solution   Referring to Example 6.3 and substituting $w = 4$ kN/m, $L = 5$ m in equations (i) and (ii) we find shearing forces of -2 kN at $x = 2$ m and $+6$ kN at

$x = 4$ m, while the bending moments are respectively -12 kNm and -8 kNm. Adding these quantities to the values in Figures 6.6c and d the shearing forces at B and C change from -20 kN to zero and from 8 kN to 18 kN respectively, and the bending moments are -28 kNm and -20 kNm.

Problems for Solution   6.3–6.6

6.2.3 *Relations Between Shearing Force, Bending Moment and Load Intensity*

Figure 6.9 represents the part of a loaded beam, such as that of Figure 6.7a, which lies between x and $(x + dx)$. A distributed load of intensity w acts down on dx. V and M are the resultants of the forces and moments acting on the

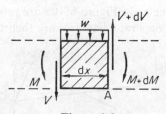

Figure 6.9

part of the beam to the left of x; $(V + dV)$ and $(M + dM)$ are the resultants of the loads acting on the right of $(x + dx)$. In Figure 6.4 the resultants to the left and right were shown to be equal and opposite. Here the possibility of a slight difference between left and right is incorporated because an infinitesimal length is being considered.

From Figure 6.9 for equilibrium of vertical forces

$$V + w\,dx - (V + dV) = 0 \qquad \text{i.e. } w = \frac{dV}{dx} \qquad (6.2)$$

For equilibrium of moments about point A

$$M + w\,dx(\tfrac{1}{2}dx) + V\,dx - (M + dM) = 0$$

Neglecting products of infinitesimals,

$$V = \frac{dM}{dx} \qquad (6.3)$$

The relations expressed by equations (6.2) and (6.3) are useful in sketching and interpreting bending moment and shearing force diagrams.

Referring to Example 6.1 (Figure 6.5) the shearing force is negative in AB and so by (6.3) the bending moment diagram has a negative slope; the reverse is true in BC. The abrupt change of shearing force at B means an abrupt change of slope of the bending moment diagram at B. Referring to Example 6.3 (Figure 6.7a) the uniformly distributed force acts down making w constant and positive in equation (6.2) so that the shearing force diagram has constant positive slope. In the left half of the beam the shearing force is negative and increasing (algebraically) so that by (6.3) the slope of the bending moment diagram is negative but increasing, i.e. becoming less negative; over the right half the shearing force is positive and increasing as is therefore the bending moment diagram slope. Note that the zero shearing force at mid-span corresponds to a stationary value of bending moment.

6.2.4 *Loads in Two Planes*

In all the preceding examples the loads acted in the xy plane. Loads acting in the xz plane can be treated in the same way to obtain separate shearing force and bending moment diagrams for this plane. Forces acting at an angle to the xz plane and the xy plane as in Figure 6.10a are resolved into their components in the two planes. Separate diagrams are then drawn for each plane.

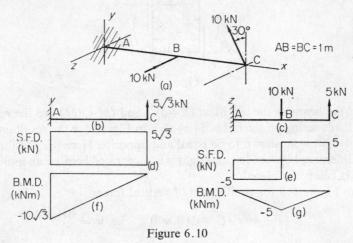

Figure 6.10

Example 6.8 Draw shearing force and bending moment diagrams for the cantilever in Figure 6.10a.

Solution The force at C is resolved into its components $10 \cos 30°$ and $10 \sin 30°$ in the xy and xz planes respectively. The loads acting in each plane are shown in Figures 6.10b and c, the xz plane being viewed from below. The shearing force and bending moment diagrams are then constructed as illustrated.

Problem for Solution 6.7

6.3 Stresses Due to Bending Moments

6.3.1 *States of Stress and Strain*

In the engineering theory of bending the stresses due to the shearing forces and bending moments are derived separately. The shearing forces are dealt with in Article 6.4. The theory for stresses due to bending moments is based on the following assumptions:

(a) During bending original plane cross-sections of the beam remain plane and perpendicular to the longitudinal (x) axis of the beam.

(b) σ_x is the only non-zero stress component relative to the x, y, z axes (Figure 6.1a).

Figure 6.11 shows (greatly exaggerated) an infinitesimal length of a beam before and after bending due to bending moments M in the xy plane. In the undeformed state nn is the edge of the xz plane, and mm is the edge of a plane parallel to the xz plane and distant y from it. The edges of these planes are mm$_1$ and nn$_1$ in the end view.

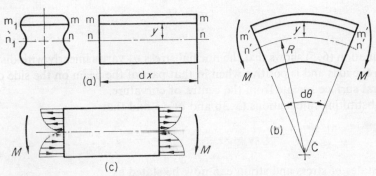

Figure 6.11 (a) Before bending. (b) After bending (greatly exaggerated). (c) Internal forces on the ends of the element

After bending, the planes nnn$_1$n$_1$ and mmm$_1$m$_1$ become curved surfaces n'n'n'$_1$n'$_1$ and m'm'm'$_1$m'$_1$. By assumption (a) the ends of the element (being cross-sections of the beam) remain plane and perpendicular to the axis, so that the lines m'n' are each on a radius of curvature. The centre of curvature is at their intersection C where the angle subtended is $d\theta$. The radius of curvature of the surface n'n'n'$_1$n'$_1$ is called R.

Since only normal stresses are assumed to act on the cross-section and these stresses are to have as resultant a bending moment, it is to be expected that part of the cross-section will be in tension and part in compression (Figure 6.11c). The surface of zero stress separating these zones is called the neutral surface and we shall consider this to be n'n'n'$_1$n'$_1$ in Figure 6.11b. Then n'n' = nn = mm = dx and from Figure 6.11b we see that n'n' = $Rd\theta$.

Now the line m'm' has length $(R+y)d\theta$ and its original length mm was, as shown above, equal to n'n' which equals $Rd\theta$. The increase of length is thus $yd\theta$ and the strain is obtained by dividing by the original length

i.e. $$\text{Strain} = \frac{yd\theta}{Rd\theta}$$

This is a normal strain in the direction of the x axis so that

$$\epsilon_x = \frac{y}{R} \tag{6.4}$$

Equation (6.4) shows that the normal strain ϵ_x is directly proportional to distance from the neutral surface of the beam. Introducing (6.4) in the strain/ stress equation (3.3a), we have, for a linear elastic material,

$$\epsilon_x = \frac{y}{R} = \frac{1}{E}(\sigma_x - \nu(\sigma_y + \sigma_z))$$

By assumption (b) $\sigma_y = \sigma_z = 0$
so that

$$\frac{\sigma_x}{y} = \frac{E}{R} \tag{6.5}$$

Equation (6.5) shows that the normal stress σ_x varies linearly with distance from the axis and is positive when in that part of the beam on the side of the neutral surface remote from the centre of curvature.

Substituting in equations (3.3b and c) we find that

$$\epsilon_y = \epsilon_z = -\frac{\nu\sigma_x}{E} = -\nu\epsilon_x \tag{6.6}$$

The states of stress and strain can now be stated as

$$\mathbf{T} = \begin{pmatrix} \sigma_x & 0 & 0 \\ 0 & 0 & 0 \\ 0 & 0 & 0 \end{pmatrix} \quad \mathbf{R} = \begin{pmatrix} \epsilon_x & 0 & 0 \\ 0 & \epsilon_y & 0 \\ 0 & 0 & \epsilon_z \end{pmatrix} \tag{A6.1, 2}$$

wherein the components are given by equations (6.5) and (6.6). This variation of σ_x with y over the ends of a beam segment is shown in Figure 6.12a. Figure 6.12b shows an end surface with dA an infinitesimal portion distant y from the z axis. The force exerted in the x direction on dA is $\sigma_x dA$.

Substituting from (6.5) the total force on the end surface is $\int_A (E/R)y\,dA$ where $\int_A$ signifies that the integration is carried over the entire area A of the end surface.

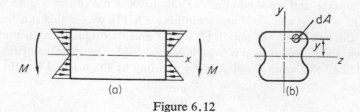

Figure 6.12

However, since there is no force in the x direction, it follows that

$$\int_A \frac{E}{R} y \, dA = 0 \qquad (6.7a)$$

If the beam is homogeneous, E is constant over the cross-section and equation (6.7a) reduces to

$$\int_A y \, dA = 0 \qquad (6.7b)$$

Since $\int_A y \, dA$ is the first moment of area of the whole cross-section about the z axis, equation (6.7b) requires that the z axis should pass through the centre of area.

Considering now the moments on the end surface, we have for area dA a moment $(\sigma_x \, dA)y$. Substituting from (6.5) the total moment for the whole surface is

$$\int_A \frac{E}{R} y^2 \, dA$$

This must equal the moment applied to the end surface, i.e. tne bending moment M.

Therefore

$$M = \int_A \frac{E}{R} y^2 \, dA \qquad (6.8a)$$

If the material is homogeneous, E is constant and (6.8a) becomes

$$M = \frac{E}{R} \int_A y^2 \, dA \qquad (6.8b)$$

The quantity $\int_A y^2 \, dA$ is known as the second moment of area of the cross-section about the z axis, and is denoted in general by the symbol I. Equation (6.8b) can then be rewritten as

$$\frac{M}{I} = \frac{E}{R} \qquad (6.9)$$

Combining equations (6.5) and (6.9) we have

$$\frac{\sigma_x}{y} = \frac{E}{R} = \frac{M}{I} \qquad (6.10)$$

Thus in general σ_x varies with y and, through M, with x.

6.3.2 *Second Moment of Area*

From the basic definition

$$I = \int_A y^2 \, dA$$

the following standard results may be derived.

Rectangular Cross-section   Take dA as a strip of width b and thickness dy as in Figure 6.13a. Then with the z axis through the centre of area

$$I = \int_{-\frac{1}{2}h}^{\frac{1}{2}h} y^2 b\,\mathrm{d}y = \frac{bh^3}{12} \qquad (6.11)$$

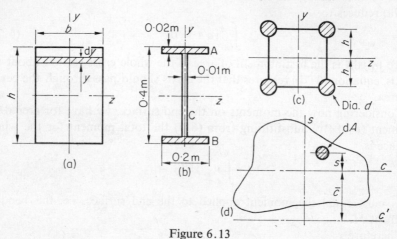

Figure 6.13

Circular Cross-section   (diameter d): Similarly it can be shown that

$$I = \frac{\pi d^4}{64} \qquad (6.12)$$

Complex Shapes   These can frequently be divided into rectangular or circular subsections. For instance the I-section in Figure 6.13b could be considered as made up of two hatched rectangles A, B (the flanges) and the unshaded part C (the web). The latter rectangle is symmetric about the z axis, and so equation (6.11) gives its contribution to I. However the z axis is not an axis of symmetry of either A or B so that equation (6.11) has to be used in conjunction with the Parallel Axes Theorem

$$I_{c'} = I_c + A\bar{c}^2 \qquad (6.13)$$

Here I_c is the second moment of area about an axis c through the centroid of the area (Figure 6.13d), A is the area and $I_{c'}$ is the second moment of the area about an axis c', parallel to c but distant $\bar{c}$ from it. From the figure

$$I_{c'} = \int_A (s+\bar{c})^2\,\mathrm{d}A$$

$$= \int_A s^2\,\mathrm{d}A + 2\bar{c}\int_A s\,\mathrm{d}A + \bar{c}^2\int_A \mathrm{d}A$$

$$= I_c + A\bar{c}^2 \qquad \text{Q.E.D.}$$

since

$$\int_A s\, \mathrm{d}A = 0$$

(first moment of area about centroid).

Example 6.9 Find I about the z axis for the cross-sections of Figures 6.13b and c.

Solution (a) Using the Parallel Axes Theorem with c the axis of symmetry of the flange we have $\bar{c} = 190$ mm and

$$(I)_{\text{flange}} = \frac{200 \times 20^3}{12} + (200 \times 20) \times 190^2 - 1{\cdot}445 \times 10^8 \text{ mm}^4$$

Using (6.11)

$$(I)_{\text{web}} = \frac{10 \times 360^3}{12} = 0{\cdot}389 \times 10^8 \text{ mm}^4$$

For the whole cross-section

$$I = 2 \times (I)_{\text{flange}} + (I)_{\text{web}} = 3{\cdot}279 \times 10^{-4} \text{ m}^4$$

(b) For Figure 6.13c we have, using (6.12) and ignoring the thin-sheet connexions,

$$I = 4 \times \left(\frac{\pi d^4}{64} + \frac{\pi d^2 h^2}{4} \right)$$

It is sometimes easier to calculate the second moment of area of a complex shape by subtracting contributions of some subdivisions.

Example 6.10 Find I for the 'box'-type section in Figure 6.14a.

Solution The section may be thought of as made up of a solid rectangle of sides 200 mm and 100 mm, less a solid rectangle of sides 180 mm and 80 mm. Then applying equation (6.11)

$$I = \left(\frac{200 \times 100^3}{12} - \frac{180 \times 80^3}{12} \right) \times 10^{-12} = 899 \times 10^{-8} \text{ m}^4$$

The second moment of the box section is $53{\cdot}9\%$ of that of a solid rectangular section 200 mm by 100 mm, while its area is just 28% of that of the complete rectangle. This is because area was multiplied by the square of distance from the z axis, when defining I. Therefore material at a small distance from the axis contributes less relatively to I than material further from the axis.

Hollow Circular Section Using the method of Example 6.10 we find for a tube with outer and inner diameters d_1 and d_2

$$I = \frac{\pi}{64} (d_1^4 - d_2^4)$$

Problems for Solution   6.8–6.11

6.3.3 *Stresses* (*Doubly Symmetric Cross-sections*)

The stress distribution in a beam is most conveniently expressed by extracting from (6.10)

$$\sigma_x = \frac{M}{I} y \qquad (6.14)$$

In equation (6.14) the stress σ_x is related to the applied loads through the bending moment M, to the shape of the cross-section through the second moment of area I, and to position within the cross-section through y. Furthermore, since M and I may vary with position along the beam axis, σ_x is dependent on x. Therefore to find the stress at any position (x, y, z) within the beam the bending moment at x and the second moment of area of the cross-section at x are determined and introduced, with the coordinate y, into equation (6.14).

For doubly symmetric cross-sections as in Figure 6.14, we have from equation (6.7b) that the neutral axis, i.e. the z axis, is at half the depth of the cross-section.

Example 6.11   A beam of uniform cross-section as in Figure 6.13b is loaded in the manner of Figure 6.6 but with loads 200 kNm, 200 kN, 100 kN. Calculate the stress at points $(2, 0.2, 0)$, $(2, -0.2, 0.1)$, $(0, -0.05, 0)$.

Solution   (a) $x = 2$ m; $y = 0.2$ m; $z = 0$. At $x = 2$ m we find $M = -160$ kNm (cf. Figure 6.6d). Taking I from Example 6.9, we have, using (6.14),

$$\sigma_x = -\frac{160 \times 10^3 \times 0.2}{3.279 \times 10^{-4}} = -97.6 \times 10^6 \text{ N/m}^2$$

(b) $x = 2$ m; $y = -0.2$ m; $z = 0.1$ m. M and I are as in case (a) and so

$$\sigma_x = 97.6 \times 10^6 \text{ N/m}^2$$

Thus, for this doubly symmetric cross-section the stresses at the upper and lower surfaces have the same magnitude but opposite signs. The sagging bending moment is consistent with compression at the top surface and tension at the bottom. The coordinate z does not enter into the calculations and this remains true as long as the loads act only in the xy plane.

(c) $x = 0$; $y = -0.05$ m; $z = 0$. M has its greatest value 200 kNm and

$$\sigma_x = \frac{200 \times 10^3 \times -0.05}{3.279 \times 10^{-4}} = -30.5 \times 10^6 \text{ N/m}^2$$

Failure Conditions   In Example 6.11 we did not consider whether the material could withstand the calculated stresses. From the stress array (A6.1) the greatest shear stress at a point in the beam is, using result (1.12), $|\frac{1}{2}\sigma_x|$.

Then if $(\sigma_x)_{max}$ is the greatest value of σ_x anywhere in the beam

$$\tau_{max} = \tfrac{1}{2}|(\sigma_x)_{max}|$$

By the maximum shear stress theory failure occurs when $\tau_{max} = \tfrac{1}{2}\sigma_0$, where σ_0 is the yield stress in a tensile test (see Article 1.12). Hence at failure $|(\sigma_x)_{max}| = \sigma_0$ and if a safety factor f_s is to be allowed

$$|(\sigma_x)_{max}| \leqslant \sigma_0/f_s \tag{6.15}$$

If the beam has uniform cross-section, I in equation (6.14) is a constant and so the maximum stress is found by substituting the greatest values of M and y. Non-uniform beams will be considered separately in Section 6.3.6.

Example 6.12 Calculate the safety factor in Example 6.11 if σ_0 is 250 MN/m².

Solution As noted before the maximum bending moment is 200 kNm and from Figure 6.13b the greatest values of y are ± 0.2 m. Substituting in equation (6.14), as in Example 6.11, $(\sigma_x)_{max} = \pm 122$ MN/m².

Applying condition (6.15)

$$f_s = \frac{250}{122} = 2.05$$

Example 6.13 Find the maximum intensity of distributed load which can be carried over BC in Figure 6.8a, if the beam has uniform cross-section as in Figure 6.14a, $b = 2$ m, $\sigma_0 = 250$ MN/m², and a safety factor of 2 is required.

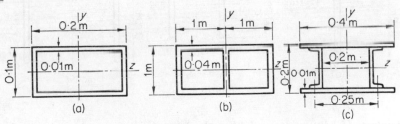

Figure 6.14

Solution From Figure 6.8a the greatest bending moment is $1.5 \, wb^2$, i.e. $6w$ (units Nm); the greatest value of y, i.e. y_{max}, is ± 0.05 m. Taking I from Example 6.10 and substituting in (6.14)

$$(\sigma_x)_{max} = \frac{6w \times 0.05 \times 10^8}{899} = 3.33w \times 10^4 \text{ N/m}^2$$

Applying condition (6.15)

$$3.33w \times 10^4 \leqslant \frac{250 \times 10^6}{2}$$

Hence

$$w \leqslant 3750 \text{ N/m}$$

Example 6.14 Find the least value for h (Figure 6.13c) when $d = 0.02$ m, and the greatest bending moment is 20 kNm. Take $\sigma_0 = 270$ MN/m^2 and $f_s = 1.5$.

Solution Following the discussion in Example 6.9 we obtain

$$I = \pi(1 + 4h^2) \times 10^{-8} \text{ m}^4$$

if h is in cm. The greatest y is $(h + 0.5d)$ and so from (6.14) and (6.15)

$$(\sigma_x)_{max} = \frac{20 \times 10^3 \times (h + 0.5d) \times 10^{-2}}{\pi(1 + 4h^2) \times 10^{-8}} \leqslant \frac{270 \times 10^6}{1.5}$$

i.e. $h \geqslant 9.73$ cm

Problems for Solution 6.12–6.17

6.3.4 *Stresses (Singly Symmetric Cross-section)*

The methods described in previous sections can be applied to beams of this type (Figure 6.15) provided the loads act in the plane of symmetry, which, for those illustrated, is the xy plane. As required by equation (6.7b) of Section 6.3.1 the z axis passes through the centroid of the cross-section.

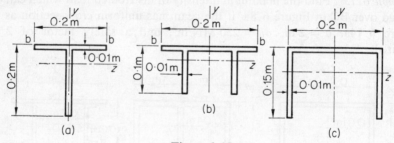

Figure 6.15

Example 6.15 Investigate whether the beam of T-type cross-section as in Figure 6.15a can support a bending moment of 30 kNm. Take $\sigma_0 = 250$ MN/m^2.

Solution To find the centroid we take first moments of area about bb in Figure 6.15a. Let $\bar{y}$ be the distance from bb to the z axis. The sum of the first moments of web and flange considered separately must be the same as the first moment of the whole cross-section.

First moment of flange is $(200 \times 10) \times 5$ mm^3 = 10,000 mm^3.

First moment of web is $(190 \times 10) \times 105$ mm^3 = 19.95×10^4 mm^3.

First moment of whole cross-section is (area) $\times \bar{y}$ = $3900 \times \bar{y}$ mm^3.

Hence

$$10,000 + 19 \cdot 95 \times 10^4 = 3900 \, \bar{y}$$

and

$$\bar{y} = 53 \cdot 7 \text{ mm} = 0 \cdot 0537 \text{ m}$$

The position of the z (neutral) axis is now known and so the second moment of area can be calculated using the Parallel Axes Theorem.

$$(I)_{\text{flange}} = (200 \times 10^3/12) + (200 \times 10) \times (\bar{y} - 5)^2 = 477 \times 10^4 \text{ mm}^4$$

where the quantity $(\bar{y} - 5)$ mm is the net distance between the z axis and a parallel axis through the centre of area of the flange.

$$(I)_{\text{web}} = (10 \times 190^3/12) + (10 \times 190) \times (105 - \bar{y})^2 = 1071 \times 10^4 \text{ mm}^4$$

Hence

$$I = 1548 \times 10^{-8} \text{ m}^4$$

The greatest value of y within the cross-section is the distance from the z axis to the bottom of the web, which is $-0 \cdot 1463$ m.

Substituting in equation (6.14).

$$(\sigma_x)_{\text{max}} = \frac{30,000 \times -0 \cdot 1463}{1548 \times 10^{-8}} = -283 \times 10^6 \text{ N/m}^2$$

Since condition (6.15) is invalidated this beam cannot support the proposed loads. It is interesting to note that at the top of the flange the stress is only $(30,000 \times 0 \cdot 0537 \times 10^8/1548)$, i.e. $104 \times 10^6 \text{ N/m}^2$.

Special Failure Conditions The onset of elastic failure of some materials cannot be described by a single characteristic stress. For these materials specified values of normal stress will be used with different values for tension and compression. For doubly symmetric cross-sections the maximum tensile and compressive stresses are equal and so comparison need only be made with the lesser magnitude of failure stress. However for cross-sections with only one axis of symmetry the greatest tensile and compressive stresses are not necessarily equal and so each failure condition must be considered separately

Example 6.16 Two T-type cross-section beams are bonded together to form a beam with cross-section as in Figure 6.15b. The material has limit stresses of $+100$ and -240 MN/m². Determine whether this beam can support a bending moment of ± 10 kNm.

Solution Take moments of area about bb to find the centroid

$$(200 \times 10) \times 5 + 2 \times (10 \times 90) \times 55 = ((200 \times 10) + 2(10 \times 90))\bar{y}$$

i.e.

$$\bar{y} = 28 \cdot 7 \text{ mm}$$

$$(I)_{\text{flange}} = (200 \times 10^3/12) + (200 \times 10)(\bar{y} - 5)^2 = 114 \cdot 1 \times 10^4 \text{ mm}^4$$

$$(I)_{\text{webs}} = 2((10 \times 90^3/12) + (10 \times 90)(55 - \bar{y})^2) = 246 \cdot 3 \times 10^4 \text{ mm}^4$$

Hence

$$I = 360 \cdot 4 \times 10^{-8} \text{ m}^4$$

The extreme values of y are $+0 \cdot 0287$ m and $-0 \cdot 0713$ m.

Substituting in equation (6.14) we have, for $M = +10$ kNm, at the upper surface a stress of $(10^4 \times 0 \cdot 0287 \times 10^8/360 \cdot 4) \text{ N/m}^2$, i.e. $79 \cdot 7 \text{ MN/m}^2$; at the lower surface the stress obtained using $y = -0 \cdot 0713$ m is -198 MN/m^2. These stresses are within the permitted range $100 \geqslant \sigma_x \geqslant -240 \text{ MN/m}^2$.

If the maximum bending moment is -10 kNm the stresses at the upper and lower surfaces are $-79 \cdot 7 \text{ MN/m}^2$ and 198 MN/m^2 respectively, the latter being outside the permitted range. However, if the beam were inverted with the same load the extreme values of y would become $+0 \cdot 0713$ m and $-0 \cdot 0287$ m and the extreme stresses -198 MN/m^2 and $79 \cdot 7 \text{ MN/m}^2$, which are acceptable. Thus we see that under these conditions both the sign of the bending moment and the physical arrangement of the beam may be critical to its safety.

Problems for Solution   6.18–6.23

6.3.5 *Shape of Cross-section*

Referring to equation (6.14) in Section 6.3.3 we see that for a given bending moment the maximum stress is determined by I and y_{max} (the greatest value of y). If these geometrical properties of the cross-section are combined as

$$Z = \frac{I}{y_{\text{max}}} \tag{6.16}$$

where Z is called the section modulus, then

$$(\sigma_x)_{\text{max}} = \frac{M}{Z} \tag{6.17}$$

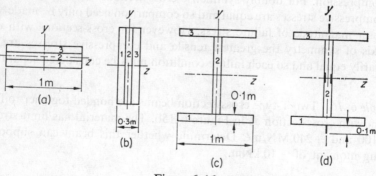

Figure 6.16

As an illustration of the relative significance of I and Z consider bending in the xy plane of the cross-sections in Figure 6.16, all of which have the same area. For example for Figure 6.16a

$$I = 1 \times 0 \cdot 3^3/12 \text{ m}^4; \quad y_{\text{max}} = 0 \cdot 15 \text{ m}; \quad Z = 0 \cdot 015 \text{ m}^3.$$

Similar calculations for the other sections lead to the summary in Table 6.1. The thin projections in Figure 6.16d have negligible area.

Table 6.1

	(a)	(b)	(c)	(d)
$I \times 10^4$ (m⁴)	22·5	250	690	690
Ratio to (a) value	1	11·1	30·7	30·7
$Z \times 10^3$ (m³)	15	50	115	98·6
Ratio to (a) value	1	3·33	7·67	6·57

From the table we see that I increases thirtyfold due to the redistribution of material further from the z axis. However Z increases less strongly between (a) and (c) due to the concomitant increase of y_{max}. For Figure 6.16d y_{max} increases but I is effectively constant so that Z decreases and by (6.17) the moment at which yielding begins is reduced compared to case (c). We therefore conclude that while it is generally desirable to dispose material as far from the neutral axis as possible when designing a cross-section, the effect of y_{max} must also be considered.

The section modulus is useful in choosing a cross-section for a given duty from a stock list of standard types. In general no stock beam will be an ideal choice for a given duty and the one selected is the nearest safe cross-section. This process is illustrated in Example 6.17 which follows.

Example 6.17 A simply supported beam carries a distributed load of linearly varying intensity. If the total load is 80 kN over a span of 4 m choose the most suitable cross-section from the stock list in Table 6.2. Take $\sigma_0 = 270$ MN/m² and allow for a safety factor of 2. (All the cross-sections listed are doubly symmetric.)

Table 6.2

Reference	$I \times 10^8$ (m⁴)	Depth (m)	Mass (kg/m)
A	2290	0·20	29
B	2220	0·15	45
C	4570	0·20	46
D	4430	0·25	31
E	5550	0·25	36

Solution This beam is loaded as in Example 6.4 (Figure 6.7b). If w_0 is the maximum intensity then

$$80 \times 10^3 = \tfrac{1}{2}w_0 L,$$

and so w_0 is 40 kN/m. From Example 6.4 in Section 6.2.2 we have that the greatest bending moment is $-w_0 L^2 / 9\sqrt{3}$, i.e. -41 kNm. From (6.17)

$$(\sigma_x)_{max} = \frac{M}{Z} = \frac{-41 \times 10^3}{Z} \tag{i}$$

But by (6.15)

$$|(\sigma_x)_{max}| \leqslant \sigma_0/f_s = 135 \times 10^6 \qquad \text{(ii)}$$

Therefore combining (i) and (ii)

$$Z \geqslant 41 \times 10^3/(135 \times 10^6) = 304 \times 10^{-6}\ m^3 \qquad \text{(iii)}$$

Since each section in Table 6.2 is doubly symmetric, y_{max} is half the depth. Then dividing each I by its y_{max} the section moduli Z are, respectively 229, 296, 457, 354, 444 all $\times 10^{-6}$ units m³.

Applying condition (iii) 'D' with $Z = 354 \times 10^{-6}\ m^3$ is the nearest safe section. The actual stress obtained by substituting this value of Z in (i) is 116 MN/m². While sections 'C' and 'E' are also safe the stresses would be lower still, representing less efficient use of material; these sections are also heavier (see Table 6.2).

Problems for Solution 6.24–6.26

6.3.6 *Non-uniform Beams*

Here we consider cases such as Figure 6.17b where the cross-section of a beam varies continuously along its longitudinal axis so that both I and y_{max} are functions of x. To find the greatest stress in beams of this type M, I and y_{max} are expressed as functions of x and equation (6.17) investigated mathematically for turning values of $(\sigma_x)_{max}$. Discontinuous cases as in Figure 6.17a can be considered as two uniform beams joined together.

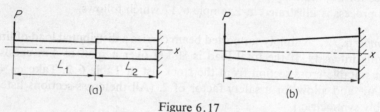

Figure 6.17

Example 6.18 A cantilever beam of square cross-section tapers from side 0.2 m at the support to 0.1 m at the free end (Figure 6.17b). Find the location of the maximum stress and its magnitude if $P = 10$ kN and $L = 4$ m.

Solution The bending moment is given by $M = 10^4 x$ Nm.

The side of the cross-section is given by $a = 0.1 (1 + (x/L))$. Now $I = a^4/12$ and $y_{max} = \frac{1}{2}a$ so that, substituting in equation (6.17), the greatest stress in the cross-section is

$$\sigma_x = 0.6 \times 10^8 \times x \left(1 + \frac{x}{L}\right)^{-3} \qquad \text{(i)}$$

Setting $d\sigma_x/dx = 0$ a stationary value is found for $x = \frac{1}{2}L$. By taking the second derivative of (i) it can be shown that this stationary value is a maximum. Substituting $x = \frac{1}{2}L = 2$ m in (i) we have $(\sigma_x)_{max} = 35.5$ MN/m².

Problems for Solution 6.27–6.29

6.3.7 *Loading in Two Planes*

When loads act in both xy and xz planes a bending moment diagram may be constructed for each plane as in Section 6.2.4. While stresses due to loads in the xy plane are obtained as before from equation (6.14) it is convenient for clarity to add the subscript z to M and I, since these quantities relate to a plane for which z is constant

i.e.
$$\frac{\sigma_x}{y} = \frac{M_z}{I_z} \tag{6.18}$$

To obtain the effect of loads in the xz plane the procedure of Section 6.3.1 is applied, with z everywhere replacing y. Then attaching subscript y to M and I we have

$$\frac{\sigma_x}{z} = \frac{M_y}{I_y} \tag{6.19}$$

Combining these equations, the total normal stress in the x direction is

$$\sigma_x = \frac{M_z y}{I_z} + \frac{M_y z}{I_y} \tag{6.20}$$

where M_z and M_y are the bending moments in the xy and xz planes.

I_z and I_y are the second moments of area of the cross-section about the z and y axes.

Example 6.19 The beam shown in Figure 6.18 is simply supported in both xy and xz planes. Find the greatest tensile and compressive stresses.

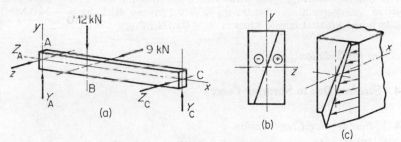

Figure 6.18 AB = 2 m, BC = 4 m; breadth of cross-section 120 mm, depth 240 mm

Solution From statics the reactive loads in the xy plane are $Y_A = 8$ kN, $Y_C = 4$ kN, and in the xz plane are $Z_A = 6$ kN, $Z_C = 3$ kN. In each plane the system is as in Example 6.1 (Figure 6.5) of Section 6.2.1, so that the greatest bending moment has magnitude Pbc/L where $b =$ AB, $c =$ BC. Hence the greatest value of M_z is -16 kNm and of M_y is 12 kNm (viewing the xz plane from below).

Here
$$I_z = 120 \times 240^3/12 \text{ mm}^4 = 1\cdot382 \times 10^{-4} \text{ m}^4$$
and
$$I_y = 240 \times 120^3/12 \text{ mm}^4 = 0\cdot346 \times 10^{-4} \text{ m}^4$$

Substituting in (6.20)

$$\sigma_x = \frac{-16,000 \times 10^4 \times y}{1 \cdot 382} + \frac{12,000 \times 10^4 \times z}{0 \cdot 346} \, \text{N/m}^2$$

$$= (-115 \cdot 5y + 346 \cdot 5z) \times 10^6 \, \text{N/m}^2 \tag{i}$$

where $-0 \cdot 12 \leqslant y \leqslant +0 \cdot 12$ and $-0 \cdot 06 \leqslant z \leqslant 0 \cdot 06$.

From equation (i) it is clear that the greatest positive value of σ_x occurs when $z > 0$ and $y < 0$. Putting $z = 0 \cdot 06$ m and $y = -0 \cdot 12$ m the greatest tensile stress is $34 \cdot 7$ MN/m^2. The maximum compressive stress occurs at $y = 0 \cdot 12$ m and $z = -0 \cdot 06$ m and is $-34 \cdot 7$ MN/m^2.

It may be noted that the line of zero stress in the cross-section is $y = 3z$ this being obtained by setting $\sigma_x = 0$ in equation (i) (see Figure 6.18b). The variation of tensile stress in the cross-section is indicated in Figure 6.18c. Under this generalized loading the greatest stresses still occur at the greatest (perpendicular) distance from the zero stress line (i.e. the generalized neutral axis.)

Example 6.20   A cantilever of length 4 m with the cross-section of Figure 6.13b is subject to a loading of the type of Figure 6.10. Find the greatest tensile and compressive stresses at B.

Solution   From Figure 6.10 the bending moment M_z in the xy plane is $-5\sqrt{3}$ kNm and in the xz plane M_y is -5 kNm. From Example 6.9 (Section 6.3.2), I_z is $3 \cdot 28 \times 10^{-4}$ m^4, and $I_y = (2 \times (20 \times 200^3) + 360 \times 10^3)/12$ mm$^4 = 0 \cdot 267 \times 10^{-4}$ m^4.

Substituting in equation (6.20)

$$\sigma_x = -26 \cdot 4y - 187z \, \text{MN/m}^2 \tag{i}$$

The maximum compressive stress will occur when y and z have their greatest positive values; putting $y = 0 \cdot 2$ m, $z = 0 \cdot 1$ m we find $\sigma_x = -24 \cdot 0$ MN/m^2. Similarly the greatest tensile stress is $+24 \cdot 0$ MN/m^2.

Problem for Solution   6.30

6.4   Stresses Due to Shearing Forces

6.4.1 *Rectangular Cross-section*

For loading in the xy plane the shearing force acts on the beam cross-section in the y direction (Figure 6.3c), i.e. a surface which is normal to the x axis is subject to forces in the y direction. From the definitions of stress components in Article 1.3 it might then be expected that there would be shear stresses τ_{xy}. The complete state of stress including the effect of bending moment is therefore of the form

$$\begin{pmatrix} \sigma_x & \tau_{xy} & 0 \\ \tau_{yx} & 0 & 0 \\ 0 & 0 & 0 \end{pmatrix} \tag{A6.3}$$

From the symmetry of the stress array it is seen that there are stresses τ_{yx} in the material, i.e. shear stresses on y surfaces in the x direction, as in Figure 6.19a. It is now assumed that the stresses τ_{xy} are not dependent on z.

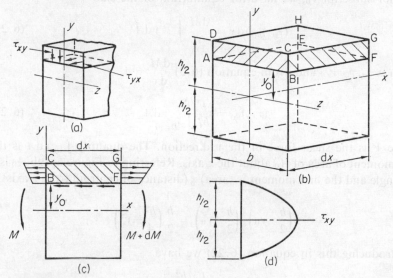

Figure 6.19

Consider a portion of a beam of rectangular cross-section subject to bending moments in the xy plane of M and $(M + dM)$ as shown in Figures 6.19b and c. The state of stress is of the form of array (A6.3). Substituting in equation (6.14) the normal stresses at x and $(x + dx)$ are

$$\frac{My}{I} \quad \text{and} \quad \frac{(M + dM)y}{I}$$

Therefore the normal forces due to these stresses on opposite ends of the portion of beam are unequal. In particular for the slab of material lying above the plane $y = y_0$ the force in the x direction on end EFGH exceeds that on the end ABCD by

$$\int_{A_0} \frac{(M + dM)y\, dA}{I} - \int_{A_0} \frac{My\, dA}{I}$$

i.e.

$$\frac{dM}{I} \int_{A_0} y\, dA$$

where A_0 is the area of ABCD.

Since the upper surface DCGH and the sides BCGF and ADHE are exterior surfaces which are observed to be force free, it follows that the slab can only be maintained in equilibrium by a tangential force on its lower surface ABFE (Figure 6.19c). This internal force is supplied by the shear stress component τ_{yx}. Assuming uniform distribution of this stress, and adding exterior subscript y_0, we have for equilibrium of the slab

$$(\tau_{yx})_{y_0} b \, dx = \frac{dM}{I} \int_{A_0} y \, dA \tag{6.21}$$

Since $\tau_{yx} = \tau_{xy}$ and, from equation (6.3) $\dfrac{dM}{dx} = V$

$$(\tau_{xy})_{y_0} = \frac{V}{bI} \int_{A_0} y \, dA \tag{6.22}$$

where V is the shear force in the y direction. The quantity $\int_{A_0} y \, dA$ is the first moment of area of A_0 about the z axis. Referring to Figure 6.19b A_0 is a rectangle and the first moment is (area) × (distance of centroid from z axis) i.e.

$$b\left(\frac{h}{2} - y_0\right) \frac{1}{2} \left(\frac{h}{2} + y_0\right) = \frac{b}{2}\left(\frac{h^2}{4} - y_0^2\right)$$

Introducing this in equation (6.22) we have

$$(\tau_{xy})_{y_0} = \frac{V}{2I}\left(\frac{h^2}{4} - y_0^2\right) \tag{6.23}$$

The shear stress is greatest when $y_0 = 0$, zero if $y_0 = \pm \frac{1}{2}h$, and varies in parabolic fashion with y_0 as in Figure 6.19d. The maximum value is

$$(\tau_{xy})_{y_0=0} = \frac{Vh^2}{8I}$$

Substituting $I = bh^3/12$ we have

$$(\tau_{xy})_{y_0=0} = \frac{3V}{2bh}$$

Noting that (bh) is the cross-sectional area, we see that the maximum stress magnitude is 1·5 times the average shear stress.

Example 6.21   A beam is formed by bonding together three identical bars of square cross-section as in Figure 6.20a. Find the shear stresses at the interfaces if the shearing force is 18 kN.

Solution   Here $I = 40 \times 120^3/12 \text{ mm}^4 = 576 \times 10^{-8} \text{ m}^4$. Substituting in equation (6.23) to find stress at the upper interface

$$(\tau_{xy})_{y_0=0\cdot02} = \frac{18{,}000}{2 \times 576 \times 10^{-8}}\left(\frac{0\cdot12^2}{4} - 0\cdot02^2\right) = 5 \times 10^6 \text{ N/m}^2$$

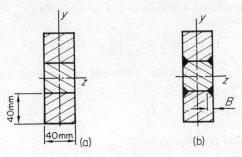

Figure 6.20

Then $(\tau_{xy})_{y_0=-0.02} = 5$ MN/m² since y_0 is squared in (6.23).

These are the stresses to be withstood by the adhesive.

It is interesting to note that the greatest shear stress τ_{xy} obtained by setting $y_0 = 0$ in (6.23), is $(\tau_{xy})_0 = 5 \cdot 63$ MN/m². If the beam were loaded as in Figure 6.6 the greatest normal stress due to the bending moment would be

$$\sigma_x = 20 \times 10^3 \times 0 \cdot 06 \times 10^8/576 \text{ N/m}^2 = 208 \text{ MN/m}^2$$

Failure Conditions The x and y axes are principal axes at $y = \pm \tfrac{1}{2}h$ since $\tau_{xy} = 0$ there. Thus at the upper and lower surfaces the failure condition remains $|(\sigma_x)_{\max}| = \sigma_0$ as discussed in Section 6.3.3.

At the neutral axis $\sigma_x = 0$ and the stress array is

$$\begin{pmatrix} 0 & (\tau_{xy})_0 & 0 \\ (\tau_{yx})_0 & 0 & 0 \\ 0 & 0 & 0 \end{pmatrix} \qquad (A6.4)$$

By drawing a Mohr Stress Circle, or using equations (1.8) it can be shown that $\tau_{\max} = (\tau_{xy})_0$ so that failure would occur for

$$(\tau_{xy})_0 = \tfrac{1}{2}\sigma_0$$

However for a typical load situation such as in Figure 6.6, we have seen in Example 6.21 that $(\tau_{xy})_0$ is only $5 \cdot 63$ MN/m² while $(\sigma_x)_{\max}$ is 208 MN/m². We therefore conclude that in homogeneous beams of rectangular cross-section the shear stresses are not important; however if the section is built up using adhesives, rivets, etc., the shear force at an interface must be borne by the bonding agent.

Example 6.22 If the bars discussed in Example 6.21 were instead welded together, giving cross-section as in Figure 6.20b, find the weld depth B if τ_{xy} in the weld is not to exceed 25 MN/m².

Solution The left side of equation (6.21) becomes $(\tau_{xy})_{y_0} \times 2B \times dx$, and then proceeding as before

$$(\tau_{xy})_{y_0} = \frac{V}{I} \times \frac{b}{4B}\left(\frac{h^2}{4} - y_0^2\right) \leqslant 25 \times 10^6$$

where b, h are width, depth of the complete section. Substituting as in Example 6.21 we find $B \geqslant 0 \cdot 004$ m.

Problems for Solution 6.31, 6.32

6.4.2 *I-sections and Hollow Sections*

I-sections A similar argument to that in 6.4.1 shows that if b_0 is the width at y_0 (Figure 6.21a)

$$(\tau_{xy})_{y_0} = \frac{V}{b_0 I}\int_{A_0} y \, dA \tag{6.24}$$

Here

$$\int_{A_0} y \, dA = b_1\left(\frac{h-h_1}{2}\right) \times \frac{1}{2}\left(\frac{h+h_1}{2}\right) + b_0\left(\frac{h_1}{2} - y_0\right) \times \frac{1}{2}\left(\frac{h_1}{2} + y_0\right)$$

so that

$$(\tau_{xy})_{y_0} = \frac{V}{b_0 I}\left(\frac{b_1}{8}(h^2 - h_1^2) + \frac{b_0}{2}\left(\frac{h_1^2}{4} - y_0^2\right)\right) \tag{6.25}$$

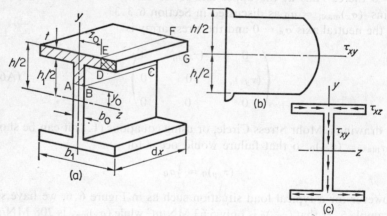

Figure 6.21

Therefore the stress is greatest at the neutral axis where $y_0 = 0$; the least value within the web occurs at $y_0 = \frac{1}{2}h_1$. Since $b_0 \ll b_1$ it is clear from (6.25) that the stress variation over the depth of the web is not large (see Figure 6.21b). If the surface $y = y_0$ lies in the flange, b_0 in equation (6.24) becomes b_1, and, evaluating A_0, we have

$$(\tau_{xy})_{y_0} = \frac{V}{2I}\left(\frac{h^2}{4} - y_0^2\right) \tag{6.26}$$

Substituting $y_0 = \frac{1}{2}h_1$ in each of (6.25) and (6.26) the stresses in the web and flange at their junction are found to be in the ratio (b_1/b_0). The variation of τ_{xy} along the y axis in Figure 6.21b shows that the shearing force is largely carried by the web. A reasonable estimate of the maximum value of τ_{xy} can be obtained by dividing the shear force by the cross-sectional area of the web only.

It is found that the shearing forces in the y direction also cause shearing stresses τ_{xz} in the flanges. By similar reasoning to that in Section 6.4.1, consideration of the forces due to the normal stresses on the ends of the portion of flange DEFG in Figure 6.21a shows that there is a net force on it in the x direction of

$$\int_{A_0} \frac{\mathrm{d}M}{I} y \, \mathrm{d}A$$

where A_0 is the area of the portion of flange lying between $z = z_0$ and $z = \frac{1}{2}b_1$. This force can only be maintained in equilibrium by a shearing force in the x direction on the surface DEF; assuming uniform distribution this force is $(\tau_{zx})_{z_0} t \, \mathrm{d}x$ where t is the flange thickness. Hence

$$(\tau_{xz})_{z_0} = \frac{V}{tI} \int_{A_0} y \, \mathrm{d}A \qquad (6.27)$$

Here

$$\int_{A_0} y \, \mathrm{d}A = t(\tfrac{1}{2}b_1 - z_0) \times \tfrac{1}{2}(h - t)$$

and so τ_{xz} is a linear function of z_0 with maximum at $z_0 = 0$; similar results could be derived for the other parts of the flanges, and the directions of τ_{xz} and τ_{xy} on the left end of element $\mathrm{d}x$ are shown in Figure 6.21c.

If the critical conditions for a beam are to be influenced by the shear stress this is likely to occur at the top of the web where the shear stress τ_{xy} is still relatively large and the normal stress σ_x is also near its greatest value.

Example 6.23 An I-section, overall size $0.2 \text{ m} \times 0.2 \text{ m}$, has both flanges and web of 10 mm thickness. The beam is subject to a bending moment of 20 kNm and a shearing force of 18 kN. Determine the state of stress at the top of the web.

Solution Here (referring to Figure 6.21a) $b_1 = 0.2 \text{ m}$, $h = 0.2 \text{ m}$, $h_1 = 0.18 \text{ m}$, $b_0 = 0.01 \text{ m}$. Then

$$I = (0.2^4 - 0.19 \times 0.18^3)/12 = 0.41 \times 10^{-4} \text{ m}^4$$

Substituting in equation (6.25) we find $\tau_{xy} = 8.35 \text{ MN/m}^2$, and using (6.14)

$$\sigma_x = 20 \times 10^3 \times 0.09 \times 10^4/0.41 \text{ N/m}^2 = 43.9 \text{ MN/m}^2$$

i.e.

$$T = \begin{pmatrix} 43 \cdot 9 & 8 \cdot 35 & 0 \\ 8 \cdot 35 & 0 & 0 \\ 0 & 0 & 0 \end{pmatrix} MN/m^2$$

The maximum shear stress at this point could be found if required by first obtaining the principal stress array.

Rectangular Hollow Sections The shear stresses τ_{xy} in the webs of a thin-walled hollow section of rectangular form (Figure 6.22a) can be found in the same way as for I-sections; equation (6.25) is applicable if b_0 is replaced

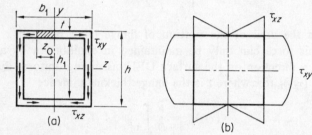

Figure 6.22

by $2t$. To determine the distribution of τ_{xz} in the flanges it is assumed that τ_{xz} is zero at the midpoint. Then (6.27) can be applied by interpreting A_0 as the area of flange between $z = 0$ and $z = z_0$, i.e.

$$(\tau_{xz})_{z_0} = \frac{V}{tI}(tz_0 \times \tfrac{1}{2}(h-t)) \tag{6.27a}$$

As for the I-section, τ_{xz} varies linearly in the flange. The distributions of τ_{xy} and τ_{xz} and their directions are shown in Figure 6.22.

Problems for Solution   6.33, 6.34

6.5   Uniform Composite Beams

This class of beam is formed from two or more prismatic bars bonded together and each extending over the entire span of the beam as in Figure 6.23a. Thus all cross-sections are the same, and have two or more parts representing different materials. As in Article 6.3 only beams loaded in a plane of symmetry will be considered. Making the same assumptions as in

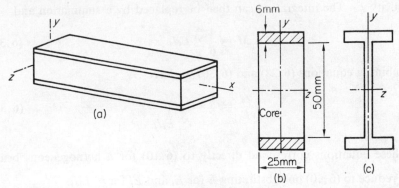

Figure 6.23

Section 6.3.1 we again obtain,

$$\epsilon_x = \frac{y}{R}$$

and applying the strain/stress equations at a point in the cross-section

$$\frac{\sigma_x}{E_i} = \epsilon_x = \frac{y}{R} \tag{6.28}$$

Here Young's Modulus has been denoted E_i as a reminder that the value to be substituted must be that for the material actually present at the point being considered. Following the arguments of 6.3.1 equilibrium of force in the x direction requires (compare equation 6.7a)

$$\int_A \sigma_x \, dA = \int_A \left(\frac{E_i}{R}\right) y \, dA = 0$$

Since F_i is a variable over A it cannot be taken outside the integration as in Section 6.3.1, and the equilibrium condition is finally

$$\int_A y(E_i \, dA) = 0 \tag{6.29}$$

Therefore the z axis must pass, not through the centre of area of the cross-section, but through the centre of weighted area (i.e. weighted by the local value of Young's Modulus). Continuing the argument of 6.3.1 the bending moment M at any position on the span is (compare equation 6.8a)

$$M = \int_A \frac{E_i}{R} y^2 \, dA = \frac{1}{R} \int_A E_i y^2 \, dA$$

The integral is a second moment of weighted area. In most cases the cross-section consists of a comparatively small number n of regions of different materials as in Figure 6.23b. The weighted second moment of area of one such region is the product of its second moment of area I_i and its modulus of elasticity E_i. The integration can then be replaced by a summation and

$$M = \frac{1}{R} \sum_1^n E_i I_i \tag{6.30}$$

Combining equations (6.28) and (6.30) we have

$$\frac{\sigma_x}{y E_i} = \frac{1}{R} = \frac{M}{\sum\limits_1^n E_i I_i} \tag{6.31}$$

These equations correspond directly to (6.10) for a homogeneous beam and reduce to (6.10) on substituting E for E_i and EI for $\sum\limits_1^n E_i I_i$.

It may be noted that when the lines of separation in the cross-section are all parallel to the z axis, a fictional equivalent section can be drawn as in Figure 6.23c, composed only of the stiffest material. The width of the ith material is scaled in the ratio E_i/E_j where E_j relates to the stiffest material. Thus the composite beam with stiff skins and light core is seen to be equivalent to an I-section in a single material.

Example 6.24 An experimental low-mass beam has cross-section as in Figure 6.23b. The core is of expanded plastic and the skins of glass-reinforced plastic, with Young's Moduli 7 GN/m² and 7 MN/m² for skin and core respectively. Find the greatest stresses in the two materials if the maximum bending moment is 110 Nm.

Solution   For core

$$EI = 7 \times 10^6 \left(\frac{25 \times 50^3}{12} \times 10^{-12} \right) = 1 \cdot 82 \text{ Nm}^2$$

For skins

$$EI = 7 \times 10^9 \left(\frac{25 \times (62^3 - 50^3)}{12} \times 10^{-12} \right) = 1647 \text{ Nm}^2$$

Therefore

$$\Sigma EI = 1649 \text{ Nm}^2$$

and so from equations (6.31)

$$\sigma_x = \frac{110 y E_i}{1649}$$

In the core the greatest y is 25 mm and $E_i = 7 \times 10^6$ N/m². Therefore the greatest σ_x is $0 \cdot 012$ MN/m². In the skins the greatest y is 31 mm and $E_i = 7 \times 10^9$ N/m². Therefore the greatest σ_x is $14 \cdot 5$ MN/m².

It is interesting to note that while the core makes a negligible direct contribution to ΣEI, its function of separating the skins by a large distance enables the composite to develop a large ΣEI.

Problem for Solution   6.35

6.6 Unsymmetrical Bending

6.6.1 *Stress Equation*

In Article 6.3 we considered the stresses due to bending moments acting in a plane of symmetry of a beam. In the case of a beam having a cross-section with two axes of symmetry, bending moments acting in another plane were treated by resolution into the planes of symmetry (see Sections 6.2.4, 6.3.7). We now seek to treat directly cases, such as in Figure 6.24, of pure bending in a plane which is not a plane of symmetry, due to moments only.

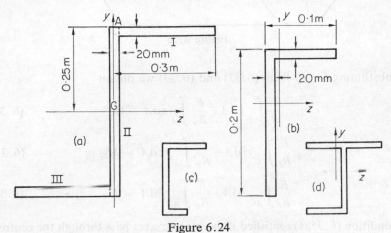

Figure 6.24

Making the same assumptions as in Section 6.3.1 and proceeding as in the derivation of equation (6.4), for both the xy and xz planes we find that the normal strain at a point (y, z) in the cross-section is given by

$$\epsilon_x = \frac{y}{R_z} + \frac{z}{R_y} \qquad (6.32)$$

where R_z and R_y are the radii of curvature of the neutral surface in the xy and xz planes respectively. Then using the equations (6.6) we have

$$\sigma_x = E\epsilon_x \qquad (6.33)$$

This normal stress represents a normal force distribution on the cross-section which must have a resultant effect equal to the action of the external forces at the cross-section. Thus (cf. Section 6.3.1) there must be zero force in the x direction, and moments M_z and M_y as in Figure 6.25. Then, considering a small element dA at position (y, z), we require that the stress distribution satisfy the following relations

$$\int_A \sigma_x \, dA = 0; \quad \int_A \sigma_x y \, dA = M_z; \quad \int_A \sigma_x z \, dA = M_y$$

$$(6.34a, b, c)$$

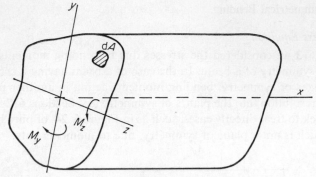

Figure 6.25

Substituting for σ_x from (6.32) and (6.33) we obtain

$$\frac{E}{R_z}\int_A y\,\mathrm{d}A + \frac{E}{R_y}\int_A z\,\mathrm{d}A = 0 \qquad (6.35a)$$

$$\frac{E}{R_z}\int_A y^2\,\mathrm{d}A + \frac{E}{R_y}\int_A yz\,\mathrm{d}A = M_z \qquad (6.35b)$$

$$\frac{E}{R_z}\int_A yz\,\mathrm{d}A + \frac{E}{R_y}\int_A z^2\,\mathrm{d}A = M_y \qquad (6.35c)$$

Condition (6.35a) is satisfied if the y and z axes pass through the centre of area of the cross-section.

The terms $\int_A y^2\,\mathrm{d}A$ and $\int_A z^2\,\mathrm{d}A$ in equations (6.35b and c) will be recognized as the second moments of area I_z and I_y of the cross-section (see Sections 6.3.2 and 6.3.7). The term $\int_A yz\,\mathrm{d}A$ is called the product moment of area of the cross-section and is denoted I_{yz}. Then equations (6.35b and c) become

$$\frac{EI_z}{R_z} + \frac{EI_{yz}}{R_y} = M_z; \quad \frac{EI_{yz}}{R_z} + \frac{EI_y}{R_y} = M_y$$

Solving these for the radii of curvature in the particular case when $M_y = 0$ and substituting the results in (6.32) and (6.33) we have

$$\sigma_x = \frac{M_z(yI_y - zI_{yz})}{(I_yI_z - I_{yz}^2)} \qquad (6.36)$$

This equation determines the stress distribution in the cross-section when the loads act only in the xy plane and the y axis is not necessarily an axis of symmetry. Equating (6.36) to zero we obtain

$$y = \frac{I_{yz}}{I_y}z$$

This is the equation of the neutral axis, which is inclined at $\tan^{-1}(I_{yz}/I_y)$ to the z axis. If $I_{yz} = 0$ the z axis is the neutral axis and (6.36) reduces to (6.18).

6.6.2 *Properties of Areas*

The product moment of area was defined in Section 6.6.1 as

$$I_{yz} = \int_A yz\,dA$$

It is usually most convenient to evaluate this quantity for axes y, z passing through the centre of area. The product moment for any other parallel set of axes can then be found using a result analogous to the Parallel Axes

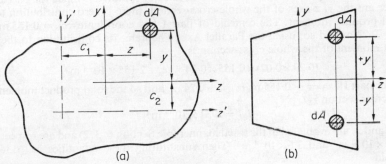

(a)   (b)

Figure 6.26

Theorem for second moments of area (see Section 6.3.2, equation 6.13). To derive this we take centroidal axes y, z and a parallel set y', z' as in Figure 6.26a. An element dA has coordinates relative to the new axes of

$$z' = z + c_1; \quad y' = y + c_2$$

By definition, the product moment for the new axes is

$$I_{y'z'} = \int_A y'z'\,dA = \int_A (y+c_2)(z+c_1)\,dA$$

$$= \int_A yz\,dA + c_1 \int_A y\,dA + c_2 \int_A z\,dA + c_1c_2 \int_A dA$$

The first term is the product moment for the centroidal axes, and the integrals in the second and third terms, being first moments about centroidal axes, are zero. Therefore

$$I_{y'z'} = I_{yz} + Ac_1c_2 \tag{6.37}$$

Thus if the product moment is known for centroidal axes it can easily be determined for parallel axes. A particularly important case of the product moment for centroidal axes occurs when one axis is an axis of symmetry

(Figure 6.26b). Since for any value of z there are equal elements dA at all values of $\pm y$, their contributions $yz\,dA$ and $(-y)z\,dA$ to the product moment cancel out. Therefore if one centroidal axis is an axis of symmetry the product moment of area is zero. It is immediately clear why for the sections considered in Article 6.3, equation (6.36) could be replaced by (6.14). Stress calculations using (6.36) can now be made.

Example 6.25 Determine the product moment of area for the cross-section in Figure 6.24a, and calculate the stress at the point A for a bending moment of 100 kNm in the xy plane.

Solution The section can be considered composed of rectangles I, II, III, each of which has zero product moment about its axes of symmetry. Since for part II the latter are the y, z axes of the whole cross-section, there is zero contribution to the total product moment. The centroid of part I has coordinates $c_1 = 0.155$ m and $c_2 = 0.24$ m and so, using the Parallel Axes Theorem, its contribution to the product moment of the whole cross-section is

$$(0.29 \times 0.02) \times 0.155 \times 0.24 \quad \text{i.e.} \quad 2.155 \times 10^{-4}\,\text{m}^4$$

For part III $c_1 = -0.155$ m, $c_2 = -0.24$ m, and so the total product moment for the cross-section is

$$I_{yz} = 4.31 \times 10^{-4}\,\text{m}^4$$

I_z and I_y are evaluated in the usual manner (see Section 6.3.2) and are respectively $8.76 \times 10^{-4}\,\text{m}^4$ and $3.6 \times 10^{-4}\,\text{m}^4$. Then substituting these quantities in (6.36) we obtain

$$\sigma_x = M_z\,(2790y - 3340z) \tag{i}$$

Putting $M_z = 10^5$ Nm, $z = 0$, $y = 0.25$ m we obtain

$$\sigma_x = 69.7\,\text{MN/m}^2$$

It is interesting to note that if the area III in Figure 6.24a were on the opposite side of the y axis, c_1 would be $+0.155$ m and its contribution to the product moment would cancel that due to part I. This is as would be expected since the section would then be a channel with one axis of symmetry. However, in the case of the section in Figure 6.24c there will clearly be a non-zero product moment.

Problems for Solution 6.36, 6.37

6.6.3 *Circle of Inertia*

We shall now develop the relationships between the properties of area measured with respect to axes z, y and those for alternative axes z', y' inclined at θ to the first set (see Figure 6.27a). Point P at the centre of element dA has coordinates (z, y) and $(z,' y')$ relative to the alternative axes. Then

$$z' = \text{GS} = \text{GT}\cos\theta + \text{TP}\sin\theta = z\cos\theta + y\sin\theta \tag{6.38a}$$

and, similarly,

$$y' = \text{SP} = -z\sin\theta + y\cos\theta \tag{6.38b}$$

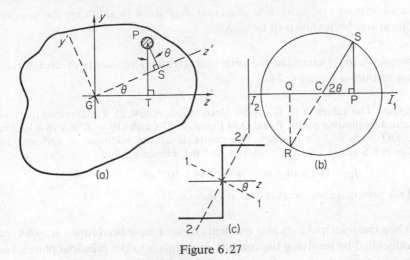

Figure 6.27

To find the second moment of area with respect to the z' axis we start from the definition (Section 6.3.2) and substitute from (6.38) as follows

$$I_{z'} = \int_A (y')^2 \, dA - \int_A (-z \sin\theta + y \cos\theta)^2 \, dA$$

$$= \sin^2\theta \int_A z^2 \, dA - 2 \sin\theta \cos\theta \int_A yz \, dA + \cos^2\theta \int_A y^2 \, dA$$

$$= I_z \cos^2\theta - 2I_{yz} \sin\theta \cos\theta + I_y \sin^2\theta \qquad (6.39a)$$

By a similar procedure we find

$$I_{y'z'} = I_{yz} (\cos^2\theta - \sin^2\theta) + (I_z - I_y) \sin\theta \cos\theta \qquad (6.39b)$$

It will be seen that equations (6.39) have the same form as equations (1.3a and 1.4a) if the quantities corresponding to σ_x, σ_y, τ_{xy} are I_z, I_y, $-I_{yz}$. We may therefore interpret for area all the properties of the stress equations derived in Articles 1.6–1.10. There is one choice of axes—the principal axes of inertia—for which the product moment is zero and the second moments have extreme values I_1 and I_2. Formal expressions for these quantities may be found by substituting the corresponding quantities in equations (1.7 and 1.8). Alternatively a Mohr Circle may be drawn as described in Article 1.10, product moments being plotted positive upwards.

This Circle of Inertia (e.g. Figure 6.27b) is always wholly to the right of the product axis since second moments are always positive. We also note that if $I_z = I_y$ the Circle reduces to a point. Consequently if the second moments of area are equal about any two mutually perpendicular axes then they are the same for any other sets of axes. Thus for example the second moment for a square of side a is $a^4/12$ for any axis through its centre of area. Lastly, we note that an axis of symmetry is a principal axis since in this case the product moment has been shown to be zero.

Example 6.26 Determine the principal axes and moments of area for the cross-section in Figure 6.24a.

Solution The values of I_z, I_y, I_{zy} are given in Example 6.25. The Circle of Inertia is obtained by plotting points P and Q in Figure 6.27b with OP $= I_z = 8{\cdot}76 \times 10^{-4}\,\mathrm{m}^4$ and OQ $= I_y = 3{\cdot}6 \times 10^{-4}\,\mathrm{m}^4$. The centre is at the midpoint of QP and the radius is CS where PS $= I_{yz} = 4{\cdot}31 \times 10^{-4}\,\mathrm{m}^4$. Hence we find

$$I_1 = 11{\cdot}2 \times 10^{-4}\,\mathrm{m}^4; \quad I_2 = 1{\cdot}16 \times 10^{-4}\,\mathrm{m}^4; \quad \theta = -29°\,33'$$

and the principal axes are as shown in Figure 6.27c.

When the principal axes and moments of area have been found stresses can be calculated by resolving the bending moments into the principal planes and then using an equation of the form of (6.20). The method described in Sections 6.6.1 and 6.6.2 usually results in less tedious calculations.

Problems for Solution 6.38–6.40

6.6.4 *Shear Centre*

In the previous sections we have considered how to calculate normal stress σ_x due to the most general case of pure bending. However if the bending moments are due to transverse forces (as in Figure 6.10) then as we have shown in Article 6.4 shear stresses τ_{xy} and τ_{xz} will be present in the cross-section, and as we shall now show, their presence may require some modification of the load system.

Consider a T-section (Figure 6.28a) subject to bending moments due to transverse forces in the xy plane. Although this is not a plane of symmetry, as was considered in the theory of simple bending in Section 6.3.1, we shall first assume that simple bending occurs with z as neutral axis. By application of equation (6.24) the shearing stresses τ_{xy} could be derived and would clearly be much greater in ABCD than in PQRS. Therefore, since most of the shearing force on the cross-section is carried by ABCD, the resultant force does not coincide with the load plane xy but passes through some point G' nearer to G_1, the centroid of ABCD. As a result it is found that twisting occurs and our assumption of simple bending due to forces in the xy plane is false.

Consider now the cantilever in Figure 6.28b with a rigid horizontal bar attached to its free end. A force at M would clearly cause anticlockwise twisting of the beam, while one at N would cause clockwise twisting. Evidently there is some intermediate position of the force for which no twisting occurs.

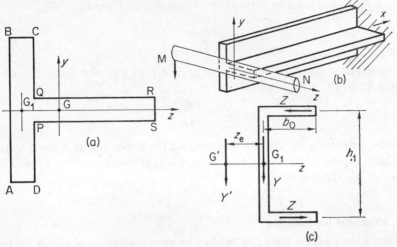

Figure 6.28

This is called the centre of flexure and for a *uniform* beam is found to lie on the line of action of the resultant of the shear stresses due to simple bending.

If the T-section were loaded in the plane of symmetry xz the shearing stresses would be symmetric about the z axis and the resultant force on the cross-section would act along this axis. The point G' at the intersection of the lines of action of both these resultant shear forces is therefore called the shear centre of the cross-section. For a uniform beam it coincides with the centre of flexure and is thus the point through which a transverse force must act if bending is to occur without twisting. In the case of a cross-section with two axes of symmetry, the shear centre is clearly at the centre of area, as was tacitly assumed in Section 6.3.7.

We now consider the location of the shear centre for a few common cross-sections. For the T-section of Figure 6.28a it is sufficiently accurate to take the shear centre at G_1. Similarly, for angle sections the resultant shear forces due to loads parallel to the y and z axes may be considered to act along the centre lines of the flanges, thus locating the shear centre at their intersection (Figure 6.24b).

The shear centre of a channel section (Figure 6.28c) must lie on the axis of symmetry z. For bending about the z axis the distribution of shear stress τ_{xz} and τ_{xy} in flanges and web respectively can be found using equations (6.24) and (6.27) and is effectively as in the left half of Figure 6.22b. There are therefore resultant shear forces Y and Z as in Figure 6.28c. This system of a force Y and a couple Zh_1 can be replaced by a force $Y' = Y$ acting at a point G', z_e to the left of G_1 such that

$$Y'z_e = Zh_1$$

Since τ_{xz} varies linearly with z the mean value is, from (6.27a)

$$\frac{Vb_0h_1}{4I_z}$$

This stress is taken to act over an area b_0t to give the force Z, and so putting $Y' = V$

$$z_e = \frac{tb_0^2h_1^2}{4I_z}$$

Lastly, extending these ideas we note that the flange forces Z for a section such as in Figure 6.24a would have zero resultant moment about G, which is therefore the shear centre.

6.7 Problems for Solution

6.1. Draw the shearing force and bending moment diagrams and sketch the form of the deflexion curve for the beams in Figure 6.29. Indicate the positions of any points of contraflexure.

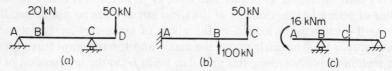

Figure 6.29 (a) AB = CD = $\frac{1}{2}$BC = 1 m. (b) AB = 2 m,
BC = 1 m. (c) AB = BC = CD = 1 m

6.2. Find the greatest concentrated load which can act vertically down at C in Figure 6.29c if there is to be no contraflexure.

6.3. Draw the shearing force and bending moment diagrams, and sketch the deflexion curve for the beams in Figure 6.30.

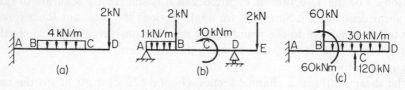

Figure 6.30 (a) AB = CD = $\frac{1}{2}$BC = 1 m.
(b) AB = BC = CD = 4 m, DE = 3 m. (c) AB = BC = CD = 2 m

6.4. Repeat Problem 6.1a and b but with uniformly distributed downward loads of intensity 20 kN/m added over parts AC of the beams.

6.5. Referring to Figure 6.7b replace the support at B by a fixed support and remove the support at A. Draw the shearing force and bending moment diagrams for this cantilever.

6.6. Find what concentrated moment must be exerted at D of Figure 6.29c to give a point of contraflexure at C.

6.7. If the inclined force in Figure 6.10a is reversed draw the shearing force and bending moment diagrams.

6.8. Calculate I about the z axis for the cross-section shown in Figure 6.14b.

6.9. A girder has the cross-section shown in Figure 6.14c, being built up from two plates and two channels. The latter have each $I = 0.195 \times 10^8$ mm⁴ about the z axis. Calculate I about the z axis for the girder.

6.10. Calculate the second moments of area about the y axis for the cross-sections in Figures 6.13a and b and Figures 6.14a and b.

6.11. Calculate the second moment of area for a solid circular cross-section of diameter 100 mm. Find dimensions for a tube to have the same second moment if the external and internal diameters are to be in the ratio 1.1.

6.12. If a beam of cross-section as Figure 6.14b is subject to a maximum bending moment of 2 MNm find the greatest stress. If the y and z axes were interchanged find the moment to produce the same stress due to bending about the new z axis.

6.13. A beam of I-shaped cross-section has depth of 300 mm, web thickness 25 mm, and has flanges 150 mm wide and 25 mm thick. If the material has $\sigma_0 = 200$ MN/m² and a safety factor of 2 is required investigate whether the beam is suitable for the duty shown in Figure 6.30c.

6.14. A beam of cross-section as in Figure 6.14c has $\sigma_0 = 240$ MN/m². Allowing a safety factor of 1.5 determine the maximum distributed load which it can carry over a span of 5 m when loaded as in Figure 6.7a. 1 (channel) = 1950 cm⁴.

6.15. Repeat Problem 6.14 for loading as in Figure 6.7b.

6.16. A beam having the cross-section described in Problem 6.13 is required to carry a uniformly distributed load of intensity 12 kN/m. If $\sigma_0 = 280$ MN/m² and a safety factor of 2 is required find the greatest simply supported span which can be used when the web is (a) vertical, (b) horizontal.

6.17. Repeat Problem 6.16 if the beam is supported as a cantilever.

6.18. Find the greatest tensile and compressive stresses in a beam of cross-section as Figure 6.15c when subject to a bending moment of −5 kNm.

6.19. A cantilever has T-type cross-section of overall dimensions 150 mm × 150 mm and both flange and web are 20 mm thick. If $\sigma_0 = 240$ MN/m² and a safety factor of 2 is necessary calculate the greatest intensity of uniformly distributed load which can be supported over a span of 2 m.

6.20. A girder has cross-section similar to that in Figure 6.14c but without the lower plate. Find the maximum simply supported span over which a uniformly distributed load of intensity 30 kN/m can be carried if $\sigma_0 = 240$ MN/m² and $f_s = 1.2$. Each channel has area of 30.3 cm². (See Problem 6.14.)

6.21. If the beam of Problem 6.20 has a span of 8 m find the greatest concentrated force which can be supported anywhere over the span for the same safety factor.

6.22. If the beam of Problem 6.19 has instead normal stress limits of −160 MN/m² and 80 MN/m² find the load intensity.

6.23. Investigate whether the beam of Problem 6.18 can be used for the duty in Figure 6.30b if it has normal stress limits of -80 MN/m^2 and 100 MN/m^2.

6.24. Find the best cross-section for the duty in Figure 6.29a, from those listed in Table 6.2. $\sigma_0 = 240$ MN/m^2.

6.25. A large road sign is supported by steel tubes. The wind loading effect for each tube can be represented as in Figure 6.8a. If $w = 1\cdot8$ kN/m and $b = 3$ m select the most suitable section from Table 6.3, allowing a minimum safety factor of 2. Take $\sigma_0 = 250$ MN/m^2.

Table 6.3

Reference	Outer diameter (m)	$I \times 10^8$ (m^4)	Mass (kg/m)
A	0·17	1000	25
B	0·17	1500	37
C	0·19	1600	29
D	0.19	2300	43
E	0·22	2400	33
F	0·22	3400	49

6.26. Find which if any of the tubes in Table 6.3 can be simply supported over a span of 25 m when used as gas pipes, while maintaining a safety factor of 2. $\sigma_0 = 250$ MN/m^2.

6.27. A cantilever of solid circular cross-section tapers from $d = 3D$ at root to $d = D$ at the free end. A concentrated force Y is exerted downwards at the free end. Find the position of the maximum stress.

6.28. A cantilever of length L tapers in width only from b at the root to zero at the free end where a concentrated force P acts. Find an expression for the maximum stress if the depth is h.

6.29. Repeat Problem 6.28 if the total load P is uniformly distributed over the whole surface of the beam.

6.30. Repeat Example 6.20 but using the cross-section of Figure 6.14c. (Centroidal second moments of channel are 1950, 152 cm^4, and its area is 30·3 cm^2.)

6.31. If the beam of Figure 6.20 were formed instead from two bars of cross-section 40 mm × 60 mm find the depth of the welds if τ_{xy} is not to exceed 25 MN/m^2.

6.32. Referring to Example 6.22 find $\tau_{\max}$ at a point in the weld material at $y_0 = 0\cdot02$ m. (*Hint* Use $\sigma_x = My/I$ to complete the stress array A6.3.)

6.33. Calculate the maximum value of τ_{xy} in Example 6.23 and compare with the quantity (V/web area).

6.34. A box section of 0·2 m × 0·2 m overall size and 10 mm wall thickness is subject to a shear force of 100 kN and bending moment of 76·3 kNm. Compare the greatest shear stresses evaluated at the top of the web and at the middle of the flange upper surface.

6.35. A composite beam has titanium skins 20 mm wide and 1 mm deep, and a core of expanded material 100 mm deep. Find the greatest uniformly distributed load which can be carried over a simply supported span of 4 m. Young's Moduli are 110 GN/m² and 200 MN/m² for skin and core respectively. The titanium yielded at 400 MN/m² in a simple tensile test. Assume failure of the composite occurs in the titanium.

6.36. A beam has an unequal angle cross-section as in Figure 6.24b and is subject to a bending moment $M_z = +20$ kNm. Find the maximum tensile and compressive stresses.

6.37. Find the greatest bending moment M_z which can be applied to the section in Figure 6.24d if the greatest shear stress is not to exceed 80 MN/m². The section consists of an unequal angle as in Figure 6.24b and a flange of width 200 mm and depth 20 mm.

6.38. Find the principal axes and moments of area for the sections of Figures 6.24b and 6.24d. (See also Problem 6.37.)

6.39. Repeat Problem 6.36 using the principal moments of area to calculate the stresses.

6.40. A beam with cross-section as in Figure 6.15b is subject to pure bending due to moments of 10 kNm in a plane xw where the w direction is inclined at $+45°$ to the positive z axis. Find the greatest normal stresses (a) using the principal moments of area, (b) by finding properties of area based on the w direction. State the equation of the neutral axis.

Beams—2 (Deflexions, and Stiffness Methods)

7.1 Slope and Deflexion

7.1.1 Differential Equation for Bending

Figure 7.1a shows part of a beam bent by loads acting in the xy plane. The x axis represents the unloaded position of the beam axis and the curved heavy line (called the deflexion curve) its bent position. The linear displacement in the y direction of a point originally on the x axis is called the deflexion v. The slope of a point on the deflexion curve relative to the x axis is called θ (Figure 7.1b). In the figure the deflexions have been greatly magnified for clarity, only very small values being acceptable in practice.

The two points m and n are an infinitesimal distance ds apart along the deflexion curve where the curvature corresponds to positive bending moment (cf. Figure 6.4). The radii of curvature R at m and n intersect at C. The slope at m is θ and at n is $(\theta - d\theta)$ where $d\theta$ is the angle subtended by ds at the centre of curvature. Then $ds = |R\,d\theta|$. Now the positive increment ds in Figure 7.1b is associated with reduction of θ by $d\theta$. Therefore ds and $d\theta$ have opposite signs, and

$$\frac{1}{R} = -\frac{d\theta}{ds}$$

As previously stated, the slopes and deflexions allowable in practice are very small, so that

$$dx = ds \cos \theta \doteqdot ds$$

and referring to Figure 7.1c

$$\frac{dv}{dx} = \tan \theta \doteqdot \theta$$

Then

$$\frac{1}{R} = -\frac{d}{dx}\left(\frac{dv}{dx}\right) = -\frac{d^2v}{dx^2}$$

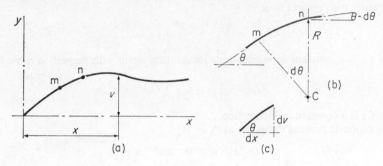

Figure 7.1

Another expression for $1/R$ is obtained from equation (6.9) of Section 6.3.1

i.e.
$$\frac{1}{R} = \frac{M}{EI}$$

Equating these results

$$EI \frac{d^2v}{dx^2} = -M \tag{7.1}$$

The quantity EI in this important differential equation is called the flexural rigidity for the plane of loading.

7.1.2 *Direct Integration for Deflexions*

Since the bending moment M is a function of x but not of v, it is possible to solve equation (7.1) by direct integration, in the course of which two constants of integration will be introduced. These constants are deduced from the boundary conditions, i.e. known values of slope and deflexion at supports or other points on the beam. It follows that at least two such quantities must be known.

Example 7.1 A simply supported beam carried distributed load of uniform intensity w as in Figure 6.7a. Find expressions for slope and deflexion and determine the greatest values.

Solution For equilibrium, from symmetry, the reactive loads are each $wL/2$. Over AB

$$M = -\frac{wLx}{2} + wx(\tfrac{1}{2}x) = -\frac{wLx}{2} + \frac{wx^2}{2} \tag{i}$$

where x is distance from A. Therefore

$$EI \frac{d^2v}{dx^2} = -M = \frac{wLx}{2} - \frac{wx^2}{2} \tag{ii}$$

Integrating with respect to x

$$EI\frac{dv}{dx} = \frac{wLx^2}{4} - \frac{wx^3}{6} + C_1 \tag{iii}$$

where C_1 is a constant of integration. Integrating again with respect to x we have

$$EIv = \frac{wLx^3}{12} - \frac{wx^4}{24} + C_1x + C_2 \tag{iv}$$

where C_2 is a constant of integration.

The supports prevent deflexion at A or B

i.e. $$v = 0 \quad \text{at} \quad x = 0 \quad \text{and} \quad x = L$$

Substituting $v = 0$ at $x = 0$ in (iv) we find $C_2 = 0$ and then using the second condition we obtain $C_1 = -wL^3/24$. Hence (iv) and (iii) become

$$EIv = \frac{wLx^3}{12} - \frac{wx^4}{24} - \frac{wL^3x}{24} \tag{v}$$

$$EI\frac{dv}{dx} = \frac{wLx^2}{4} - \frac{wx^3}{6} - \frac{wL^3}{24} \tag{vi}$$

Equation (v) gives the vertical displacement of the beam at any position while equation (vi) gives the rate of change of displacement with respect to x, i.e. the slope of the deflexion curve.

From symmetry the greatest deflexion will be at $x = \frac{1}{2}L$. Substituting in (v)

$$(v)_{x=\frac{1}{2}L} = -\frac{5wL^4}{384EI}$$

As expected the deflexion is negative, i.e. downwards. Substituting in (vi) it is found that the slope is zero at $x = \frac{1}{2}L$ and is greatest at the supports where

$$\theta_A = \left(\frac{dv}{dx}\right)_{x=0} = -\frac{wL^3}{24EI} \quad \text{and} \quad \theta_B = \left(\frac{dv}{dx}\right)_{x=L} = \frac{wL^3}{24EI}$$

Example 7.2 Find expressions for deflexion and slope at the free end of the beam of Figure 7.2a. Determine the values of the loads to induce unit slope and zero deflexion at B.

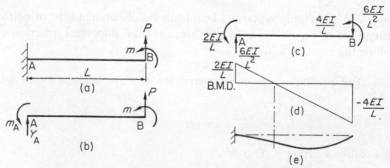

Figure 7.2 (a) Beam. (b) Arbitrary reactive loads. (c) Actual loads. (d) Bending moment diagram. (e) Form of deflexion curve

Solution Taking moments about A

$$m_A + m + PL = 0 \qquad \text{i.e.} \qquad m_A = -m - PL$$

For $\Sigma Y = 0$

$$Y_A = -P$$

Hence, over AB

$$M = m_A - Y_A x = -m - PL + Px$$

Substituting in (7.1) and integrating

$$EI \frac{dv}{dx} = mx + PLx - \frac{Px^2}{2} + C_1 \tag{i}$$

and

$$EIv = \frac{mx^2}{2} + \frac{PLx^2}{2} - \frac{Px^3}{6} + C_1 x + C_2 \tag{ii}$$

where C_1 and C_2 are constants of integration. Support conditions are

$$v = 0 \quad \text{at} \quad x = 0 \tag{iii}$$

and

$$\frac{dv}{dx} = 0 \quad \text{at} \quad x = 0 \tag{iv}$$

Putting (iii) and (iv) in (i) and (ii)

$$C_1 - C_2 = 0$$

and so

$$\frac{dv}{dx} = \frac{1}{EI}\left(mx + PLx - \frac{Px^2}{2}\right) \tag{v}$$

$$v = \frac{1}{EI}\left(\frac{mx^2}{2} + \frac{PLx^2}{2} - \frac{Px^3}{6}\right) \tag{vi}$$

To find slope at D put $x = L$ in (v)

$$\left(\frac{dv}{dx}\right)_{x=L} = \frac{mL}{EI} + \frac{PL^2}{2EI} \tag{vii}$$

The deflexion at B is found by putting $x = L$ in (vi)

$$(v)_{x=L} = \frac{mL^2}{2EI} + \frac{PL^3}{3EI} \tag{viii}$$

For $(v)_{x=L} = 0$ we have from (viii)

$$P = -\frac{3m}{2L} \tag{ix}$$

Setting $(dv/dx)_{x=L}$ equal to unity in (vii) and substituting from (ix)

$$m = \frac{4EI}{L} \tag{x}$$

This is a result of some importance in later work. It states that the moment to be exerted at B to produce unit rotation (slope) at B when deflexion is prevented at B is $4EI/L$. The deflected form of the beam is sketched in Figure 7.2e. Substituting (x) in (ix) $P = -6EI/L^2$. Hence using the equilibrium equations $m_A = 2EI/L$ and $Y_A = 6EI/L^2$ (Figure 7.2c).

Problems for Solution   7.1–7.3

7.1.3 *Modified (Macauley) Integration Method*

In the examples of the previous section bending moment was a continuous function of x. For most loading systems this will not be true. In Figure 7.3 the bending moment function is given by the following relationships:—

Over AB $M = -Y_A x$ for $0 \leqslant x \leqslant x_1$ (7.2a)

Over BC $M = -Y_A x + P_1(x - x_1)$ for $x_1 < x \leqslant x_2$ (7.2b)

Over CD $M = -Y_A x + P_1(x - x_1) + P_2(x - x_2)$ for $x_2 < x \leqslant L$ (7.2c)

To apply the direct integration method to this problem would therefore require integration of the three differential equations obtained by substituting equations (7.2) in (7.1), viz.

$$EI \frac{d^2v}{dx^2} = Y_A x \tag{7.3a}$$

$$EI \frac{d^2v}{dx^2} = Y_A x - P_1(x - x_1) \tag{7.3b}$$

$$EI \frac{d^2v}{dx^2} = Y_A x - P_1(x - x_1) - P_2(x - x_2) \tag{7.3c}$$

There would be 6 constants of integration with only two support conditions viz. $v = 0$ at $x = 0$ and $x = L$. The other four conditions needed are obtained from compatibility of deformation, i.e. slope and deflexion at B must be the same for both AB and BC and slope and deflexion at C must be the same for BC and CD.

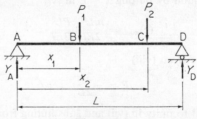

Figure 7.3

This method tends to be tedious. An easier mathematical procedure is based on the observation that equations (7.2a) and (7.2b) could be derived in turn from (7.2c) by striking out $P_2(x - x_2)$ and $P_1(x - x_1)$ when x is such that the bracket term is negative. We now adopt the convention of using square brackets for terms which have zero value if they are not positive. Thus (7.2c) written as

$$M = -Y_A x + P_1[x-x_1] + P_2[x-x_2]$$

would represent the bending moment for the whole beam. Within CD the square bracket terms are both positive so that the bending moment function is as (7.2c).

In BC $[x-x_2]$ would be negative and therefore zero leaving M as in (7.2b). Similarly over AB $[x-x_1]$ and $[x-x_2]$ are both zero, and $M = -Y_A x$ as in (7.2a).

Equations (7.3) are now replaced by

$$EI\frac{d^2v}{dx^2} = -M = Y_A x - P_1[x-x_1] + P_2[x-x_2] \tag{7.4}$$

Integration of (7.4) introduces only two constants of integration, to be determined from support conditions in the usual way. The general procedure is as follows:—

(a) Write the bending moment equation for the whole beam, using square brackets for terms which are zero over part of the span.

(b) Substitute this expression in (7.1) and integrate in the usual way keeping the bracket terms as entities throughout.

(c) Substitute the support conditions to find constants of integration.

When values of x are introduced the square bracket terms are retained only if positive.

Example 7.3 Find slope and deflexion equations for the beam of Figure 7.3 when $P_2 = 0$. Determine the maximum deflexion. Let $AB = x_1 = a$ and $BD = b$.

Solution From equilibrium the reactive loads are $Y_A = Pb/L$ and $Y_D = Pa/L$. Since the most general position for bending moment is between B and D

$$M = -Y_A x + P[x-a] \tag{i}$$

The square bracket term is zero over AB, i.e. when $x \leqslant a$. Substituting in (7.1) and introducing the reactive loads

$$EI\frac{d^2v}{dx^2} = \frac{Pbx}{L} - P[x-a]$$

Integrating

$$EI\frac{dv}{dx} = \frac{Pbx^2}{2L} - P\frac{[x-a]^2}{2} + C_1 \tag{ii}$$

and

$$EIv = \frac{Pbx^3}{6L} - P\frac{[x-a]^3}{6} + C_1 x + C_2 \tag{iii}$$

where C_1 and C_2 are constants of integration. Substituting the support conditions $v = 0$ at $x = 0$ and $x = L$ into (iii) we have $C_2 = 0$ and $C_1 = Pb(b^2-L^2)/6L$. Hence

$$\frac{dv}{dx} = \frac{1}{EI}\left(\frac{Pbx^2}{2L} - \frac{P[x-a]^2}{2} + \frac{Pb}{6L}(b^2-L^2)\right) \qquad \text{(iv)}$$

and

$$v = \frac{1}{EI}\left(\frac{Pbx^3}{6L} - \frac{P[x-a]^3}{6} + \frac{Pb(b^2-L^2)x}{6L}\right) \qquad \text{(v)}$$

To determine the position of maximum deflexion it is first necessary to find whether it lies on AB or BD. This is done by determining from (iv) the slopes at A, $x = 0$, and B, $x = a$. Since these are found to be negative and positive respectively there is evidently a point of zero slope between A and B giving a stationary value of v. Setting $(dv/dx) = 0$ in (iv) and noting that $[x-a]$ is negative for AB we find $x = \sqrt{(L^2-b^2)/3}$. Substituting this value of x in (v)

$$v_{max} = -\frac{Pb(L^2-b^2)^{3/2}}{9\sqrt{3}\,EIL}$$

Other Loads   The beam in Figure 7.4 is supported by reactive forces Y_B and Y_E as shown, and the bending moment equation is

$$M = Px - Y_B[x-x_1] + w[x-x_1] \times \tfrac{1}{2}[x-x_1]$$

$$-w[x-x_2] \times \tfrac{1}{2}[x-x_2] - m[x-x_3]^0 \quad (7.5)$$

The third term generates bending moment due to a uniformly distributed load which acts downwards from $x = x_1$ to position x. Since this load actually terminates at $x = x_2$, the fourth term is added generating bending moment due to a uniformly distributed load acting upwards from $x = x_2$ to

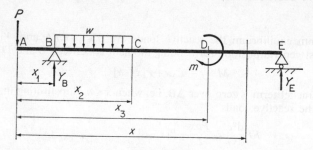

Figure 7.4

x. This has the effect of cancelling the bending moment generated by the third term for $x > x_2$. The fifth term generates the bending moment due to the concentrated moment m. The multiplying square bracket $[x-x_3]^0 = 1$ incorporates the moment m only if $x > x_3$.

Example 7.4 Find deflexion at $x = 0$ and $x = 3$ m and slope at $x = 7$ m for the beam of Figure 7.4. Take $EI = 4000$ kNm², $P = 10$ kN, $w = 4$ kN/m, $m = 8$ kNm (clockwise), $x_1 = 1$ m, $x_2 = 3$ m, $x_3 = 5$ m, AE $= 7$ m.

Taking moments about B

$$10 \times 1 - 2 \times 4 \times 1 - 8 + Y_E \times 6 = 0$$

Hence

$$Y_E = 1 \text{ kN} \quad \text{and from} \quad \Sigma Y = 0 \quad Y_B = 17 \text{ kN}$$

Hence

$$M = 10x - 17[x-1] + \tfrac{1}{2}w[x-1][x-1] - \tfrac{1}{2}w[x-3][x-3] - 8[x-5]^0$$

Substituting in (7.1) and integrating twice we find

$$EIv = -\frac{10x^3}{6} + \frac{17[x-1]^3}{6} - \frac{4[x-1]^4}{24} + \frac{4[x-3]^4}{24} + \frac{8[x-5]^2}{2} + C_1 x + C_2 \quad \text{(i)}$$

where C_1 and C_2 are constants of integration. Substituting the support conditions $v = 0$ at $x = 1$ m and $x = 7$ m in (i) we obtain

$$C_1 + C_2 = 1.67 \quad \text{and} \quad 7C_1 + C_2 = 117$$

Solving

$$C_1 = 19.22 \quad \text{and} \quad C_2 = -17.55$$

Introducing these in (i) we can find $(v)_{x=0} = 4.39 \times 10^{-3}$ m and $(v)_{x=3m} = 3.78 \times 10^{-3}$ m. Differentiating (i) we get $(dv/dx)_{x=7m} = -1.28 \times 10^{-3}$.

Problems for Solution 7.4–7.9

7.1.4 *Superposition*

In the bending moment equations (7.5) for the arbitrary load system in Figure 7.4 the loads are not functions of x. Therefore the differential equation of bending (7.1) is also linear in the loads, and will remain so after integration with respect to x. Thus the slope (dv/dx) and deflexion v will be linear functions of the loads and we may therefore state the following Superposition Principle.

Slope and deflexion at a point on a beam subject to several loads are the algebraic sums of the slopes and deflexions developed at that point when each load acts separately on the beam.

A list of particular values of deflexions and slopes for simple load systems is given in Figure 7.5. These 'standard results' can often be utilized in conjunction with the Superposition Principle to obtain solutions for combinations of loads.

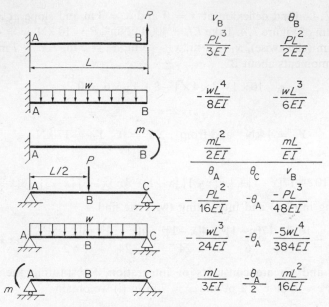

Figure 7.5

Example 7.5 Find the loads to produce unit slope and zero deflexion at B in Figure 7.2a using Superposition and standard results.

Solution From Figure 7.5 the deflexions at B due to P and m acting separately are $PL^3/3EI$ and $mL^2/2EI$ respectively. However the total deflexion at B is zero and therefore by Superposition

$$v_B = \frac{PL^3}{3EI} + \frac{mL^2}{2EI} = 0 \tag{i}$$

Obtaining the end slopes from Figure 7.5 we have

$$\theta_B = \frac{mL}{EI} + \frac{PL^2}{2EI} = 1 \tag{ii}$$

(i) and (ii) can be solved for P and m. Comparison with Example 7.2 shows considerable economy of effort.

Problems for Solution 7.10–7.12

7.1.5 Non-uniform Beams

When the cross-section varies continuously in size and/or shape, the beam is said to be non-uniform and I in equation (7.1) is a function of x. Here we shall consider only one simple case. Many of the more complex cases may require numerical evaluation; a more detailed investigation may be found in for example reference 1.

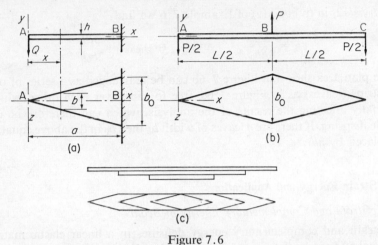

Figure 7.6

Example 7.6 Find the greatest stress and deflexion for a beam tapering in breadth and loaded as in Figure 7.6a.

Solution Here

$$M = Qx \qquad \text{(i)}$$

and

$$b = b_0 \left(\frac{x}{a}\right) \qquad \text{(ii)}$$

where b_0 is the breadth at $x = a$. Then

$$I = \frac{bh^3}{12} = \frac{b_0 h^3 x}{12a} = I_0 \left(\frac{x}{a}\right) \qquad \text{(iii)}$$

Greatest stress at any particular x is (equation 6.17)

$$(\sigma_x)_{\text{max}} = \frac{My_{\text{max}}}{I} = \frac{6Qa}{b_0 h^2} \qquad \text{(iv)}$$

and is thus independent of x. Substituting (i) and (iii) in (7.1)

$$EI_0 \frac{\mathrm{d}^2 v}{\mathrm{d}x^2} = -Qa$$

This equation is equivalent to that for a uniform beam of second moment I_0 subject to end moment Qa (anticlockwise) and so from Figure 7.5 we have

$$(v)_{x=0} = -\frac{Qa^3}{2EI_0} \qquad \text{(v)}$$

The beam in Figure 7.6b is from symmetry equivalent to two spans of Figure 7.6a placed back to back at B. Then substituting $Q = 0.5P$, $a = 0.5L$,

$I_0 = b_0 h^3/12$, in (iv) and (v) of Example 7.6 we find

$$v_B = \frac{3PL^3}{8Eb_0h^3} \quad \text{and} \quad (\sigma_x)_{max} = \frac{3PL}{2b_0h^2}$$

The plan area shown in Figure 7.6b can be rearranged as a series of over-laid strips (leaves) as in Figure 7.6c, free to slide over each other, without materially affecting the results of the analysis, which can therefore be used for a leaf spring. If there are n leaves of width b_1 then b_0 in the above equations is replaced by nb_1.

7.2 Strain Energy and Applications

7.2.1 *Strain and Complementary Energy in Beams*

The strain and complementary energy densities in a linear elastic material are equal and can be determined from equation (4.12) of Section 4.1.2. The states of stress and strain in an infinitesimal length dx of a beam (as in Figure 6.12) were found in Section 6.3.1 and expressed there by arrays (A6.1) and (A6.2). Using this information equation (4.12) reduces to

$$U_0 = \tfrac{1}{2}\sigma_x \epsilon_x$$

Substituting for ϵ_x using equation (6.6) of Section 6.3.1 we obtain $U_0 = \sigma_x^2/2E$ in which, from equation (6.14), we have $\sigma_x = My/I$ so that finally

$$U_0 = \frac{M^2 y^2}{2EI^2}$$

Referring to Figure 6.12 the volume having this strain energy density is a prism of length dx and cross-section dA. Therefore the strain energy of the length dx of the beam is, using (4.13)

$$dU = \int_A U_0 dA dx = \frac{M^2 dx}{2EI^2} \int_A y^2 dA$$

But

$$I = \int_A y^2 dA$$

so that

$$dU = \frac{M^2 dx}{2EI} \tag{7.6}$$

For a beam of length L for which M is a continuous function of x the total strain energy is

$$U = \int_0^L \frac{M^2 dx}{2EI} \tag{7.7}$$

In equations (7.6) and (7.7) the strain energy has been expressed in terms of the applied loads through the bending moment M. If desired the strain energy can be related to deflexion v by substituting for M from equation (7.1).

Example 7.7 Obtain an expression for the strain energy of a cantilever with a concentrated force P at the free end. Evaluate the strain energy when $L = 4$ m, $EI = 8$ MNm^2 and P is (a) 6 kN, (b) 12 kN.

Solution Referring to the beam of this description illustrated in Figure 7.5 and measuring x from the right-hand end B for convenience, then $M = -Px$ and from (7.7)

$$U = \int_0^L \frac{P^2 x^2 dx}{2EI} = \frac{P^2 L^3}{6EI}$$

Substituting for P we find strain energies of 48 Nm and 192 Nm respectively, so that doubling the load quadruples the strain energy.

Example 7.8 Find the strain energy of a simply supported beam subject to a concentrated downward force at $x = b$ as in Figure 6.5.

Solution Since the bending moment diagram is discontinuous (Figure 6.5) there is no single continuous expression for M in terms of x for substitution in equation (7.7). However since energy is a scalar quantity the total strain energy of the beam may be taken as the sum of the energies of any number of sub-divisions.

Over AB

$$M = -\frac{Pcx}{L}$$

so that

$$U_{AB} = \int_0^b \frac{P^2 c^2 x^2 dx}{2L^2 EI} \tag{i}$$

Over BC it is convenient to measure x as a local variable from C to B, giving $M = -Pbx/L$. Then

$$U_{BC} = \int_0^c \frac{P^2 b^2 x^2 dx}{2L^2 EI} \tag{ii}$$

Evaluating (i) and (ii) and adding we find

$$U = \frac{P^2 b^2 c^2}{6EIL}$$

It is interesting to note that the Conservation of Energy Principle can be used to find the displacement of a load. Referring to Example 7.7 the strain energy found there must be equal to the work done by the load P. Since $P \propto v_B$ it follows that the work done by the load is $\frac{1}{2}Pv_B$ (compare Figure 4.1b), and so equating to the strain energy we find the familiar $v_B = PL^3/3EI$.

The sign of a displacement determined from energy considerations is positive or negative according as it has the same or opposite direction to the corresponding load. If the load is a concentrated moment then the corresponding displacement is a slope.

Problems for Solution   7.13–7.15

7.2.2 *Beam Displacements (Castigliano Method)*

Castigliano's Theorem: Part 2 (see Article 4.4) provides a useful method for finding displacements of the loads on a beam. Partial derivatives are taken of the complementary energy $\bar{U}$ (expressed in terms of the applied loads) with respect to the loads. Then if F is any concentrated force or moment the corresponding displacement is by equation (4.26)

$$\delta = \frac{\partial \bar{U}}{\partial F} \tag{7.8}$$

Noting that for a linear elastic material $U = \bar{U}$ in equation (7.7) we can write

$$\frac{\partial \bar{U}}{\partial F} = \frac{\partial}{\partial F}\left(\int_0^L \frac{M^2 dx}{2EI}\right) = \int_0^L \frac{2M\left(\frac{\partial M}{\partial F}\right) dx}{2EI}$$

$$= \int_0^L \frac{M\frac{\partial M}{\partial F} dx}{EI} \tag{7.9}$$

For instance in the case of Example 7.7 where $M = -Px$

$$v_B = \frac{\partial \bar{U}}{\partial P} = \int_0^L \frac{(-Px)(-x)\,dx}{EI} = \frac{PL^3}{3EI}$$

Example 7.9   Find the slope at A for the beam in Figure 7.7a.

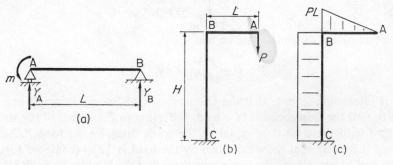

Figure 7.7

Solution   Taking moments about A

$$m + Y_B L = 0 \quad \text{and} \quad Y_B = -\frac{m}{L} = -Y_A$$

Hence $M = m - mx/L$ over AB. The slope θ_A at A is the displacement of the moment m and so applying (7.8) and (7.9)

$$\theta_A = \frac{\partial \bar{U}}{\partial m} = \frac{1}{EI} \int_0^L M \frac{\partial M}{\partial m} dx$$

Here $(\partial M/\partial m) = 1 - (x/L)$ and, evaluating the integral,

$$\theta_A = \frac{mL}{3EI}$$

Example 7.10   Find the vertical deflexion at A in Figure 7.7b.

Solution   The bending moment diagram is shown in Figure 7.7c. The required deflexion is the displacement of the load P. Over AB, taking x from A, $M = Px$ and $(\partial/\partial MP) = x$.

Hence

$$\left(\frac{\partial \bar{U}}{\partial P}\right)_{AB} = \frac{1}{EI} \int_0^L (Px)x \, dx = \frac{PL^3}{3EI} \tag{i}$$

Over BC, taking x from B, $M = PL$ and $(\partial M/\partial P) = L$. Hence

$$\left(\frac{\partial \bar{U}}{\partial P}\right)_{BC} = \int_0^H \frac{(PL)L \, dx}{EI} = \frac{PL^2 H}{EI} \tag{ii}$$

Adding (i) and (ii)

$$v_A = \frac{\partial \bar{U}}{\partial P} = \frac{PL^2}{EI}\left(\frac{L}{3} + H\right)$$

It should be noted from the above examples that only the displacement of a concentrated load can be determined by this method. To find displacements at other points or due to distributed loads we add dummy (fictitious) concentrated forces or moments corresponding to the required displacements. The complementary energy is obtained in the usual way for the real and dummy loads, and partial derivatives are taken with respect to the dummy loads which are then given their true zero value.

Example 7.11   Continuing Example 7.7 find the slope at the free end B.

Solution   Referring to Figure 7.2a let m be a dummy moment the displacement of which is the required slope. Then at any position x from B the bending moment is

$$M = -(m + Px)$$

Therefore

$$\theta_B = \frac{\partial \bar{U}}{\partial m} = \int_0^L \frac{M}{EI} \frac{\partial M}{\partial m} dx$$

Now $(\partial M/\partial m) = -1$

so that

$$\theta_B = \int_0^L \frac{(m+Px)\,(1)\,\mathrm{d}x}{EI} \qquad \text{(i)}$$

$$= \frac{mL}{EI} + \frac{PL^2}{2EI}$$

But $m = 0$ for the dummy load

therefore

$$\theta_B = \frac{PL^2}{2EI}$$

The sign is positive since it is apparent that P alone will cause a slope of the same sign as would be caused by m. Had P acted downwards the slope would have been negative. It should be noted that $m = 0$ can be introduced directly into equation (i) so reducing the labour of integration.

Example 7.12 Find the end deflexion of the partly loaded cantilever in Figure 6.8a.

Solution A dummy concentrated force P is added at C to provide a load corresponding to the required deflexion. Considering P to act upwards, and measuring distances from C for convenience we find bending moment over CB as

$$M = -Px + \tfrac{1}{2}wx^2 \quad \text{so that} \quad \frac{\partial M}{\partial P} = -x$$

and

$$\left(\frac{\partial \bar{U}}{\partial P}\right)_{CB} = \frac{1}{EI}\int_0^b (-Px + \tfrac{1}{2}wx^2)\,(-x)\,\mathrm{d}x$$

Putting $P = 0$ we obtain

$$\left(\frac{\partial \bar{U}}{\partial P}\right)_{CB} = -\frac{wb^4}{8EI} \qquad \text{(i)}$$

Over BA

$$M = -Px + wb(x - \tfrac{1}{2}b)$$

and a similar procedure leads to

$$\left(\frac{\partial \bar{U}}{\partial P}\right)_{BA} = -\frac{19wb^4}{12EI} \qquad \text{(ii)}$$

Hence adding (i) and (ii)

$$v_B = \frac{\partial \bar{U}}{\partial P} = \left(\frac{\partial \bar{U}}{\partial P}\right)_{BA} + \left(\frac{\partial \bar{U}}{\partial P}\right)_{CB} = -\frac{41wb^4}{24EI}$$

The negative sign indicates that the displacement of the dummy load P is opposite to its assumed direction, i.e. B moves down due to the distributed load w, as would be expected.

Problems for Solution 7.16–7.19

7.2.3 *Curved Beams*

The Castigliano method is particularly convenient for the determination of displacements of slender curved beams such as that lying in the xy plane

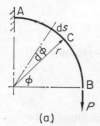 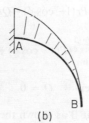 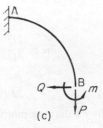

| (a) | (b) | (c) |

Figure 7.8

in Figure 7.8. Taking an element of length ds along the axis, we can replace dx by ds in equation (7.7) to obtain the strain energy as

$$U = \int_s dU = \int_s \frac{M^2 ds}{2EI} \tag{7.10}$$

where s is the total length of the axis of the curved beam. In the particular case when the beam is an arc of a circle of radius r then $ds = r d\phi$ and (7.10) becomes

$$U = \bar{U} = r \int_0^{\phi_0} \frac{M^2 d\phi}{2EI} \tag{7.11}$$

where $d\phi$ and ϕ_0 are the angles subtended by ds and s at the centre of curvature.

Example 7.13 Find the vertical and horizontal deflexions and the slope at B in Figure 7.8a. Sketch the bending moment diagram.

Solution Here $\phi_0 = \pi/2$. At a point C (Figure 7.8a) the bending moment is the moment of the force P about C.

i.e. $$M = P(r - r \cos \phi)$$

The variation of bending moment is plotted in Figure 7.8b. Here $(\partial M/\partial P) = r(1 - \cos \phi)$ and the vertical deflexion δ_V at B is the displacement of force P

i.e.
$$\delta_V = \frac{\partial \bar{U}}{\partial P} = \frac{1}{EI} \int_0^{\pi/2} M \frac{\partial M}{\partial P} r d\phi$$

$$= \frac{Pr^3}{EI} \int_0^{\pi/2} (1 - \cos \phi)^2 d\phi$$

$$= \frac{Pr^3}{EI} \left(\frac{3\pi}{4} - 2 \right)$$

To find the horizontal deflexion δ_H a dummy force Q is added (Figure 7.8c). Then
$$M = Pr(1 - \cos \phi) + Qr \sin \phi$$
and
$$\partial_H = \frac{\partial \bar{U}}{\partial Q} = \frac{r^3}{EI} \int_0^{\pi/2} (P(1 - \cos \phi) + Q \sin \phi) \sin \phi d\phi$$

$$= \frac{Pr^3}{2EI} \qquad \text{when} \qquad Q = 0$$

Adding a dummy moment m at B as shown the slope is found to be
$$\frac{\partial \bar{U}}{\partial m} = -\frac{Pr^2}{EI} \left(\frac{\pi}{2} - 1 \right)$$

and is therefore opposite in sense to that assumed for m.

Example 7.14   Find the vertical deflexion at B in Figure 7.9a.

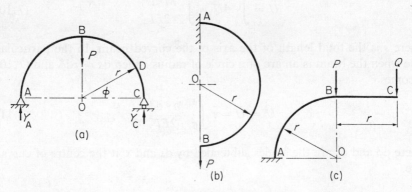

Figure 7.9

Solution   Taking moments about A
$$Y_C \times 2r = Pr \quad \text{and so} \quad Y_A = Y_C = \tfrac{1}{2}P$$

At a point D on CB the bending moment is
$$M = Y_C(r - r \cos \phi) = \tfrac{1}{2}Pr(1 - \cos \phi)$$

Then

$$\left(\frac{\partial \bar{U}}{\partial P}\right)_{CB} = \frac{Pr^3}{EI} \int_0^{\pi/2} \tfrac{1}{4}(1 - \cos \phi)^2 \, d\phi$$

Measuring ϕ from OA over AB a similar result is obtained and the deflexion of B is found to be half of δ_V in Example 7.13.

Problems for Solution   7.20–7.22

7.3 Statically Indeterminate Beams

7.3.1 *Introduction*

As discussed in Article 6.1 a beam requires either one fixed support, or a hinge and a roller, to maintain equilibrium and prevent rigid body displacements (Figures 6.2d and e). Frequently however, as in Figure 7.10, additional

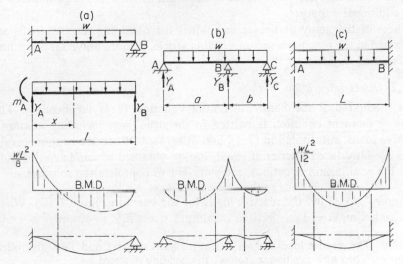

Figure 7.10   In each case the bottom diagram shows the form of the deflexion curve

supports are provided, and the total number of reactive loads then exceeds the number of equilibrium equations. For instance in Figure 7.10a the reactive loads are Y_A, Y_B, m_A and the equilibrium equations are (taking moments about A)

$$m_A + Y_B L = \tfrac{1}{2} w L^2; \quad Y_A + Y_B = wL \tag{7.12}$$

These equations only permit any two of the reactions to be expressed in terms of the third and we say that the system has one degree of (static) indeterminacy. The beam in Figure 7.10b also has one degree of indeterminacy, while that in Figure 7.10c has two degrees of indeterminacy.

Until the reactive loads are determined the bending moments cannot be calculated, and so stress, deflexions and loads cannot be evaluated (refer to equations 6.14 and 7.1). A great many techniques have been devised to derive sufficient additional relations between the reactive loads to complete the solution for these loads. These techniques all depend ultimately on the fact that each support imposes some restriction on the deformation of the beam; in particular if the supports are not themselves deformable each imposes a known deflexion and in the case of a fixed support a known slope as well. For example the beam in Figure 7.10a may be thought of as a cantilever which has been required to have zero deflexion at B. Alternatively we could think of the beam as simply-supported but required to have zero slope at A. Similarly the beam in Figure 7.10c might be interpreted as a cantilever subject to requirements of zero deflexion and slope at B. It should also be noted that non-zero values of deflexion or slope could be imposed at the additional supports.

Three of the many particular techniques for obtaining reactive loads are described in the remaining sections of this article. A more universal procedure is discussed in Article 7.4.

7.3.2 *Direct Integration Method*

This method uses the basic differential equation (7.1) for bending. The bending moment equation is written in the usual way including unknown reactive loads, substituted in (7.1), and integrated. When the support conditions are introduced a series of equations are obtained which, in conjunction with the equilibrium equations, are sufficient to complete the solution.

Example 7.15 Find the reactive loads for the beam in Figure 7.10a. When $w = 10 \text{ kN/m}$, $L = 3 \text{ m}$, find the maximum stress if $Z = 80 \text{ cm}^3$.

Solution The reactive loads are m_A, Y_A at the fixed support, and Y_B at the roller, as shown. Then at a position x from A the bending moment is

$$M = m_A - Y_A x + \tfrac{1}{2}wx^2 \tag{i}$$

Substituting in (7.1) and integrating, with C_1, C_2 constants, we have

$$EI\frac{dv}{dx} = -m_A x + \tfrac{1}{2}Y_A x^2 - \frac{wx^3}{6} + C_1 \tag{ii}$$

$$EIv = -\tfrac{1}{2}m_A x^2 + \frac{Y_A x^3}{6} - \frac{wx^4}{24} + C_1 x + C_2 \tag{iii}$$

The support conditions are $v = 0$ at $x = 0$ and $x = L$, and $(dv/dx) = 0$ at $x = 0$. Using the first and last of these $C_1 = C_2 = 0$; using $v = 0$ at $x = L$ we find from (iii)

$$12m_A - 4Y_A L + wL^2 = 0 \tag{iv}$$

Equilibrium equations for this problem were previously written as (7.12); solving simultaneously with (iv) we find

$$Y_A = \frac{5wL}{8}; \quad Y_B = \frac{3wL}{8}; \quad m_A = \frac{wL^2}{8} \tag{v}$$

Equations (i), (ii) and (iii) can now be written in terms of the load w only; hence bending moment, slope and deflexion can be evaluated, and using (6.14) stress can be calculated. The bending moment diagram and the deflexion curve are sketched in the figure. The magnitude of the greatest bending moment is $wL^2/8$, as in a uniformly loaded cantilever, but occurs at the root. Then

$$(\sigma_x)_{max} = \frac{M_{max}}{Z} = \frac{10^4 \times 9 \times 10^6}{8 \times 80} = 140 \times 10^6 \text{ N/m}^2.$$

Problems for Solution 7.23–7.25

7.3.3 *Superposition Method*

In this method the indeterminate beam is considered as the combination of two statically determinate beams subject to one or more conditions of compatibility of deformation. For example in Figure 7.11 the two simple cantilevers can be combined to form Figure 7.11a by imposing the compatibility condition of zero deflexion at C required by the roller support.

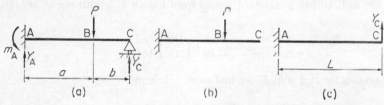

(a) (b) (c)

Figure 7.11

The routine for this method consists first of removing from the indeterminate beam sufficient of the supports to make it determinate. As pointed out in Section 7.3.1 this can be done in more than one way; in Figure 7.11 the support at C has been removed to give the *primary* structure in Figure 7.11b. A *secondary* structure is formed of the same beam as the primary, but subject only to the reactive load(s) corresponding to the support(s) removed from the actual beam; for example the beam in Figure 7.11c is subject to the reactive load Y_C treated as if it were an applied load. The deflexions at C for the primary and secondary systems can be determined separately and then superposed to give the total deflexion at C. However, this deflexion is known to be zero and so from compatibility of deformation between the primary and secondary systems a relation between the applied and reactive loads is obtained. As many such relations can be derived as there are compatibility conditions.

Example 7.16 Find the reactive loads in Figure 7.11a if $a = b = L/2$. Evaluate the central deflexion if $EI = 800$ kNm^2, $P = 16$ kN and $L = 2$ m.

Solution The primary and secondary structures are chosen as described above. To find the deflexion v_{C1} of the primary at end C, we note that since BC is unloaded it must remain straight and so has slope equal to that at B throughout its length. Consequently, using standard results from Figure 7.5,

$$v_{C1} = v_B + \theta_B \times BC = -\frac{Pa^3}{3EI} - \frac{Pa^2b}{2EI} = -\frac{5PL^3}{48EI} \tag{i}$$

For the secondary

$$v_{C2} = \frac{Y_C L^3}{3EI} \tag{ii}$$

But for compatibility

$$v_{C1} + v_{C2} = 0 \tag{iii}$$

Combining (i)–(iii) we obtain $Y_C = 5P/16$, and from equilibrium $Y_A = 11P/16$ and $m_A = 3PL/16$.

An expression for the central deflexion is easily obtained using the Superposition method (Section 7.1.4). The system is considered as the combination of two simply supported beams, one with central load P, and the other with moment m_A at the left end. Utilizing standard results from Figure 7.5, with $m_A = 3PL/16$ we obtain

$$v_B = -\frac{PL^3}{48EI} + \left(\frac{3PL}{16}\right)\frac{L^2}{16EI}$$

Substituting for P, L and EI we find $v_B = -1 \cdot 46$ mm.

Example 7.17 Find the reactive loads for the beam in Figure 7.10b when $a = b = L/2$ and the central support is set δ above the other supports.

Solution This system is considered as the combination of two simply supported beams, the primary being uniformly loaded and the secondary bearing a central upward concentrated force Y_B. Then, using standard results from Figure 7.5,

$$v_B = v_{B1} + v_{B2} = -\frac{5wL^4}{384EI} + \frac{Y_B L^3}{48EI} \tag{i}$$

But compatibility of deformation makes $v_B = \delta$. Hence

$$Y_B = \frac{48EI\delta}{L^3} + \frac{5wL}{8}$$

It is interesting to examine the alternative case in which the supports are all at the same level before loading but the central support can yield under load. If the latter is elastic with stiffness k (force/unit deflexion) then we put $v_B = -Y_B/k$ in (i) (negative sign since the force opposes deflexion).

Problems for Solution 7.26–7.29

7.3.4 *Castigliano Method*

The Castigliano method for deflexions can be adapted to establish relations between the reactive loads for a statically indeterminate beam. Certain of the reactive loads are selected to be considered as if they were applied loads; the number so chosen is equal to the number of degrees of indeterminacy. For example the system in Figure 7.10a could be treated as a simple cantilever subject to the distributed load and an apparent applied load Y_B. The Castigliano Theorem for deflexions can then be used to find the displacement of Y_B. Thus measuring x from B towards A in Figure 7.10a

$$M = -Y_B x + \frac{wx^2}{2}$$

and

$$v_B = \frac{\partial \bar{U}}{\partial Y_B} = \frac{1}{EI} \int_0^L M \frac{\partial M}{\partial Y_B} \, dx = \frac{L^3}{EI} \left(+\frac{Y_B}{3} - \frac{wL}{8} \right)$$

However if this system is to be compatible with the actual structure, we require $v_B = 0$, and so Y_B is $3wL/8$ as found in Example 7.15. The other reactive loads are then deduced using the equilibrium equations.

This application of the Castigliano Theorem is sometimes called Least Work, sometimes the Compatibility Theorem; the latter title seems a better description.

Problem for Solution 7.30

7.4 Stiffness Methods

7.4.1 *Introduction*

In the previous articles of this chapter various methods have been developed for determination of the deformation of a beam, and, what is really more important, for the evaluation of the reactive loads of statically indeterminate beams. However, none of these methods is very easy to apply for beams with a large number of loads and supports. Alternative methods have been developed which aim to express the response of the system by linear algebraic equations relating loads on, and displacements of, the beam at a number of chosen points called nodes. When the nodal loads are expressed in terms of the nodal displacements we have a set of stiffness equations (compare Article 5.3). Since these are comparatively easy to form and solve using a digital computer, very complex systems can be dealt with.

We begin our discussion by obtaining stiffness equations for a portion of a beam subject only to loads at its ends as in Figure 7.12a (compare Section 5.3.1 and Figure 5.2). Since this portion of beam is not required to be infinitesimal we shall call it a finite element of the beam or simply, beam element. It may be part of a single span as between 1 and 2 in Figure 7.12b, or it could be one complete span from a continuous beam as between 1 and 2 in Figure 7.12c. In either case the complete beam is regarded as a collinear assembly of finite elements which are numbered in sequence using bold type **1, 2 . . . n**; the nodes of the assembly are numbered in sequence using ordinary type 1, 2, . . . n. Then as in Section 5.3.2 we build up stiffness equations for the assembly from those for its elements.

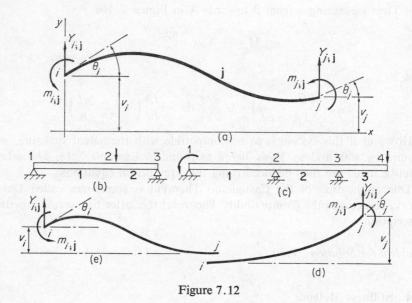

Figure 7.12

In this chapter we shall for convenience discuss only loading of the assembly in the *xy* plane. However all the results obtained could easily be adapted to loading in the *xz* plane. The discussion is of course limited to beams for which the *y* direction is an axis of symmetry of the cross-section.

7.4.2 *Element Stiffness Equations*

The loads acting in the *xy* plane on the ends of the beam element lying between nodes *i* and *j* are shown in Figure 7.12a. At node *i* there is a force $Y_{i;j}$ in the *y* direction. The first subscript *i* is the nodal number and the second **j** is the number of the element. Also at node *i* a moment $m_{i;j}$ can act as shown. Similarly a force $Y_{j;j}$ and a moment $m_{j;j}$ can act on the element at node *j*. The displacements of the nodes due to bending consist of deflexions v_i and v_j, and of rotations, i.e. slopes, θ_i and θ_j, as shown. (At this point the reader is reminded of the general discussions in Article 6.1 and Section 7.1.1 relating to loads on, and displacements of, a beam.)

We now write the following set of linear algebraic equations relating each of the element end loads to the end displacements.

$$Y_{i;j} = a_{11}v_i + a_{12}\theta_i + a_{13}v_j + a_{14}\theta_j$$
$$m_{i;j} = a_{21}v_i + a_{22}\theta_i + a_{23}v_j + a_{24}\theta_j$$
$$Y_{j;j} = a_{31}v_i + a_{32}\theta_i + a_{33}v_j + a_{34}\theta_j$$
$$m_{j;j} = a_{41}v_i + a_{42}\theta_i + a_{43}v_j + a_{44}\theta_j$$

$$(7.13)$$

where the quantities a_{11} etc. are stiffness coefficients. In matrix form these become

$$
\begin{bmatrix} Y_{i;j} \\ m_{i;j} \\ Y_{j;j} \\ m_{j;j} \end{bmatrix} =
\begin{bmatrix}
a_{11} & a_{12} & a_{13} & a_{14} \\
a_{21} & a_{22} & a_{23} & a_{24} \\
a_{31} & a_{32} & a_{33} & a_{34} \\
a_{41} & a_{42} & a_{43} & a_{44}
\end{bmatrix}
\begin{bmatrix} v_i \\ \theta_i \\ v_j \\ \theta_j \end{bmatrix}
\tag{7.14}
$$

In these equations the end loads and displacements of the element form column matrices or vectors which we write for reference as

$$\mathbf{F}_j = \{Y_{i;j} \quad m_{i;j} \quad Y_{j;j} \quad m_{j;j}\} \tag{7.15}$$
$$\mathbf{\delta}_j = \{v_i \quad \theta_i \quad v_j \quad \theta_j\} \tag{7.16}$$

(Note the use of braces { } to indicate a column matrix written as a row matrix to save space—see also Appendix II.)

The square matrix of stiffness coefficients in equation (7.14) is called the stiffness matrix of the beam element and is denoted by $\mathbf{k}_j$. Then equations (7.14) can be written in the shorthand form

$$\mathbf{F}_j = \mathbf{k}_j\mathbf{\delta}_j \tag{7.14a}$$

A general expression for the element stiffness matrix can be deduced, as will be shown in Section 7.4.3, using methods already described in Articles 7.1 and 7.3. Using this expression the stiffness matrix for a particular beam element can be obtained as a set of numbers. Then for a given set of nodal displacements the element end loads can be found by the arithmetic operations corresponding to the matrix product $\mathbf{k}_j\mathbf{\delta}_j$.

7.4.3 *Stiffness Matrix of a Beam Element*

Referring to Figure 7.12a, the bending moment at a position x from node i is

$$M = m_{i;j} - Y_{i;j}x$$

Then substituting in the differential equation of flexure (7.1) we have

$$EI\frac{d^2v}{dx^2} = -m_{i;j} + Y_{i;j}x$$

Integrating,

$$EI \frac{dv}{dx} = -m_{i;j}x + \tfrac{1}{2} Y_{i;j} x^2 + C_1 \qquad (7.17)$$

and

$$EIv = -\tfrac{1}{2} m_{i;j} x^2 + \tfrac{1}{6} Y_{i;j} x^3 + C_1 x + C_2 \qquad (7.18)$$

where C_1 and C_2 are constants of integration to be determined from the support conditions.

As a first step let displacements of the element be prevented at node i. Thus the element nodal displacement vector (7.16) is (Figure 7.12d)

$$\delta_j = \{0 \quad 0 \quad v_j \quad \theta_j\}$$

The zero displacements at node i are equivalent to $v = 0$ and $\theta = (dv/dx) = 0$ at $x = 0$. Substituting these values in equations (7.18) and (7.17) respectively we find $C_1 = C_2 = 0$. Then, if L is the length of the element, the displacements at node j can be found by substituting $x = L$ in equations (7.17) and (7.18) as follows,

$$EI\theta_j = EI \left(\frac{dv}{dx} \right)_{x=L} = -m_{i;j}L + \tfrac{1}{2} Y_{i;j} L^2$$

$$EIv_j = EI(v)_{x=L} = -\tfrac{1}{2} m_{i;j} L^2 + \tfrac{1}{6} Y_{i;j} L^3$$

Solving these for $Y_{i;j}$ and $m_{i;j}$ we have

$$Y_{i;j} = -\frac{12EI}{L^3} v_j + \frac{6EI}{L^2} \theta_j$$

and   (7.19)

$$m_{i;j} = -\frac{6EI}{L^2} v_j + \frac{2EI}{L} \theta_j$$

Now, applying the equilibrium equations (6.1a) for a beam to the element of Figure 7.12a, we can write

$$Y_{i;j} + Y_{j;j} = 0; \qquad m_{i;j} + m_{j;j} + Y_{j;j} L = 0 \qquad (7.20)$$

where the second of these equations was obtained by taking moments about node i. Then substituting from (7.19) in (7.20)

$$Y_{j;j} = \frac{12EI}{L^3} v_j - \frac{6EI}{L^2} \theta_j$$

                                                                                                       (7.21)

$$m_{j;j} = -\frac{6EI}{L^2} v_j + \frac{4EI}{L} \theta_j$$

Now if the known displacements v_i and θ_i equal to zero are substituted in equations (7.13) we have

$$Y_{i;j} = a_{13}v_j + a_{14}\theta_j$$

$$m_{i;j} = a_{23}v_j + a_{24}\theta_j$$

$$Y_{j;j} = a_{33}v_j + a_{34}\theta_j \qquad (7.22)$$

$$m_{j;j} = a_{43}v_j + a_{44}\theta_j$$

Comparing the first two of these with equations (7.19) we see that

$$a_{13} = -12EI/L^3 \qquad a_{14} = 6EI/L^2$$

$$a_{23} = -6EI/L^2 \qquad a_{24} = 2EI/L \qquad (7.23)$$

These constitute a block of four coefficients in the stiffness matrix of equations (7.14)

i.e.
$$\mathbf{k_j} = \begin{bmatrix} a_{11} & a_{12} & -\dfrac{12EI}{L^3} & \dfrac{6EI}{L^2} \\[2mm] a_{21} & a_{22} & -\dfrac{6EI}{L^2} & \dfrac{2EI}{L} \\[2mm] a_{31} & a_{32} & a_{33} & a_{34} \\[2mm] a_{41} & a_{42} & a_{43} & a_{44} \end{bmatrix} \qquad (7.24)$$

It is worth noting that this block of coefficients in $\mathbf{k_j}$ was determined by considering end loads at node i due to displacements at node j only.

If we now compare equations (7.21) and (7.22) we are able to write

$$a_{33} = 12EI/L^3 \qquad a_{34} = -6EI/L^2$$

$$a_{43} = -6EI/L^2 \qquad a_{44} = 4EI/L \qquad (7.25)$$

which constitute the lower right-hand block of coefficients in (7.24), and are those which relate end loads at node j to displacements at node j.

The above discussion has obtained (equations 7.23 and 7.25) the stiffness coefficients which relate end loads on the element to displacements at node j. To find the remaining coefficients we proceed in a similar manner.

Let displacements be prevented at node j so that (Figure 7.12e)

$$\boldsymbol{\delta_j} = \{v_i \quad \theta_i \quad 0 \quad 0\}$$

Then substituting v and (dv/dx) equal to zero at $x = L$ in equations (7.18) and (7.17), and proceeding as before we obtain

$$Y_{i;j} = \frac{12EI}{L^3} v_i + \frac{6EI}{L^2} \theta_i$$

$$m_{i;j} = \frac{6EI}{L^2} v_i + \frac{4EI}{L} \theta_i \qquad (7.26)$$

and, using the equilibrium equations (7.20)

$$Y_{j;j} = -\frac{12EI}{L^3} v_i - \frac{6EI}{L^2} \theta_i$$

$$m_{j;j} = \frac{6EI}{L^2} v_i + \frac{2EI}{L} \theta_i \tag{7.27}$$

Comparing equations (7.26) and (7.27) with those obtained by putting v_j and θ_j equal to zero in (7.13) we find the remaining coefficients and can complete (7.24) giving the element stiffness matrix as

$$
\mathbf{k_j} = \begin{bmatrix}
\dfrac{12EI}{L^3} & \dfrac{6EI}{L^2} & -\dfrac{12EI}{L^3} & \dfrac{6EI}{L^2} \\[2mm]
\dfrac{6EI}{L^2} & \dfrac{4EI}{L} & -\dfrac{6EI}{L^2} & \dfrac{2EI}{L} \\[2mm]
-\dfrac{12EI}{L^3} & -\dfrac{6EI}{L^2} & \dfrac{12EI}{L^3} & -\dfrac{6EI}{L^2} \\[2mm]
\dfrac{6EI}{L^2} & \dfrac{2EI}{L} & -\dfrac{6EI}{L^2} & \dfrac{4EI}{L}
\end{bmatrix} \tag{7.28}
$$

On inspection it is seen that the matrix is symmetric since

$$a_{pq} = a_{qp} \quad \text{for} \quad p, q = 1, 2, 3, 4 \quad (p \neq q) \quad \text{(compare equation 5.10)}$$

A more convenient form of (7.28) for calculations is

$$
\mathbf{k_j} = \frac{EI}{L^3} \begin{bmatrix}
12 & 6L & -12 & 6L \\
6L & 4L^2 & -6L & 2L^2 \\
-12 & -6L & 12 & -6L \\
6L & 2L^2 & -6L & 4L^2
\end{bmatrix} \tag{7.29}
$$

This relatively simple matrix contains all the information needed to calculate the element end loads for any given set of element nodal displacements δ_j. It is clear from (7.29) that the stiffness coefficients do not all have the same physical dimensions. In this respect a stiffness matrix differs from a stress array, all the components of which have the same dimensions, e.g. N/m^2. To avoid errors in calculations with stiffness equations it is recommended that all quantities are expressed in terms of newtons and metres.

Example 7.18 Write down the stiffness equations and find the end loads for the beam of length 4 m shown in Figures 7.13a and b where the right-hand support has been given an upward displacement of 0·01 m.

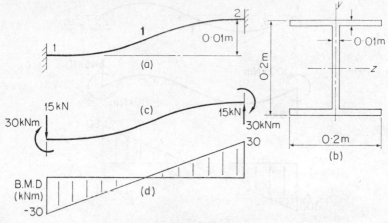

Figure 7.13

Solution An exact calculation makes $I = 0.41 \times 10^{-4}$ m⁴. For numerical convenience we shall take E to be 195×10^9 N/m² so that $EI = 8 \times 10^6$ Nm². Then the quantity EI/L^3 is 0.125×10^6 N/m and, substituting in (7.29), the stiffness matrix is

$$\mathbf{k}_1 = 0.125 \times 10^6 \begin{bmatrix} 12 & 24 & -12 & 24 \\ 24 & 64 & -24 & 32 \\ -12 & -24 & 12 & 24 \\ 24 & 32 & -24 & 64 \end{bmatrix} \tag{i}$$

where **1**, the number of the element, replaces **j**. Now the fixed supports prevent rotation at the ends, nodes 1 and 2, so that $\theta_i = \theta_1 = 0$ and $\theta_j = \theta_2 = 0$. Deflexion is prevented at node 1 $(v_i = v_1 = 0)$ and is 0.01 m at node 2 $(v_j = v_2 = 0.01)$. The element nodal displacement vector is therefore

$$\delta_1 = \{0 \quad 0 \quad 0.01 \quad 0\}$$

and the stiffness equations (7.14a) are

$$\mathbf{F}_1 = \begin{bmatrix} Y_{1;1} \\ m_{1;1} \\ Y_{2;1} \\ m_{2;1} \end{bmatrix} = 0.125 \times 10^6 \begin{bmatrix} 12 & 24 & -12 & 24 \\ 24 & 64 & -24 & 32 \\ -12 & -24 & 12 & -24 \\ 24 & 32 & -24 & 64 \end{bmatrix} \begin{bmatrix} 0 \\ 0 \\ 0.01 \\ 0 \end{bmatrix} = 10^3 \begin{bmatrix} -15 \\ -30 \\ 15 \\ -30 \end{bmatrix}$$

The end loads at node 1 for example are a down force of 15 kN and a clockwise moment of 30 kNm as shown in Figure 7.13c; the corresponding bending moment diagram is also shown. It should be noted how easily this statically indeterminate problem has been solved.

Example 7.19 A beam with deflexion curve as in Figure 7.14a has the same length and properties as that in Example 7.18. Find the end loads if the end slopes shown are 0.01 and 0.005 radian in magnitude.

Solution The stiffness matrix is the same as (i) of Example 7.18. For this simply supported system the end deflexions are zero. Therefore

$$\delta_1 = \{0 \quad -0.01 \quad 0 \quad -0.005\}$$

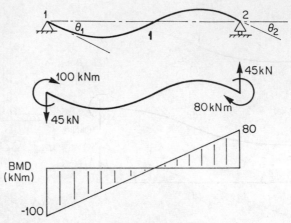

Figure 7.14

Evaluating the matrix product $k_1\delta_1$ we find

$$F_1 = 10^3\{-45 \quad -100 \quad 45 \quad -80\}$$

The force $Y_{1;2} = -45 \times 10^3$ N acts down at node 1. The end loads and the corresponding bending moment diagram are shown in Figures 17.14b and c.

As an alternative method of obtaining the general expression for the element stiffness matrix we could use the Superposition technique described in Section 7.1.4. For example when v_i and θ_i are zero as in Figure 7.12d the beam element is in the same condition as a cantilever with support at node i and applied loads $Y_{j;j}$, $m_{j;j}$ at node j. Using standard results from Figure 7.5 we can write

$$EI\theta_j = m_{j;j}L + \tfrac{1}{2}Y_{j;j}L^2$$

Similarly we obtain an expression for v_j and then, solving, equations (7.21) can be found. The procedure for finding the stiffness coefficients is as before.

Another possibility is to use Castigliano's Theorem: Part 2 as discussed in Section 7.2.2. For example referring to Figure 7.12e the system is again equivalent to a cantilever, and the bending moment at x from node i is

$$M = m_{i;j} - Y_{i;j}x$$

To find θ_i we take the appropriate partial derivative of the complementary energy. Thus

$$\theta_i = \frac{\partial \bar{U}}{\partial m_{i;j}} = \frac{1}{EI}\int_0^L M \frac{\partial M}{\partial m_{i;j}}\,dx$$

$$= \frac{1}{EI}\int_0^L (m_{i;j} - Y_{i;j}x)\,dx$$

and

$$EI\theta_i = m_{i;j}L - \tfrac{1}{2}Y_{i;j}L^2$$

Then v_i is found from $(\partial \bar{U}/\partial Y_{i;j})$ and, solving, equations (7.26) are obtained.

Problems for Solution 7.31, 7.32

7.4.4 *Structure of the Element Stiffness Equations*

After substituting from (7.29) the element stiffness equations (7.14) can be written in partitioned form as

$$
\begin{bmatrix} Y_{i;j} \\ m_{i;j} \\ \hline Y_{j;j} \\ m_{j;j} \end{bmatrix} = \frac{EI}{L^3} \left[\begin{array}{cc|cc} 12 & 6L & -12 & 6L \\ 6L & 4L^2 & -6L & 2L^2 \\ \hline -12 & -6L & 12 & -6L \\ 6L & 2L^2 & -6L & 4L^2 \end{array} \right] \begin{bmatrix} v_i \\ \theta_i \\ \hline v_j \\ \theta_j \end{bmatrix} \tag{7.30}
$$

Introducing

$$\mathbf{F}_{i;j} = \{Y_{i;j} \quad m_{i;j}\}$$

and

$$\mathbf{F}_{j;j} = \{Y_{j;j} \quad m_{j;j}\} \tag{7.31}$$

as vectors of element end loads at nodes i and j respectively, we can write the element end load vector as

$$\mathbf{F}_j = \{\mathbf{F}_{i;j} \quad \mathbf{F}_{j;j}\} \tag{7.32}$$

Similarly, we can write the element nodal displacement vector as

$$\boldsymbol{\delta}_j = \{\boldsymbol{\delta}_i \quad \boldsymbol{\delta}_j\} \tag{7.33}$$

where

$$\boldsymbol{\delta}_i = \{v_i \quad \theta_i\} \quad \text{and} \quad \boldsymbol{\delta}_j = \{v_j \quad \theta_j\} \tag{7.34}$$

Finally, the partitioned element stiffness matrix can be represented as

$$
\begin{bmatrix} \mathbf{k}_{ii;j} & \mathbf{k}_{ij;j} \\ \mathbf{k}_{ji;j} & \mathbf{k}_{jj;j} \end{bmatrix} \tag{7.35}
$$

where a typical sub-matrix is

$$\mathbf{k}_{ij;j} = \frac{EI}{L^3} \begin{bmatrix} -12 & 6L \\ -6L & 2L^2 \end{bmatrix} \tag{7.36}$$

When this block of coefficients was first determined in Section 7.4.3 (equations 7.23) it was pointed out that they related end loads at node i to displacements at node j. The nodal subscripts ij in (7.36) indicate this connexion; the third subscript j denotes the number of the beam element.

The element stiffness equations (7.30) can now be expressed in terms of the sub-matrices defined above, as

$$
\begin{bmatrix} \mathbf{F}_{i;\mathrm{j}} \\ \mathbf{F}_{j;\mathrm{j}} \end{bmatrix} = \begin{bmatrix} \mathbf{k}_{ii;\mathrm{j}} & \mathbf{k}_{ij;\mathrm{j}} \\ \mathbf{k}_{ji;\mathrm{j}} & \mathbf{k}_{jj;\mathrm{j}} \end{bmatrix} \begin{bmatrix} \boldsymbol{\delta}_i \\ \boldsymbol{\delta}_j \end{bmatrix}
\tag{7.37}
$$

These equations show clearly the relation between end load vectors and nodal displacement vectors for an element. They will be seen to correspond in form to the equations (5.17) for a bar subject to axial forces; indeed the latter are merely a particularly simple case for which the end load and nodal displacement vectors consist of single quantities (e.g. $\mathbf{F}_{i;\mathrm{j}} = \{X_{i;\mathrm{j}}\}$ and $\boldsymbol{\delta}_j = \{u_j\}$).

7.4.5 *Stiffness Equations for a Collinear Assembly*

When we consider an assembly of elements as in Figures 7.12b and c external loads are applied only at the nodes. At the ith node these loads are a force Y_i and a moment m_i as shown in Figure 7.15a. These together constitute a vector of nodal loads

$$
\mathbf{F}_i = \{Y_i \quad m_i\}
\tag{7.38}
$$

If all the nodal loads are listed as a column matrix we obtain the assembly load vector

$$
\begin{aligned}
\mathbf{F} &= \{\mathbf{F}_1 \quad \mathbf{F}_2 \ldots \mathbf{F}_i \ldots \mathbf{F}_n\} \\
&= \{Y_1 \quad m_1 \quad Y_2 \quad m_2 \ldots Y_i \quad m_i \ldots Y_n \quad m_n\}
\end{aligned}
\tag{7.39}
$$

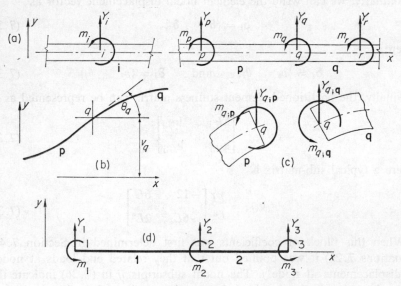

Figure 7.15

Similarly, using the nodal displacement vector of equation (7.34), the assembly displacement vector is

$$\delta = \{\delta_1 \quad \delta_2 \dots \delta_i \dots \delta_n\}$$
$$= \{v_1 \quad \theta_1 \quad v_2 \quad \theta_2 \dots v_i \quad \theta_i \dots v_n \quad \theta_n\} \tag{7.40}$$

Following the same procedure for the assembly as for an element we write a set of linear equations expressing the nodal loads in terms of the nodal displacements, as follows:

$$Y_1 = s_{11}v_1 + s_{12}\theta_1 + \dots + s_{1,(2n-1)}v_n + s_{1,2n}\theta_n$$
$$m_1 = s_{21}v_1 + s_{22}\theta_1 + \dots + s_{2,(2n-1)}v_n + s_{2,2n}\theta_n$$

.
.
.

$$Y_n = s_{(2n-1),1}v_1 + s_{(2n-1),2}\theta_1 + \dots + s_{(2n-1),(2n-1)}v_n + s_{(2n-1),2n}\theta_n$$
$$m_n = s_{2n,1}v_1 + s_{2n,2}\theta_1 + \dots + s_{2n,(2n-1)}v_n + s_{2n,2n}\theta_n \tag{7.41}$$

where s_{11} etc. are stiffness coefficients.

Equations (7.41) may be written in matrix form as

$$
\begin{bmatrix} Y_1 \\ m_1 \\ \cdot \\ \cdot \\ \cdot \\ m_n \end{bmatrix}
=
\begin{bmatrix}
s_{11} & s_{12} & \dots & s_{1,2n} \\
s_{21} & s_{22} & \dots & s_{2,2n} \\
& & & \\
& & & \\
& & & \\
s_{2n,1} & s_{2n,2} & \dots & s_{2n,2n}
\end{bmatrix}
\begin{bmatrix} v_1 \\ \theta_1 \\ \\ \\ \\ \theta_n \end{bmatrix}
\tag{7.42}
$$

or in shorthand form

$$\mathbf{F} = \mathbf{K}\boldsymbol{\delta} \tag{7.42a}$$

where $\mathbf{K}$ is a square matrix of size $2n \times 2n$, known as the assembly stiffness matrix, and $\mathbf{F}$ and $\boldsymbol{\delta}$ are assembly load and displacement vectors as defined by equations (7.39) and (7.40).

As in Section 5.3.2 these assembly stiffness equations can be derived from the stiffness equations of the elements of the assembly, by applying twin conditions of compatibility of displacement, and of equilibrium, at each node. The first of these conditions requires that the deflexion curve is continuous at a node, as for example at q in Figure 7.15b. This condition has in fact already been satisfied by using nodal displacements to define the element end displacements (Figure 7.12a).

The second, or equilibrium, condition is satisfied by requiring that the external loads at a node must be equal to the resultants of the end loads exerted at that node on the elements meeting at it (compare Section 5.3.2). For example at node q of the assembly of Figure 7.15a the external force Y_q must be the resultant of the end forces on the elements (Figure 7.15c).

i.e. $\qquad\qquad\qquad\qquad Y_q = Y_{q;\mathbf{p}} + Y_{q;\mathbf{q}}$

In a similar manner a relation can be obtained between the external moment m_q and the element end moments, and we can write the two relations in matrix form as

$$\begin{bmatrix} Y_q \\ m_q \end{bmatrix} = \begin{bmatrix} Y_{q;\mathbf{p}} \\ m_{q;\mathbf{p}} \end{bmatrix} + \begin{bmatrix} Y_{q;\mathbf{q}} \\ m_{q;\mathbf{q}} \end{bmatrix} \qquad (7.43)$$

Then using the definitions (7.31) and (7.38) we obtain the vector form of the equilibrium condition

$$\mathbf{F}_q = \mathbf{F}_{q;\mathbf{p}} + \mathbf{F}_{q;\mathbf{q}} \qquad\qquad (7.43a)$$

where $\mathbf{p}$ and $\mathbf{q}$ are elements meeting at node q. We observe that if there is no external load the element end loads must be equal and opposite.

We shall now illustrate the development of the assembly stiffness equations by reference to the two-element assembly in Figure 7.15d. Since there are three nodes the assembly load and displacement vectors are

$$\mathbf{F} = \{ Y_1 \quad m_1 \quad Y_2 \quad m_2 \quad Y_3 \quad m_3 \}$$

and

$$\boldsymbol{\delta} = \{ v_1 \quad \theta_1 \quad v_2 \quad \theta_2 \quad v_3 \quad \theta_3 \}$$

At node 1 the element end loads are $Y_{1;1}$ and $m_{1;1}$ and so for equilibrium $Y_1 = Y_{1;1}$ and $m = m_{1;1}$. Node 2 is at the junction of elements **1** and **2** and so we could obtain the equilibrium condition by substituting $q = 2$, $\mathbf{p} = \mathbf{1}$ and $\mathbf{q} = \mathbf{2}$ in (7.43). In summary the equilibrium conditions (7.43) require

at node 1 $\qquad Y_1 = Y_{1;1}; \quad m_1 = m_{1;1}$

at node 2 $\qquad Y_2 = Y_{2;1} + Y_{2;2}; \quad m_2 = m_{2;1} + m_{2;2}$ $\qquad (7.44)$

at node 3 $\qquad Y_3 = Y_{3;2}; \quad m_3 = m_{3;2}$

Now the element end loads can be expressed in terms of nodal displacements by means of the element stiffness equations (7.14). Element **1** has its ends at nodes 1 and 2, therefore, putting $i = 1$, $j = 2$ and $\mathbf{j} = \mathbf{1}$ in (7.14) we obtain the stiffness equations as shown below. Similarly the equations shown for element **2** were obtained using $i = 2$, $j = 3$, $\mathbf{j} = \mathbf{2}$.

$$
\begin{bmatrix} Y_{1;1} \\ m_{1;1} \\ Y_{2;1} \\ m_{2;1} \end{bmatrix} = \begin{bmatrix} a_{11} & a_{12} & a_{13} & a_{14} \\ a_{21} & a_{22} & a_{23} & a_{24} \\ a_{31} & a_{32} & a_{33} & a_{34} \\ a_{41} & a_{42} & a_{43} & a_{44} \end{bmatrix}_1 \begin{bmatrix} v_1 \\ \theta_1 \\ v_2 \\ \theta_2 \end{bmatrix}
\tag{7.45}
$$

$$
\begin{bmatrix} Y_{2;2} \\ m_{2;2} \\ Y_{3;2} \\ m_{3;2} \end{bmatrix} = \begin{bmatrix} a_{11} & a_{12} & a_{13} & a_{14} \\ a_{21} & a_{22} & a_{23} & a_{24} \\ a_{31} & a_{32} & a_{33} & a_{34} \\ a_{41} & a_{42} & a_{43} & a_{44} \end{bmatrix}_2 \begin{bmatrix} v_2 \\ \theta_2 \\ v_3 \\ \theta_3 \end{bmatrix}
\tag{7.46}
$$

Using these relations we can substitute for element end loads in the nodal equilibrium equations (7.44), obtaining, for example

$$Y_1 = Y_{1;1} = (a_{11})_1 v_1 + (a_{12})_1 \theta_1 + (a_{13})_1 v_2 + (a_{14})_1 \theta_2$$

$$m_2 = m_{2;1} + m_{2;2} = (a_{41})_1 v_1 + (a_{42})_1 \theta_1 + ((a_{43})_1 + (a_{21})_2) v_2$$
$$+ ((a_{44})_1 + (a_{22})_2) \theta_2 + (a_{23})_2 v_3 + (a_{24})_2 \theta_3$$

The complete set of equations can be written in matrix form as follows:

$$
\begin{bmatrix} Y_1 \\ m_1 \\ Y_2 \\ m_2 \\ Y_3 \\ m_3 \end{bmatrix} = \begin{bmatrix} (a_{11})_1 & (a_{12})_1 & (a_{13})_1 & (a_{14})_1 & 0 & 0 \\ (a_{21})_1 & (a_{22})_1 & (a_{23})_1 & (a_{24})_1 & 0 & 0 \\ (a_{31})_1 & (a_{32})_1 & ((a_{33})_1 + (a_{11})_2) & ((a_{34})_1 + (a_{12})_2) & (a_{13})_2 & (a_{14})_2 \\ (a_{41})_1 & (a_{42})_1 & ((a_{43})_1 + (a_{21})_2) & ((a_{44})_1 + (a_{22})_2) & (a_{23})_2 & (a_{24})_2 \\ 0 & 0 & (a_{31})_2 & (a_{32})_2 & (a_{33})_2 & (a_{34})_2 \\ 0 & 0 & (a_{41})_2 & (a_{42})_2 & (a_{43})_2 & (a_{44})_2 \end{bmatrix} \begin{bmatrix} v_1 \\ \theta_1 \\ v_2 \\ \theta_2 \\ v_3 \\ \theta_3 \end{bmatrix}
\tag{7.47}
$$

These equations express nodal loads in terms of nodal displacements and are the required stiffness equations. The square matrix is the assembly stiffness matrix K (equations 7.42a) and, comparing with the formal expression in equations (7.42), it is seen that, for example,

$$s_{11} = (a_{11})_1; \quad s_{33} = ((a_{33})_1 + (a_{11})_2); \quad s_{66} = (a_{44})_2$$

Now, referring to equations (7.45)–(7.47), s_{11} is seen to relate nodal force Y_1 to nodal displacement v_1; $(a_{11})_1$ is the only element stiffness coefficient relating end force at node 1 to deflexion v_1 at node 1, and so s_{11} is as given. However s_{33}, which relates force Y_2 to deflexion v_2, has contributions from both element stiffness matrices; $(a_{33})_1$ relates end force $Y_{2;1}$ to v_2 and $(a_{11})_2$ relates end force $Y_{2;2}$ to v_2. Thus we see that the process of obtaining the assembly stiffness matrix from the components of the element stiffness matrices is related to that described in Section 5.3.2. The difference is due to there being two components of load and of displacement at each node so that there are $2n$ stiffness equations and $\mathbf{K}$ is $2n \times 2n$. Consequently, for example, although $(a_{21})_1$ is associated with a load at node 1 it appears in row 2 of $\mathbf{K}$.

The general procedure for obtaining $\mathbf{K}$ can now be inferred. A square null matrix $\mathbf{K}$ is formed of size $2n \times 2n$ where n is the number of nodes. The components of the various element stiffness matrices $\mathbf{k}_j$ are then assigned to a location in $\mathbf{K}$ according to the following rules for rows and columns:

Stiffness coefficient a_{gh} of an element which has left end at node number i and right end at node number j

	Address in $\mathbf{K}$		Address in $\mathbf{K}$
$g = 1$	Row $2i - 1$	$h = 1$	Column $2i - 1$
$g = 2$	Row $2i$	$h = 2$	Column $2i$
$g = 3$	Row $2j - 1$	$h = 3$	Column $2j - 1$
$g = 4$	Row $2j$	$h = 4$	Column $2j$ $\qquad$ (7.48)

This procedure is especially easy to carry out with a digital computer. Some locations in $\mathbf{K}$ receive contributions from more than one $\mathbf{k}_j$; these contributions are accumulated as in (7.47). Other locations will have received no contribution after all the element matrices have been incorporated in $\mathbf{K}$; these locations remain zero and may be very numerous in larger structures since each element is connected only to a small number of other elements.

For example in the case of element **1** of the assembly discussed above (equations 7.45–7.47), nodes i and j are 1 and 2. Then the coefficient a_{14} has $g = 1$, $h = 4$ and from rules (7.48) should be placed in row $2i - 1$, column $2j$, that is to say it contributes to s_{14}. Coefficient a_{32} has $g = 3$, $h = 2$ and is placed in row $2j - 1$, column $2i$ (row 3, column 2). Similarly coefficient a_{23} of element **2** $(i = 2, j = 3)$ is placed in row $2i$, column $2j - 1$ (row 4, column 5). Lastly we note that some locations such as row 6 column 1 of (7.47) remain unoccupied.

Example 7.20 A beam assembly as in Figure 7.15d is formed from two elements each of length 4 m. Element **1** has $EI = 8$ MNm^2, element **2** has $EI = 16$ MNm^2. Form the assembly stiffness equations, find nodal loads, and sketch the bending moment diagram if $\delta = \{0\ 0\ 0\cdot01\ 0\ 0\ 0\}$.

Solution Element **1** has the same properties as the beam element of Example 7.18 and so

$$\mathbf{k}_1 = 0\cdot125 \times 10^6 \begin{bmatrix} 12 & 24 & -12 & 24 \\ 24 & 64 & -24 & 32 \\ -12 & -24 & 12 & -24 \\ 24 & 32 & -24 & 64 \end{bmatrix} \qquad \text{(i)}$$

Element **2** has *EI* twice that of element **1** but the same length, so that from equation (7.29) of Section 7.4.3 $k_2 = 2 \times k_1$. **K** is of size $2n \times 2n$, i.e. 6×6. We now introduce components of k^1, k_2 in **K** using the rules (7.48). Element **1** has ends at nodes 1 and 2 so that, for example, $(a_{14})_1 = 0 \cdot 125 \times 10^6 \times 24$ is placed in row $(2 \times 1) - 1$ and column 2×2 of **K**.

Element **2** joins nodes 2 and 3 so that as an example $(a_{23})_2$, which is $0 \cdot 125 \times 10^6 \times -48$, is placed in row 2×2 and column $(2 \times 3) - 1$ of **K**. When this process has been carried out for all the components of k_1 and k_2 the complete stiffness equations are:

$$
\begin{bmatrix} Y_1 \\ m_1 \\ Y_2 \\ m_2 \\ Y_3 \\ m_3 \end{bmatrix} = \frac{10^6}{8}
\begin{bmatrix}
12 & 24 & -12 & 24 & 0 & 0 \\
24 & 64 & -24 & 32 & 0 & 0 \\
-12 & -24 & (12+24) & (-24+48) & -24 & 48 \\
24 & 32 & (-24+48) & (64+128) & -48 & 64 \\
0 & 0 & -24 & -48 & 24 & -48 \\
0 & 0 & 48 & 64 & -48 & 128
\end{bmatrix}
\begin{bmatrix} 0 \\ 0 \\ 0 \cdot 01 \\ 0 \\ 0 \\ 0 \end{bmatrix}
$$

$$
= 10^6
\begin{bmatrix}
-0 \cdot 015 \\
-0 \cdot 03 \\
0 \cdot 045 \\
0 \cdot 03 \\
-0 \cdot 03 \\
0 \cdot 06
\end{bmatrix}
$$

The nodal loads and bending moment diagram are shown in Figures 7.16a and b. The system is equivalent to that in Figure 7.16c.

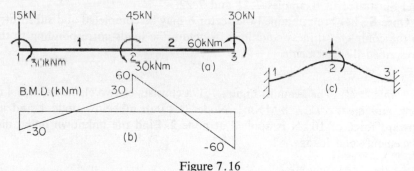

Figure 7.16

Problems for Solution   7.33, 7.34

7.4.6 *Assembly Stiffness Equations* (*Application*)

When the nodal displacements of an assembly are all known the nodal loads are found by direct substitution in the assembly stiffness equations as shown in Example 7.20. However it is more often necessary to determine the nodal displacements for a particular set of nodal loads. Now a beam structure has two degrees of freedom of rigid body displacement, comprising one translation (deflection v in the y direction) and one rotation (slope θ in the xy plane). Therefore a discussion similar to that in Section 5.3.3 would show that at least two nodal displacements must be known. These may be either (a) one deflexion and one slope, or, (b) two deflexions (which being at different nodes effectively present rigid body rotation as well as translation).

In matrix terms what this means is that the complete stiffness matrix $\mathbf{K}$ in equations (7.42a) is singular. When at least two nodal displacements are known we can proceed, by a similar argument to that of equations (5.24) and (5.25) (Section 5.3.3), to show that the stiffness equations can be reduced in size to give

$$\mathbf{F}'_R = \mathbf{K}_{RR}\boldsymbol{\delta}_R \qquad (7.49)$$

These equations can be solved for the unknown displacements $\boldsymbol{\delta}_R$. Thus

$$\boldsymbol{\delta}_R = \mathbf{K}_{RR}^{-1}\mathbf{F}'_R \qquad (7.50)$$

The reduction process was described formally in Section 5.3.3 by partitioning the assembly stiffness equations (see equations 5.28–5.30). This shows that

$$\mathbf{F}'_R = \mathbf{F}_R - \mathbf{K}_{RP}\boldsymbol{\delta}_P \qquad (7.51)$$

where $\boldsymbol{\delta}_P$ are the known, or prescribed, nodal displacements.

In practice the reduction expressed by equations (7.49) and (7.51) is given effect in the manner described in the discussion of equation (5.25), and illustrated in Examples 7.21 and 7.22.

Once $\boldsymbol{\delta}_R$ has been obtained the vector $\boldsymbol{\delta}$ may be completed and substituted in the complete stiffness equations to obtain the loads corresponding to the prescribed displacements.

Example 7.21 The beam in Figure 7.17a consists of two elements each 4 m long and having $EI = 8$ MNm². Node 3 is 0·01 m below node 1 and an upward force of 10 kN is applied at node 2. Find the unknown nodal displacements and loads.

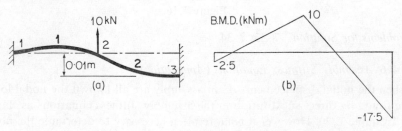

Figure 7.17

Solution The nodal load vector is

$$\mathbf{F} = \{Y_1 \quad m_1 \quad 10^4 \quad 0 \quad Y_3 \quad m_3\}$$

Here $v_1 = 0$, $v_3 = -0.01$ m, and since fixed supports have been used θ_1 and θ_3 are zero. Hence the nodal displacement vector is

$$\boldsymbol{\delta} = \{0 \quad 0 \quad v_2 \quad \theta_2 \quad -0.01 \quad 0\}$$

Both elements have the same properties as element **1** in Example 7.20. Therefore using the element stiffness matrix from Example 7.20 (equation i) for both $\mathbf{k}_1$ and $\mathbf{k}_2$ and applying rules (7.48) we obtain

$$
\begin{bmatrix} Y_1 \\ m_1 \\ 10^4 \\ 0 \\ Y_3 \\ m_3 \end{bmatrix} = \frac{10^6}{8}
\begin{bmatrix}
12 & 24 & -12 & 24 & 0 & 0 \\
24 & 64 & -24 & 32 & 0 & 0 \\
-12 & -24 & (12+12) & (-24+24) & -12 & 24 \\
24 & 32 & (-24+24) & (64+64) & -24 & 32 \\
0 & 0 & -12 & -24 & 12 & -24 \\
0 & 0 & 24 & 32 & -24 & 64
\end{bmatrix}
\begin{bmatrix} 0 \\ 0 \\ v_2 \\ \theta_2 \\ -0.01 \\ 0 \end{bmatrix}
\tag{i}
$$

Since there are sufficient prescribed displacements to eliminate rigid body displacements we can form the reduced stiffness equations (7.49). Deleting rows 1, 2, 5, 6 of $\mathbf{F}$ and $\mathbf{K}$ (rows corresponding to prescribed displacements) we have

$$
\begin{bmatrix} 10^4 \\ 0 \end{bmatrix} = \frac{10^6}{8}
\begin{bmatrix}
-12 & -24 & 24 & 0 & -12 & 24 \\
24 & 32 & 0 & 128 & -24 & 32
\end{bmatrix}
\begin{bmatrix} 0 \\ 0 \\ v_2 \\ \theta_2 \\ -0.01 \\ 0 \end{bmatrix}
$$

which may be simplified to

$$
\begin{bmatrix} 10^4 \\ 0 \end{bmatrix} = \frac{10^6}{8} \times -0.01 \begin{bmatrix} -12 \\ -24 \end{bmatrix} + \frac{10^6}{8} \begin{bmatrix} 24 & 0 \\ 0 & 128 \end{bmatrix} \begin{bmatrix} v_2 \\ \theta_2 \end{bmatrix}
$$

and finally

$$
10^3 \begin{bmatrix} -5 \\ -30 \end{bmatrix} = \frac{10^6}{8} \begin{bmatrix} 24 & 0 \\ 0 & 128 \end{bmatrix} \begin{bmatrix} v_2 \\ \theta_2 \end{bmatrix}
$$

After solving for v_2 and θ_2 the nodal displacement vector

$$\boldsymbol{\delta} = 10^{-3} \{0 \quad 0 \quad 1.67 \quad -1.875 \quad -10 \quad 0\}$$

can be substituted in (i) to obtain the nodal load vector

$$\mathbf{F} = 10^3 \{-3.125 \quad -2.5 \quad 10 \quad 0 \quad -6.875 \quad 17.5\}$$

The bending moment diagram can now be constructed (Figure 7.17b), and using it stresses could be determined in the usual way (see Article 6.3).

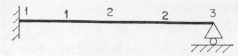

Figure 7.18

Example 7.22 Let the beam assembly of the previous example be supported instead as in Figure 7.18 and subjected to a moment of -12 kNm at node 3. Find the unknown nodal loads and displacements.

Solution Here

$$\mathbf{F} = \{Y_1 \quad m_1 \quad 0 \quad 0 \quad Y_3 \quad -12 \times 10^3\}$$

and

$$\delta = \{0 \quad 0 \quad v_2 \quad \theta_2 \quad 0 \quad \theta_3\}$$

After substituting these in (i) of Example 7.21 we first eliminate rows 1, 2 and 5 and after simplification obtain reduced stiffness equations

$$\begin{bmatrix} 0 \\ 0 \\ -12 \times 10^3 \end{bmatrix} = 10^6 \begin{bmatrix} 3 & 0 & 3 \\ 0 & 16 & 4 \\ 3 & 4 & 8 \end{bmatrix} \begin{bmatrix} v_2 \\ \theta_2 \\ \theta_3 \end{bmatrix}$$

Hence

$$\delta = 10^{-3}\{0 \quad 0 \quad 3 \quad 0.75 \quad 0 \quad -3\}$$

and

$$\mathbf{F} = 10^3\{-2.25 \quad -6 \quad 0 \quad 0 \quad +2.25 \quad -12\}$$

In this example we have obtained the displacements at the centre (node 2) and the reactive loads at the supports (Y_1, m_1, Y_3), for this statically indeterminate system.

Problems for Solution   7.35–7.39

7.4.7 *Intermediate Loads*

Since beams are frequently subject to distributed forces acting between nodes it is necessary to find some means of incorporating these effects in the stiffness method of analysis. In fact the procedure will also provide an alternative procedure for cases of concentrated loads acting between supports which were dealt with previously (see Example 7.21) by introducing a node at the point of application.

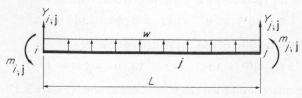

Figure 7.19

Consider the usual beam element with the addition of intermediate uniformly distributed load as in Figure 7.19. Making v_j and θ_j zero the system would be equivalent to a cantilever with support at node j. Using the Superposition method and with information from Figure 7.5 we can write

$$EIv_i = \tfrac{1}{3}Y_{i;j}L^3 - \tfrac{1}{2}m_{i;j}L^2 + \tfrac{1}{8}wL^4$$

$$EI\theta_i = -\tfrac{1}{2}Y_{i;j}L^2 + m_{i;j}L - \tfrac{1}{6}wL^3$$

Then, solving for end loads, we have

$$Y_{i;j} = \frac{12EI}{L^3}v_i + \frac{6EI}{L^2}\theta_i - \tfrac{1}{2}wL$$

$$m_{i;j} = \frac{6EI}{L^2}v_i + \frac{4EI}{L}\theta_i - \frac{wL^2}{12}$$

(7.52)

Comparing these with equations (7.26) we see that the effect of the distributed load has been to add the quantities $-\tfrac{1}{2}wL$ and $-wL^2/12$ to the right-hand sides. If v_i and θ_i are zero in equations (7.52) the beam element has then, in effect, fixed supports at each end; the end loads are the reactive loads and are then known as fixed end actions (see the Moment Distribution method in Article 14.2).

Alternatively we can interpret quantities $\tfrac{1}{2}wL$ and $wL^2/12$ subtracted from the right-hand sides of (7.52) as being the end loads which would have the same effect on the rest of the structure as the distributed load. These quantities, which are called equivalent end actions, are opposite in sign to the fixed end actions.

The fixed end actions for node j in Figure 7.19 may be found by a similar process and are $-\tfrac{1}{2}wL$ and $+wL^2/12$. Other forms of intermediate loading (Figure 7.20) could also be considered in this manner, and the element

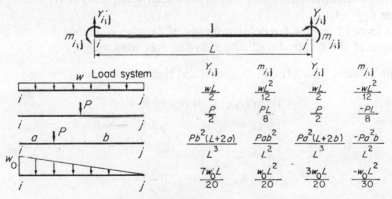

Figure 7.20

stiffness equations (7.14) can now be written in the more comprehensive form

$$
\begin{bmatrix} Y_{i;j} \\ m_{i;j} \\ Y_{j;j} \\ m_{j;j} \end{bmatrix} = \begin{bmatrix} a_{11} & a_{12} & a_{13} & a_{14} \\ a_{21} & a_{22} & a_{23} & a_{24} \\ a_{31} & a_{32} & a_{33} & a_{34} \\ a_{41} & a_{42} & a_{43} & a_{44} \end{bmatrix} \begin{bmatrix} v_1 \\ \theta_1 \\ v_2 \\ \theta_2 \end{bmatrix} + \begin{bmatrix} c_1 \\ c_2 \\ c_3 \\ c_4 \end{bmatrix} \tag{7.53}
$$

i.e.
$$
\mathbf{F_j} = \mathbf{k_j}\boldsymbol{\delta_j} + \mathbf{f_j} \tag{7.53a}
$$

where $\mathbf{f_j}$ is a vector of fixed end actions for the jth element.

Tables of these quantities are to be found in books on Moment Distribution, e.g. reference 1, and a few examples are given in Figure 7.20. For the case in Figure 7.19 we have already found

$$
\mathbf{f_j} = \left\{ -\tfrac{1}{2}wL \quad \frac{-wL^2}{12} \quad -\tfrac{1}{2}wL \quad \frac{wL^2}{12} \right\}
$$

When a collinear assembly of intermediately loaded elements is considered the equilibrium conditions (7.43) are unchanged. Therefore when the particular example of a two-element assembly is discussed, as in Section 7.4.5, substitution of element stiffness equations (7.53) will produce all the terms on the right-hand side of (7.47) with in addition a column matrix

$$
\mathbf{f} = \{(c_1)_1 \quad (c_2)_1 \quad ((c_3)_1 + (c_1)_2) \quad ((c_4)_1 + (c_2)_2) \quad (c_3)_2 \quad (c_4)_2\} \tag{7.54}
$$

The assembly stiffness equations are then of the form

$$
\mathbf{F} = \mathbf{K}\boldsymbol{\delta} + \mathbf{f} \tag{7.55}
$$

where $\mathbf{f}$ is the assembly matrix of fixed end actions. (Compare discussion of equation 5.39 in Section 5.4.3.) It is worth noting that if we formulate the system in terms of equivalent end actions then $\mathbf{f}$ in (7.55) is replaced by $\mathbf{f_e}$ where $\mathbf{f} = -\mathbf{f_e}$.

To form $\mathbf{f}$ from the $\mathbf{f_j}$ a column matrix of $2n$ rows is formed and the components of the $\mathbf{f_j}$ introduced into $\mathbf{f}$ by the following rules:

for jth element with left end at node i and right end at node j

Term $(c_g)_j$	is placed in Row
$g = 1$	$2i-1$
$g = 2$	$2i$
$g = 3$	$2j-1$
$g = 4$	$2j$

$$(7.56)$$

Example 7.23 Find the nodal displacements when a downward distributed load of uniform intensity 3 kN/m also acts on the beam of Figure 7.18 (Example 7.22).

Solution Using fixed end actions from Figure 7.20 for w acting down

$$\mathbf{f}_1 = 10^3\{6 \quad 4 \quad 6 \quad -4\} = \mathbf{f}_2$$

The matrix $\mathbf{f}$ for the structure will have $2n$, i.e. 6 rows. Applying rules (7.56) the assembly fixed end actions are

$$\mathbf{f} = 10^3\{6 \quad 4 \quad (6+6) \quad (-4+4) \quad 6 \quad -4\}$$

Adding this vector to the assembly stiffness equations previously considered and reducing by eliminating rows 1, 2 and 5 corresponding to prescribed displacements we obtain

$$\begin{bmatrix} 0 \\ 0 \\ -12 \times 10^3 \end{bmatrix} = \frac{10^6}{8} \begin{bmatrix} 24 & 0 & 24 \\ 0 & 128 & 32 \\ 24 & 32 & 64 \end{bmatrix} \begin{bmatrix} v_2 \\ \theta_2 \\ \theta_3 \end{bmatrix} + 10^3 \begin{bmatrix} 12 \\ 0 \\ -4 \end{bmatrix}$$

Solving these we can write

$$\delta = 10^{-3}\{0 \quad 0 \quad -5 \quad -0.25 \quad 0 \quad 1\}$$

Problem for Solution 7.40

7.4.8 *Assembly Stiffness Equations (Vector Form)*

Assembly stiffness equations are found by substitution for the end loads in the nodal equilibrium equations (7.43a). The qth node of the assembly in Figure 7.15a lies at the junction of elements $\mathbf{p}$ and $\mathbf{q}$ and the nodal equilibrium equation is

$$\mathbf{F}_q = \mathbf{F}_{q;\mathbf{p}} + \mathbf{F}_{q;\mathbf{q}} \tag{7.43a}$$

The generalized element stiffness equations are given by (7.37). Applying this for the $\mathbf{p}$th element which has ends at nodes p and q we have

$$\mathbf{F}_{q;\mathbf{p}} = \mathbf{k}_{qp;\mathbf{p}}\delta_p + \mathbf{k}_{qq;\mathbf{p}}\delta_q$$

Similarly

$$\mathbf{F}_{q;\mathbf{q}} = \mathbf{k}_{qq;\mathbf{q}}\delta_q + \mathbf{k}_{qr;\mathbf{q}}\delta_r$$

Substituting in (7.43a) we have

$$\mathbf{F}_q = \mathbf{k}_{qp;\mathbf{p}}\delta_p + (\mathbf{k}_{qq;\mathbf{p}} + \mathbf{k}_{qq;\mathbf{q}})\delta_q + \mathbf{k}_{qr;\mathbf{q}}\delta_r \tag{7.57}$$

This vector equation for the qth node is analogous to the scalar equations obtained at the end of Section 5.3.2. By similar arguments to Section 5.3.2 each sub-matrix $\mathbf{K}_{ij}$ of the assembly stiffness matrix consists of the algebraic sum of all element stiffness sub-matrices $k_{ij;i}$ having the same nodal subscripts. Therefore the typical assembly stiffness equation in vector form is (compare equations 5.23)

$$F_q = \sum_{p=1}^{n} \left(\sum_{\mathbf{p}=1}^{\mathbf{n}} k_{qp;\mathbf{p}} \right) \delta_p \qquad (7.58)$$

where q and p take all values from 1 to n and $\mathbf{p}$ takes all values from $\mathbf{1}$ to $\mathbf{n}$.

7.4.9 *Summary of Operations for Continuous Beam*

The following summary of operations also corresponds to the general stages of a computer program.

1. Description of system by numbering of nodes and elements; recording coordinates of nodes, end nodes of elements, and length and flexural rigidity of elements.

2. Calculation of element stiffness matrices k_j.

3. Assembly of complete structure stiffness matrix K by introducing the components of the k_j according to the rules (7.48).

4. Formation of the nodal load vector F for the structure, introducing values of all prescribed loads.

5. Formation of fixed end action matrices f_j and their assembly by rules (7.56) to form f.

6. Formation of the nodal displacement vector δ for the structure, introducing values of all prescribed displacements.

7. Reduction of the complete matrix equations $F = K\delta + f$ by elimination of rows corresponding to prescribed displacements, giving

$$F'_R = K_{RR}\delta_R + f_R$$

8. Inversion of K_{RR} and calculation of $\delta_R = K_{RR}^{-1}(F'_R - f_R)$ which is then used to complete δ.

9. Use of $F = K\delta$ to compute loads corresponding to prescribed displacements.

10. Calculations of stresses by drawing the bending moment diagram and using methods of Article 6.3.

7.5   Elastic/Plastic Analysis of Beams

7.5.1 *Pure Bending*

The stress distribution due to pure bending of a beam was discussed in Section 6.3.1. The state of stress has the form of (array A6.1) and σ_x varies with position in the cross-section according to equation (6.14) for loads in the xy plane. If y_{max} is the greatest distance from the neutral axis of the cross-section we have

$$(\sigma_x)_{\max} = \frac{My_{\max}}{I}$$

and following condition (6.15) of Section 6.3.3 yielding begins when

$$\sigma_x = \pm \sigma_0$$

Then if M_0 is the bending moment at which yielding begins

$$M_0 = \frac{\sigma_0 I}{y_{\max}} \tag{7.59}$$

The beam does not collapse at M_0 since only surface material has attained the yield condition. The bending moment can therefore be increased beyond the value M_0 the additional load bringing more of the material to the yield condition.

In the particular case of the beam of rectangular cross-section in Figure 7.21a yielding begins at upper and lower surfaces ($y_{\max} = \pm \frac{1}{2}h$) when

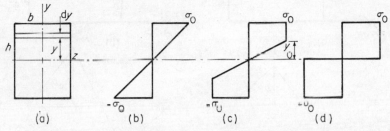

Figure 7.21   (b) $M = M_0$. (c) $M_0 < M < \hat{M}$. (d) $M = \hat{M}$

$M = +M_0$ As the numerical value of the bending moment is further increased (i.e. $|M| > M_0$) more material is brought to the yield condition, as shown, with the boundary of the remaining elastic material at $y = \pm y_0$. Considering successive elements of the cross-section the resultant moment due to the stress distribution (Figure 7.21c) is

$$M = \int_{-y_0}^{y_0} ((b \mathrm{d}y)\, \sigma_x)\, y + 2 \int_{y_0}^{\frac{1}{2}h} (b \mathrm{d}y)\, \sigma_0 y$$

From Figure 7.21c we see that $\sigma_x = (y/y_0)\sigma_0$ and therefore

$$M = b\sigma_0 \left(\frac{h^2}{4} - \frac{y_0^2}{3} \right) \tag{7.60}$$

When $y_0 = \frac{1}{2}h$ (Figure 7.21b)

$$M = M_0 = b\sigma_0 \frac{h^2}{6} \tag{7.61}$$

When $y_0 = 0$ yielding has penetrated throughout the cross-section (Figure 7.21d) so that the beam can offer no further resistance. We therefore call the bending moment obtained by putting $y_0 = 0$ in equation (7.60) the limit or fully plastic moment $\hat{M}$

i.e.

$$\hat{M} = \sigma_0 b \frac{h^2}{4} \tag{7.62}$$

The ratio of the limit moment $\hat{M}$ to the yield moment M_0 is called the shape factor as it is found to depend on the shape of the cross-section. Using equations (7.61) and (7.62) we have for the rectangular cross-section $\hat{M}/M_0 = 1\cdot5$. This result, originating from the non-uniform elastic stress distribution (6.14), may be compared with direct loading (Section 5.6.2) for which the shape factor is unity.

In the elastic design of a beam the maximum stress was required not to exceed a stress obtained by dividing the yield stress by a factor of safety f_s. For plastic design such stress based considerations are meaningless. Instead

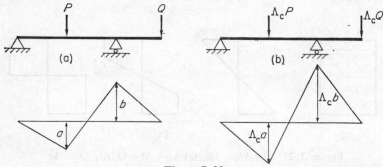

Figure 7.22

the working load must not exceed that obtained by dividing the collapse load by a load factor Λ_c. When more than one load acts all are assumed to bear a constant proportion to each other for this purpose as for example in Figure 7.22, where (a) represents design (b) collapse loads.

Before calculating some limit moments and shape factors we note a more rapid method of calculation. Using Figure 7.21d we can write

$$\hat{M} = \int_0^{\frac{1}{2}h} \sigma_0(b\,dy)\,y + \int_{-\frac{1}{2}h}^0 (-\sigma_0)\,(b\,dy)y \tag{7.63}$$

The right-hand side of this expression comprises products of first moments of area and stress. Because the stress has opposite sign on opposite sides of the neutral axis both integrals have the same sign. Therefore (7.63) is equivalent to the product of σ_0 with the sum of the magnitudes of the first moments of area of the two parts into which the cross-section is divided by the neutral axis. For the simple rectangle

$$\hat{M} = \sigma_0(2 \times (b \times \tfrac{1}{2}h \times \tfrac{1}{4}h)) = \sigma_0 \frac{bh^2}{4}$$

This procedure is also easy to perform when the cross-section can be represented as a combination of rectangles as in the following example.

Example 7.24 Determine the shape factor for the cross-section shown in Figure 7.13b.

Solution From Example 7.18 $I = 0.41 \times 10^{-4}$ m⁴, from equation (7.59) $M_0 = \sigma_0 \times 0.41 \times 10^{-4}/0.1 = 4.1 \sigma_0 \times 10^{-4}$ Nm. Treating flanges and web as three rectangles we find

$$\hat{M} = \sigma_0(2(0.2 \times 0.01 \times 0.095) + 2(0.01 \times 0.09 \times 0.045))$$

$$= 4.61 \sigma_0 \times 10^{-4} \text{ Nm}$$

The shape factor is $\hat{M}/M_0 = 1.12$.

The above example shows the I-section to have a much smaller reserve of strength beyond the yield moment than the rectangular cross-section. The reason is that at the yield moment there is proportionately less material at low stress due to the small width of the web.

For a solid circular cross-section this situation is obviously reversed and the shape factor is 1.70.

For beam cross-sections having only one axis of symmetry, e.g. Figure 7.23a, (loaded in the plane of symmetry), the elastic stress distribution is linear over the depth of the cross-section but is not symmetrical with respect to the neutral axis (Figure 7.23b). Then if y_1 and y_2 ($y_1 > y_2$) are the distances from the extreme surfaces to the neutral axis, $y_{max} = y_1$ in equation (7.59) and yielding begins at this surface only.

As the bending moment is further increased the additional moment is borne by the remaining material. When the y_2 surface reaches the yield

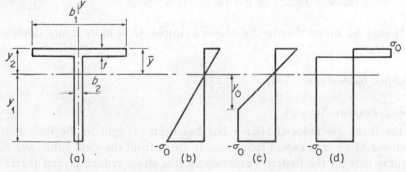

Figure 7.23   (b) $M = M_0$. (c) $M_0 < M < \hat{M}$. (d) $M = \hat{M}$

point yielding will have already penetrated to a finite depth from the y_1 surface as shown in Figure 7.23c. The elastic material in the zone $y_2 > y > y_0$ has a linear stress distribution. Comparing Figures 7.23b and 7.23c we see that the line of zero stress (neutral axis) has moved. Henceforth the term neutral axis will be reserved for the zero stress line under purely elastic conditions.

Increasing the bending moment until yielding is complete, the stress distribution of Figure 7.23d is obtained. The zero stress line is located using the condition that the resultant force due to the normal stresses on the cross-section is zero in the absence of direct loading of the beam. The forces on each part are opposite in direction and are equal to the products of the yield stress and area. Therefore for zero resultant force the zero stress line must divide the cross-section into two equal areas. Since this is also the case for a doubly symmetric cross-section (Figure 7.21) the zero stress line can be used as the fully plastic equivalent of the elastic neutral axis.

Example 7.25 Find the shape factor for the T-section of Figure 7.23a when $(y_1 + y_2) = 0.4$ m, $b_1 = 0.2$ m, $b_2 = t = 0.02$ m.

Solution Taking first moments of area about the top of the flange we find the neutral axis at $\bar{y} = 1636 \times 10^3/11,600 = 141$ mm. Then

$$y_1 = 259 \text{ mm}; \quad y_2 = \bar{y}$$

$$I = \frac{180 \times 20^3}{12} + 180 \times 20 \times 131^2 + \frac{20 \times 400^3}{12} + 20 \times 400 \times 59^2$$

$$= 1.97 \times 10^8 \text{ mm}^4 = 1.97 \times 10^{-4} \text{ m}^4$$

From equation (7.59) $M_0 = \sigma_0 \times 1.97 \times 10^{-4}/0.259 = 7.59\sigma_0 \times 10^{-4}$ Nm. The zero stress line, which is parallel to the neutral axis, is situated 110 mm from the top of the flange, and divides the cross-section into two areas of 5800 mm². Evaluating the limit moment we find

$$\hat{M} = \sigma_0(0.2 \times 0.02 \times 0.1) + (0.09 \times 0.02 \times 0.045) + (0.29 \times 0.02 \times 0.145)$$

$$= 13.22\sigma_0 \times 10^{-4} \text{ Nm}.$$

The shape factor $= 13.22/7.59 = 1.74$.

It may be noted that in the above example $\hat{M}$ is more easily calculated than M_0.

Problem for Solution 7.41

7.5.2 Residual Stresses

If the loads are reduced after a bar has been brought to the fully plastic moment $\hat{M}$ we may expect reductions in stress from the yield value. We now assume that all the material experiences this stress reduction and therefore behaves as an elastic material. Following the discussion in Section 5.6.2 we restate the elastic stress/moment equation (6.14) in terms of increments as

$$\Delta\sigma_x = \frac{\Delta M\, y}{I} \qquad (7.64)$$

Then if $\hat{M}$ is completely removed $\Delta M = -\hat{M}$ and so $\Delta\sigma_x = -\hat{M}y/I$. Consider now the rectangular section of Figure 7.21a. Substituting for $\hat{M}$ from equation (7.62) and setting $I = bh^3/12$ we find $\Delta\sigma_x = -3\sigma_0(y/h)$. The original stress distribution and $\Delta\sigma_x$ are plotted in Figures 7.24a and b

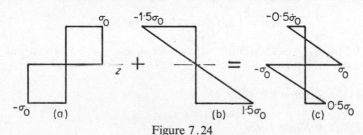

Figure 7.24

respectively. The stress after unloading is the algebraic sum of these (Figure 7.24c). For example

at $y = +\tfrac{1}{2}h$ final stress $= \sigma_0 - \dfrac{3\sigma_0}{2} = -\tfrac{1}{2}\sigma_0$

at $y = 0$ final stress $= \pm\sigma_0 + 0 = \pm\sigma_0$

It is easily verified that these residual stresses exert zero resultant force and moment on the cross-section, as is required by the absence of external loads.

If the beam is reloaded in the same way it is elastic up to the moment $\hat{M}$, since this process is just a reversal of that in Figure 7.24. If part of a beam with residual stresses is removed, say in a machining process, then the resultant force and moment may cease to be zero. In this event additional bending takes place spontaneously until the stress distribution again has the necessary properties of zero resultant force and moment.

If the load is only partially removed, or is reversed, the same methods can be applied to find the stress distribution, the appropriate value of ΔM being used. In the case of load reversal, study of Figure 7.24 shows that there will be some load change for which the upper and lower surfaces attain the yield stress. This reverse yielding has serious consequences if the loads alternate frequently, and may cause failure of the material.[2]

Example 7.26 Find an expression for the residual stress distribution in the beam of Example 7.24 (Figure 7.13b) when loaded to the limit and released. State the values at the top and bottom of the upper flange and at neutral axis. $\sigma_0 = 200$ MN/m².

Solution From Example 7.24 $\hat{M} = 4\cdot61\sigma_0 \times 10^{-4} = 92{,}200$ Nm. At the limit the stress distribution (compare Figure 7.21d) can be expressed as

$$\sigma_x = \frac{\sigma_0 y}{|y|} = 200 \times 10^6 \times \frac{y}{|y|} \tag{i}$$

During unloading we have from (7.64) the change of stress distribution

$$\Delta\sigma_x = \frac{(0 - \hat{M})y}{I} = \frac{-92{,}200y}{0\cdot41 \times 10^{-4}} = -2\cdot25y \times 10^9 \tag{ii}$$

The residual stress distribution is $(\sigma_x + \Delta\sigma_x)$ which from (i) and (ii) is

$$\sigma_{\text{res}} = \left(200 \frac{y}{|y|} - 2250y\right) \times 10^6 \text{ N/m}^2$$

At $y = 0\cdot1$ m, $\sigma_{\text{res}} = -25$ MN/m²; at $y = 0\cdot09$ m, $\sigma_{\text{res}} = -2\cdot5$ MN/m²; at $y = 0$, $\sigma_{\text{res}} = \pm200$ MN/m².

Problems for Solution   7.42, 7.43

7.5.3 *Effect of Shear Force*

In general shearing forces are transmitted by the cross-section in addition to the bending moment. In Article 6.4 the distribution of shear stress due to the shear forces was discussed and reasonable approximations were deduced for rectangular and I-sections. For the former the shear stresses are small enough compared with the normal stresses for their effect on yielding to be neglected. This is not always so for I-sections for which a satisfactory approximation (see Section 6.4.2) is that the shear stress is uniform over the web and zero in the flanges. The state of stress in the web is

$$\begin{pmatrix} \sigma_x & \tau_{xy} & 0 \\ \tau_{yx} & 0 & 0 \\ 0 & 0 & 0 \end{pmatrix} \tag{A7.1}$$

where

$$\tau_{xy} = \frac{\text{shear force}}{\text{web area}} \tag{7.65}$$

Using the von Mises yield condition (equation 5.52) and equations (1.8) to find the principal stresses we find for the web

$$\sigma_0^2 = \sigma_x^2 + 3\tau_{xy}^2 \tag{7.66}$$

whereas in the flanges the yield condition is $\sigma_0^2 = \sigma_x^2$. Therefore yielding begins in the web when the normal stress there is less than σ_0 (Figure 7.25a), resulting in a smaller contribution to the limit moment.

Example 7.27 Find the limit moment and shape factor for the I-section considered in Example 7.24 taking into account a shear force of 100 kN. $\sigma_0 = 200$ MN/m².

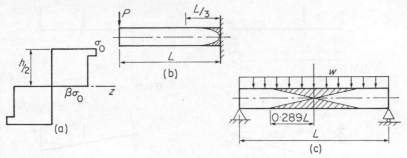

Figure 7.25

Solution   From (7.65)

$$\tau_{xy} = \frac{100,000}{0.18 \times 0.01} = 55.5 \times 10^6 \text{ N/m}^2$$

From equation (7·66)

$$\sigma_x = 10^6 \sqrt{200^2 - 3 \times (55.5)^2} = 175 \times 10^6 \text{ N/m}^2$$

Referring to Figure 7.25a $\beta = \sigma_x/\sigma_0 = 0.875$ and

$$\hat{M} = \sigma_0(2 \times (0.2 \times 0.01 \times 0.095) + 2\beta(0.01 \times 0.09 \times 0.045))$$

$$= 4.51\sigma_0 \times 10^{-4} = 90,200 \text{ Nm}$$

This relatively severe shear stress has reduced $\hat{M}$ by 2·3%. The shape factor is reduced to 1·1.

The presence of shear force means that the bending moment is varying over the span of the beam (Section 6.2.3). In the case of the cantilever of Figure 7.25b we have $M = Px$. The fully plastic moment is first developed at the root the load then having the value $\hat{P}$; at any other position the bending moment is $M = \hat{P}x = \hat{P}L(x/L) = \hat{M}(x/L)$. If the cantilever has rectangular cross-section, M and $\hat{M}$ are given by equations (7.60) and (7.62) respectively. Substituting we find $x/L = 1 - (4y_0^2/3h^2)$ as the equation determining the boundary of the yielded material, shown by cross hatching in Figure 7.25b. The result of a similar analysis for a uniformly distributed loading on a simply supported beam is shown in Figure 7.25c.

When the fully plastic moment has been developed at one section no further load can be sustained; continuous deformation is possible at that section and it is said that a plastic hinge has developed, e.g. at the root in Figure 7.25b.

7.5.4 *Collapse Loads*

As pointed out in Section 7.5.1 plastic structural analysis is concerned with the load to cause collapse rather than that to initiate yielding as in elastic analysis. Determination of collapse loads of frames and continuous beams is beyond the scope of this work (see for example references 3 and 4) but, as an introduction, single span beams will be considered.

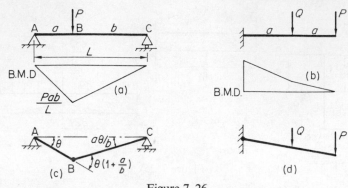

Figure 7.26

In the case of statically determinate beams (Figure 7.26a and b) there are just sufficient constraints to enable the loads at the supports to be determined using only the equilibrium equations. As the load(s) are increased the bending moment at the critical position eventually attains the limit value $\hat{M}$ and a plastic hinge is thereby created at this point. Since this permits relative rotation of the two parts of the beam it is obvious from Figures 7.26c and d that the systems are then mechanisms. No further load can be sustained. Denoting by $\hat{P}$ and P_0 the values of load causing collapse and yield respectively we have for example from the bending moment diagram in Figure 7.26a

$$P_0 = \frac{M_0 L}{ab} \quad \text{and} \quad \hat{P} = \frac{\hat{M}L}{ab}$$

Thus the gain in load capacity between elastic and plastic design of a statically determinate beam is determined by the shape factor of the cross-section.

It is instructive to determine $\hat{P}$ for the simply supported beam by considering a small movement of the mechanism. During this the loads are constant so that the external work is $\hat{P}a\theta$; the internal work is that to rotate AB and BC about the plastic hinge which exerts as it were a frictional moment $\hat{M}$. From geometry the relative rotation of AB and BC is $\theta L/b$ and so equating internal and external work

$$\hat{P}a\theta = \hat{M}\frac{\theta L}{b}$$

i.e.

$$\hat{P} = \frac{\hat{M}L}{ab}$$

This result is trivial but establishes a method which is useful for the study of indeterminate structures.

A statically indeterminate beam, e.g. Figure 7.27a, has more constraints than the number of equilibrium equations. In elastic analysis further equations

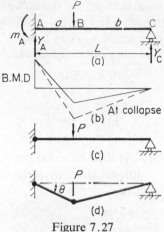

Figure 7.27

are obtained by considering the deformation of the material in conjunction with the prescribed displacements imposed by the additional constraints.

To simplify the analysis we assume a shape factor of 1, i.e. $\hat{M} = M_0$.

The system of Figure 7.27a can be analysed by superposition of deflexion patterns (see Section 7.3.3) for a cantilever length L with down force P at C, and a cantilever length L with up force Y_C at C subject to the displacement condition $v_C = 0$.

Then

$$v_C = -\left(\frac{Pa^3}{3EI} + \frac{Pa^2b}{2EI}\right) + \frac{Y_CL^3}{3EI} = 0$$

and

$$Y_C = \frac{Pa^2}{L^3}\left(a + \frac{3b}{2}\right)$$

Using $\Sigma Y = 0$ and $\Sigma m = 0$ we find

$$m_A = Pa\left(1 - \frac{a^2}{L^2} - \frac{3ab}{2L^2}\right)$$

The bending moment diagram (Figure 7.27b) has two peaks whose relative magnitude depends on the ratio a/b. The values shown are for $b = 2a$ and make A the position of greatest bending moment. If P is increased until $m_A = 5PL/27 = \hat{M}$ a plastic hinge forms at A. The beam is then effectively supported by a hinge at A and a roller at B (Figure 7.27c) so that it is statically determinate and can support additional load. However since the moment at A is constant at $\hat{M}$ the reactive forces Y_A and Y_C can be obtained from static equilibrium. As P increases so does Y_A and hence M_B (bending moment at B) until $|M_B| = \hat{M}$. A plastic hinge forms at B (Figure 7.27d) putting the beam in the condition of Figure 7.26c which is known to constitute a mechanism. The load is therefore the collapse load $\hat{P}$, and is determined using the work method. The external work during a small rotation θ of AB is $\hat{P}a = \hat{P}L\theta/3$. The internal work at hinge A is $\hat{M}\theta$ and at hinge B is $\hat{M}\theta(1+(a/b)) = 1\cdot5\hat{M}\theta$. Equating internal and external work

$$\frac{\hat{P}L\theta}{3} = \frac{5\hat{M}\theta}{2}$$

i.e.

$$\hat{P} = 7\cdot5\,\frac{\hat{M}}{L}$$

For a shape factor of unity the load to cause the first plastic hinge is also the yield load and so the additional strength of the indeterminate beam on a plastic rather an elastic basis is

$$\frac{7\cdot5\hat{M}\times5L}{L\times27\hat{M}} = 1\cdot39$$

or 39%. Note that the work method did not require a knowledge of the order of development of the plastic hinges, but only of their location.

Problem for Solution 7.44

7.6 References

1. Gere, J. M., *Moment Distribution*, Van Nostrand, Princeton, N.J., 1963.
2. Neal, B. G., *The Plastic Methods of Structural Analysis*, Chapman and Hall, London, 1959.
3. Baker, Sir J. and Heyman, J., *Plastic Design of Frames*, Vol. 1, Cambridge, 1969.
4. Horne, M. R., *Plastic Theory of Structures*, Nelson, 1971.

7.7 Problems for Solution

E(steel) = 210 GN/m² unless otherwise indicated.

7.1. A cantilever carries a uniformly distributed load of intensity w and is subject to an anticlockwise concentrated moment at the right-hand (free) end. Find slope and deflexion at the free end.

7.2. A simply supported beam is subject to an anticlockwise concentrated moment at the left end. Find expressions for the slopes at the ends and compare the maximum and central deflexions.

7.3. For the beam of Example 7.2 (Figure 7.2a) find the end loads and the shearing force and bending moment diagrams for displacements $v = 1$ and $(dv/dx) = 0$ at B.

7.4. If $P_1 = P_2 = P$, AB = BC = a, and CD = $2a$, in Figure 7.3 find the central deflexion and slope, and the end slopes. Determine the location and magnitude of the greatest deflexion.

7.5. Find the central deflexion and end slopes for the beam in Figure 7.28a if $b = 2a = $ BC and AB = CD = a.

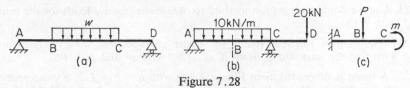

Figure 7.28

7.6. A simply supported beam of length 4 m has an anticlockwise concentrated moment 10 kN/m acting at mid-span. Calculate the slopes at the ends and the mid-point if $EI = 1 \cdot 5$ MNm².

7.7. Calculate deflexion and slope at D in Figure 7.28b if $I = 4000$ cm⁴. Find the least value for I if the deflexions at B and D are not to exceed 0·002 of AD. AB = BC = CD = 2 m.

7.8. If $w = 30$ kN/m, AB = CD = 1 m, BC = 3 m and a clockwise moment of 85 kNm acts at B find the mid-span deflexion in Figure 7.28a. $EI = 33 \cdot 6$ MNm².

7.9. Find the value of m in Figure 7.28c for which deflexion at C is half that at B. Sketch the shearing force and bending moment diagrams. AB = BC = b.

7.10. Rework Problem 7.3 using Superposition.

7.11. A cantilever AB is subject to a uniformly distributed downward load of intensity w and a concentrated force P acting up at the free end B. Find P for $v_B = 0$ and $v_B = -1$.

7.12. An anticlockwise concentrated moment m is added to the system of Problem 7.11. Determine P and m in terms of w and L for zero slope and deflexion at B.

7.13. Find expressions for the strain energy in a cantilever of length L when subject to (a) concentrated end moment, (b) uniformly distributed load of intensity w.

7.14. Calculate the strain energy of the beam in Figure 6.29b if $EI = 5$ MNm².

7.15. Find the end slope in Problem 7.13 (Part a).

7.16. Use the Castigliano Method to find the deflexion of force P in Example 7.8.

7.17. Find the end deflexion and slope of a cantilever subject to a uniformly distributed load.

7.18. Find the end deflexion and slope for a cantilever carrying a distributed load with linearly varying intensity with maximum value w_0 at the root.

7.19. If in Figure 7.7b the force P were to act horizontally to the right determine the vertical and horizontal deflexions at A and the slope at A.

7.20. Find the vertical deflexions at B and C and the slope at B in Figure 7.9c if $Q = 0$.

7.21. Find the horizontal deflexion at C in Figure 7.9a.

7.22. Find the horizontal and vertical deflexion at B in Figure 7.9b. Hence calculate the wire diameter for a vertical stiffness of 1 kN/m if $r = 0 \cdot 1$ m. If $\sigma_0 = 400$ MN/m² find the maximum value of P for a safety factor of 2.

7.23. Use the double integration method to obtain the reactive loads for the beam in Figure 7.10c.

7.24. The beam of Figure 7.10a is of square cross-section, $w = 10\,\text{kN/m}$ and $L = 4\,\text{m}$; find the least size of beam if $\sigma_0 = 240\,\text{MN/m}^2$ and $f_s = 1\cdot5$.

7.25. A beam is supported as in Figure 7.10b with $a = b = L/2$; a concentrated force P acts down at the midpoint of AB. Find the reactive loads and sketch the bending moment diagram and the deflexion curve.

7.26. Solve for the reactive loads in Figure 7.10a and c using the Superposition method.

7.27. Find the reactive loads for the beam in Figure 7.11a when $a = b = L/2$ and the roller support is set δ below the support at A.

7.28. Find an expression for the reactive load at B in Figure 7.10a if the roller support is elastic of stiffness k.

7.29. If $P = 27\,\text{kN}$, $a = 2\,\text{m}$, $b = 1\,\text{m}$ in Figure 7.11a find the maximum stress, and the deflexion at B if the cross-section is as in Figure 6.14a.
(*Hint* Once the reactive loads are determined use equation (7.1) to find an expression for v_B).

7.30. If the roller at C in Figure 7.9a is replaced by a hinge find the reactive loads.

7.31. For the system of Example 7.18 find the end loads when the right-hand support is observed to have deflexion of $0\cdot02\,\text{m}$ and slope $0\cdot006$ radian.

7.32. A 1 m long steel rod of $0\cdot05\,\text{m}$ diameter is simply supported and is observed to have end slopes of $0\cdot002$ and $0\cdot001$. Find the end loads.

7.33. A continuous beam consists of two 2 m spans with EI values of 1 and 3 MNm^2. Find the nodal loads if the nodal displacements are $\delta = \{0 \quad 0 \quad 0\cdot01 \quad 0 \quad 0 \quad 0\}$. Draw the bending moment diagram and interpret the system.

7.34. When the assembly of Example 7.20 has $\delta = \{0 \quad 0 \quad 0 \quad 0\cdot01 \quad 0 \quad 0\cdot01\}$ find the nodal loads and sketch the system and its bending moment diagram.

7.35. If the beam of Example 7.21 were supported as in Figure 7.18 find nodal loads and displacements when a force of 10 kN acts downwards at node 2.

7.36. Repeat Problem 7.35 but using a beam with the properties described in Problem 7.33.

7.37. The beam described in Example 7.21 has $v_1 = v_2 = v_3 = 0$ and a moment of 12 kNm is imposed at node 2. Sketch the bending moment diagram and the support system.

7.38. A $0\cdot05\,\text{m}$ diameter steel rod has a fixed support at node 1, a roller support at node 2, and is subject to a downward force of 4 kN at node 3. If the elements are 2 m and 1 m long find the nodal displacements and sketch the bending moment diagram.

7.39. Rework Problem 7.6 using stiffness methods. (*Hint* $\theta_1 = \theta_3$ from symmetry.)

7.40. Rework Problem 7.7 using stiffness methods but with fixed support at A (Figure 7.28b).

7.41. Determine the shape factor and fully plastic moment for the following cross-sections. Take $\sigma_0 = 250 \text{ MN/m}^2$. (a) Structural hollow section in Figure 6.14b. (b) Tube of inner and outer diameters 0·18 and 0·2 m. (c) I-Section of Figure 7.13b loaded on the z axis. (d) Square cross-section, side 20 mm, loads acting along one diagonal. (e) Section of Figure 6.15b.

7.42. Repeat Example 7.26 with the limit load only half removed.

7.43. A rectangular cross-section width b, depth h is brought to $+\hat{M}$. Find the change of moment at which the stresses at the upper and lower surfaces are sufficient to initiate reverse yielding.

7.44. Find expressions for the collapse loads. (a) The support at C of Figure 7.27a is moved to B and the load to C. (b) The roller support at C of Figure 7.27a is replaced by a fixed support ($b = 2a$).

8
Torsion

8.1 Torsion Moment

In this chapter we consider the effects of moments exerted in cross-sectional planes of a bar as in Figure 8.1c; such moments are usually called *twisting moments* or *torques* and will be denoted m_x where x is the longitudinal axis of the bar. Twisting moments may be caused by static couples as at A in Figure 8.1a or by the effects of power transmission as in Figure 8.2a. If in the latter case the angular velocity ω is constant then

$$\text{Power} = m_x \omega \qquad (8.1)$$

If the bar is to be at rest or in uniform motion the equilibrium equation must be satisfied

i.e. $$\sum m_x = 0 \qquad (8.2)$$

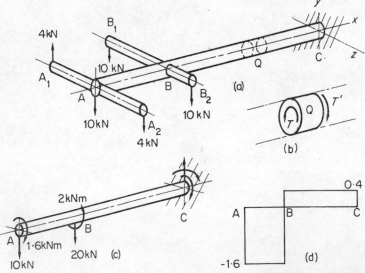

Figure 8.1 Not to scale. (a) $AA_1 = AA_2 = 0.2$ m, $BB_1 = 0.25$ m, $BB_2 = 0.05$ m, $AB = 1$ m, $BC = 2$ m. (d) Torsion moment diagram (units kNm)

The effects of twisting moments are transmitted by the material of the bar between their planes of action. If an element as in Figure 8.1b has resultant twisting moment T acting to the left, then to the right there must be a moment T' such that $T+T' = 0$. The opposite faces of the element are therefore subjected to equal but opposite moments. These resultant moments, which will be called *torsion moments*, tend to cause the ends of the element to rotate with respect to each other. We now define the torsion moment T at any position along a bar as the algebraic sum of the twisting moments acting to one side of that position. A sign convention is useful and here we shall consider moments to one side of a position on the bar to be positive if clockwise when viewed in that direction. For example referring to Figure 8.1c the moment at A is anticlockwise (i.e. negative) when viewed from B, as is that at B when viewed from A.

Example 8.1 Draw the torsion moment diagram for the tube ABC in Figure 8.1a.

Solution The forces acting on the cross-bar at A exert a couple of magnitude 1·6 kNm on the tube in the plane of its cross-section (Figure 8.1c). For all positions between A and B the resultant moment to the left of any cross-section, i.e. the torsion moment, is then by the sign convention $-1·6$ kNm. The twisting moment exerted at B is $(10 \times 0·25 - 10 \times 0·05)$, i.e. 2 kNm. Hence between B and C the torsion moment is $(-1·6+2)$, i.e. 0·4 kNm. The torsion moment diagram is plotted in Figure 8.1d. Note that to maintain equilibrium the support at C must exert a vertical force of 30 kN, a moment of 70 kNm in the longitudinal plane xy, and a moment in the yz plane of 0·4 kNm (Figure 8.1c).

Situations such as that dealt with in the above example might arise for instance due to unsymmetrical loading of a bridge deck, an aeroplane wing, etc.

Example 8.2 A turbine develops 1000 kW at 3000 r.p.m. If the driving torque is effectively applied uniformly over the length AB (Figure 8.2a), and is transmitted to the load through a gear wheel at C, draw the torsion moment diagram.

Solution Using equation (8.1) torque $= 10^6/\omega$ Nm where $\omega = 2\pi \times (3000/60)$ rad/sec. Hence the total torque applied over a length of 0·4 m is 3190 Nm, i.e. a loading intensity of 7970 Nm/m.

Then the resultant moment to the left of any position between A and B is $7970x$ Nm where x is the distance from A. Thus the torsion moment diagram increases linearly from A to B and is then constant between B and C (Figure 8.2b).

Problems for Solution 8.1, 8.2

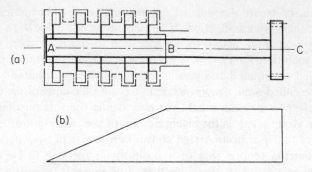

(a)

(b)

Figure 8.2   AB = BC = 0·4 m. (b) Torsion moment diagram

8.2   Bars of Circular Cross-section (Shafts)

8.2.1 *Polar Coordinate System*

The natural symmetry of a bar of circular cross-section (i.e. a rod) suggests the use of polar coordinates r, θ to describe position in the cross-section, retaining x as the longitudinal axis as in Chapters 5, 6, 7. Figure 8.3 shows an infinitesimal element in this x, θ, r coordinate system. Following the

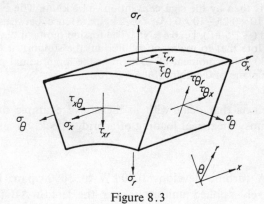

Figure 8.3

arguments of Articles 1.2 and 1.3 average forces acting on the faces of such an element can, in the limit, be expressed by the stress components shown. For example on the surfaces perpendicular to the circumferential direction the stress components are σ_θ, $\tau_{\theta x}$, $\tau_{\theta r}$. The positive directions are defined by the same convention as in Article 1.3. The stress array is then

$$\begin{pmatrix} \sigma_x & \tau_{x\theta} & \tau_{xr} \\ \tau_{\theta x} & \sigma_\theta & \tau_{\theta r} \\ \tau_{rx} & \tau_{r\theta} & \sigma_r \end{pmatrix} \tag{A8.1}$$

The method of Article 1.4 may be used to prove the symmetry of the stress array, i.e. $\tau_{\theta x} = \tau_{x\theta}$ etc. The corresponding strain array is

$$
\begin{pmatrix}
\epsilon_x & \tfrac{1}{2}\gamma_{x\theta} & \tfrac{1}{2}\gamma_{xr} \\
\tfrac{1}{2}\gamma_{x\theta} & \epsilon_\theta & \tfrac{1}{2}\gamma_{r\theta} \\
\tfrac{1}{2}\gamma_{xr} & \tfrac{1}{2}\gamma_{r\theta} & \epsilon_r
\end{pmatrix}
\tag{A8.2}
$$

The strain/stress equations may be derived from those of Article 3.3 by replacing y by θ and z by r. For example

$$
\epsilon_x = \frac{1}{E}\left(\sigma_x - \nu(\sigma_\theta + \sigma_r)\right) \quad \text{and} \quad \gamma_{x\theta} = \frac{\tau_{x\theta}}{G}
$$

8.2.2 *Stress and Twist*

The elementary theory of torsion of a rod is based on assumptions analogous to those used in bending of a bar (see Section 6.3.1). It is assumed that after torsion has taken place

(a) original plane cross-sections remain plane and that original radii remain straight.

(b) The only non-zero stress component is $\tau_{x\theta}$ for axes x, θ, r. Figure 8.4a shows a thin disk of length dx and radius r isolated from a rod and subject to torsion moment T. The deformation allowed by the above assumptions

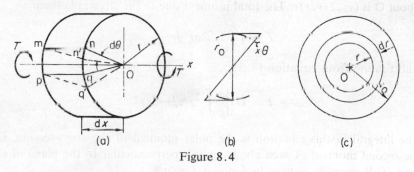

(a)   (b)   (c)

Figure 8.4

is a relative rotation of the end planes of the element. Then two parallel lines mn and pq in the cylindrical surface are rotated through a small angle to positions mn' and pq'. It is seen that the unloaded right angle nm̂p becomes after loading

$$
\text{n'}\hat{\text{m}}\text{p} = \frac{\pi}{2} - \text{nm̂n'}
$$

so that nm̂n', the change of an original right angle, is (see Article 2.3) the shear strain $\gamma_{x\theta}$ of the cylindrical surface at radius r. Then

$$
\text{nn'} = \text{qq'} = \gamma_{x\theta}dx
$$

By assumption (a) the end planes are undistorted and therefore the radii On and Oq in Figure 8.4a rotate through a small angle $d\theta$ to positions On' and Oq'. Then

$$nn' = qq' = rd\theta$$

Combining expressions for nn' and qq' we have

$$\gamma_{x\theta} = r\frac{d\theta}{dx}$$

Since the same reasoning would be true for any cylindrical surface intermediate between the outer boundary and the x axis this is a general result. Using the appropriate strain/stress equation quoted in Section 8.2.1 we have

$$\frac{\tau_{x\theta}}{r} = G\frac{d\theta}{dx} \tag{8.3}$$

These stresses act tangentially in the plane of the cross-section (Figure 8.3) are greatest at the outer surface radius r_0 and zero at the centre of the rod (Figure 8.4b). The moment in the cross-section due to the shear stresses $\tau_{x\theta}$ must be equal to the total moment exerted on this plane, i.e. the torsion moment T. The total tangential force on the ring element of Figure 8.4c due to the shear stress at radius r is $\tau_{x\theta}2\pi rdr$ and the moment of this force about O is $(\tau_{x\theta}2\pi rdr)r$. The total moment due to the stresses is then

$$T = \int_0^{r_0} \tau_{x\theta}2\pi r^2dr$$

Substituting from equation (8.3)

$$T = G\frac{d\theta}{dx}\int_0^{r_0} r^2(2\pi rdr)$$

The integral in this equation is the polar moment of the cross-section, i.e. the second moment of area about an axis perpendicular to the plane of the area. This quantity is usually denoted J giving

$$\frac{T}{J} = G\frac{d\theta}{dx} \tag{8.4}$$

Combining equations (8.3) and (8.4) we have

$$\frac{\tau_{x\theta}}{r} = \frac{T}{J} = G\frac{d\theta}{dx} \tag{8.5}$$

The similarity between the form of these equations and those (6.10) for bending of a bar should be noted.

Failure Conditions By assumption (b) the state of stress at radius r is

$$\begin{pmatrix} 0 & \tau_{x\theta} & 0 \\ \tau_{\theta x} & 0 & 0 \\ 0 & 0 & 0 \end{pmatrix} \qquad (A8.3)$$

The principal stresses can be found by a Mohr Stress Circle (see Article 1.10) and are

$$\sigma_1 = \tau_{x\theta}; \quad \sigma_2 = -\tau_{x\theta}; \quad \sigma_3 = 0$$

Then the greatest shear stress at radius r is (see equation 1.12)

$$\tfrac{1}{2}|\sigma_1 - \sigma_2| \qquad \text{i.e. } \tau_{x\theta}$$

The greatest shear stress in the rod is therefore at r_0 and when failure occurs (see Article 1.12)

$$\tau_0 = \tau_{\max} = |(\tau_{x\theta})_{r_0}| \qquad (8.6)$$

where τ_0 is the shear stress at the yield point for loading in simple shear.

Polar Moment of Area For a solid rod, radius r_0

$$J = \int_0^{r_0} 2\pi r^3 \mathrm{d}r = \tfrac{1}{2}\pi r_0^4 \qquad (8.7)$$

For a tube of inner and outer radii r_i and r_0

$$J = \tfrac{1}{2}\pi(r_0^4 - r_i^4) \qquad (8.8)$$

Example 8.3 Using a safety factor of 2 find the minimum diameter of solid shaft required for the turbine of Example 8.2. In a tensile test σ_0 was 240 MN/m^2,

Solution From Example 8.2 the greatest torsion moment is 3190 Nm. The polar moment is given by (8.7) and so from equation (8.5)

$$(\tau_{x\theta})_{r_0} = \frac{Tr_0}{J} = \frac{2 \times 3190 \times r_0}{\pi r_0^4}$$

By the maximum shear stress theory (Article 1.12) we have from the tensile test $\tau_0 = \tfrac{1}{2}\sigma_0 = 120$ MN/m^2 and so the allowable stress is $\tau_0/f_s = 120/2 = 60$ MN/m^2.
Hence $(\tau_{x\theta})_{r_0} \leqslant 60 \times 10^6$ and so $r_0 \geqslant 32 \cdot 35$ mm. Therefore the minimum diameter required is 64·7 mm.

Displacements The angle of twist θ of a rod is the relative rotation of its end planes. From equation (8.4)

$$\theta = \int_0^L \frac{T\mathrm{d}x}{GJ} \qquad (8.9)$$

where T may be a function of x, and L is the length of the shaft. This result is true whether the shaft is at rest or in uniform rotation. In particular if T is constant over L then

$$\theta = \frac{TL}{GJ} \tag{8.10}$$

Where a shaft is made up of several different parts subject to known torsion moments, the angle of twist for each part may be calculated separately using (8.9) or (8.10) and the results added.

Example 8.4   Find the total angle of twist of the turbine shaft of Example 8.2 if $G = 80 \text{ GN/m}^2$. Use the diameter found in Example 8.3.

Solution   Over AB (see Figure 8.2a) $T = 7970x$ and so using (8.9)

$$\theta_{AB} = \frac{1}{GJ} \int_0^{0.4} 7970x\,dx = 0\!\cdot\!00462 \text{ rad.}$$

Over BC $T = 3190 \text{ Nm}$ and from (8.10) $\theta_{BC} = 0\!\cdot\!00924$ rad. Then total twist $= \theta_{AB} + \theta_{BC} = 0\!\cdot\!01386$ rad.

Problems for Solution   8.3–8.6

8.2.3 *Material Economy*

The fourth power dependence of J on r means that hollow shafts may offer substantial savings in mass for the same duty as a solid shaft. From (8.5) for a given torsion moment, the greatest shear stress is proportional to r_0/J. For a hollow shaft with $r_i = nr_0$ the quantity is $1/(r_0^3(1-n^4))$ and for a solid shaft, radius r_s, it is $1/r_s^3$. Equating these we find that for the same maximum stress and torsion moment

$$\frac{r_s}{r_0} = \sqrt[3]{1-n^4}$$

Mass is proportional to cross-sectional area and so the ratio of masses of hollow to solid is $r_0^2(1-n^2)/r_s^2$. For example, if $n=0\!\cdot\!9$ the mass ratio at constant strength is $0\!\cdot\!388 : 1$. Again, if the shaft of Example 8.3 were bored out to half its diameter the safety factor would still be $1\!\cdot\!88$.

At constant power transmission torque is inversely proportional to speed of rotation (equation 8.1). It follows that a smaller shaft diameter should be possible if the speed is increased. For the same maximum stress at speeds ω_1 and ω_2 let the solid shaft radii be r_1 and r_2. Then using equations (8.5) and (8.7) we obtain

$$\frac{r_2}{r_1} = \sqrt[3]{\frac{\omega_1}{\omega_2}}$$

The ratio of masses for the two solid shafts is $(r_2/r_1)^2$. For example if $\omega_2 = 10\omega_1$ then $r_2/r_1 = 0.465$ and the masses of the shafts are in the ratio $0.215 : 1$. However it should be noted that at higher speeds other problems may arise, for instance with bearings and vibration.

Problems for Solution 8.7, 8.8

8.2.4 *Statically Indeterminate Problems*

Figure 8.5a illustrates a situation in which a twisting moment is applied to a rod at an intermediate point B, and the ends A and C are clamped. There are therefore reactive twisting moments m_A and m_C at A and C. As these

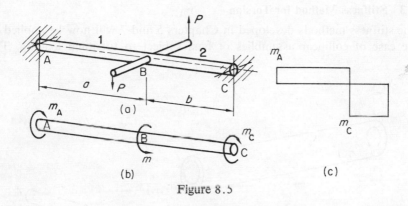

Figure 8.5

cannot be determined using only the equilibrium equation (8.2) for twisting moments, it follows that this is a statically indeterminate problem (cf. Figure 5.3b).

A second equation can be obtained by considering compatibility of deformation at B. Referring to the figure it is seen that the angles of twist of AB and BC have the same magnitude but opposite signs. Alternatively we could say that the total angle of twist over AC is zero. Using equation (8.10) this relationship can be expressed in terms of the moments.

Example 8.5 Find expressions for the reactive moments at A and C in Figure 8.5a. ABC is uniform.

Solution Referring to Figure 8.5b the reactive moments must act as shown and for equilibrium

$$m = m_A + m_C \tag{i}$$

Using the sign convention described in Article 8.1 we find that for any position between A and B the resultant twisting moment to the left is $+ m_A$. For any position between B and C the resultant moment to the left is $m_A - m$, and the torsion moment diagram is Figure 8.5c. Then using (8.10)

$$\theta_{AB} = \frac{m_A a}{GJ} \quad \text{and} \quad \theta_{BC} = \frac{(m_A - m) b}{GJ}$$

Then since $\theta_{AB} + \theta_{BC} = 0$ we find

$$m_A \left(1 + \frac{a}{b}\right) = m \tag{ii}$$

Using (i) and (ii) an expression for m_C can be obtained.

Problems for Solution   8.9–8.11

8.3   Stiffness Method for Torsion

The stiffness methods developed in Chapters 5 and 7 will now be applied to the case of collinear assemblies of bars subject to torsional loading. The

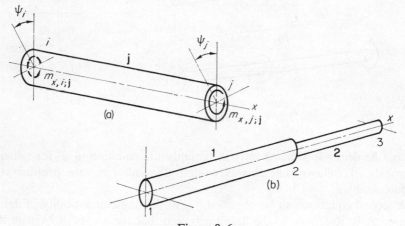

Figure 8.6

end loads and displacements for the **j**th bar of such an assembly are shown in Figure 8.6a, and the corresponding vectors are

$$\mathbf{F_j} = \{m_{x,i;j} \quad m_{x,j;j}\} \qquad \mathbf{\delta_j} = \{\psi_i \quad \psi_j\} \tag{8.11a, b}$$

The element stiffness equations then have the matrix form

$$\begin{bmatrix} m_{x,i;j} \\ m_{x,j;j} \end{bmatrix} = \begin{bmatrix} k_{ii;j} & k_{ij;j} \\ k_{ji;j} & k_{jj;j} \end{bmatrix} \begin{bmatrix} \psi_i \\ \psi_j \end{bmatrix} \tag{8.12}$$

which may be compared with equations (5.7) and (5.10). The components of the element stiffness matrix are evaluated in a similar manner to that used in Section 5.3.1. For example if ψ_i is zero the angle of twist is ψ_j, the torsion moment is $m_{x,j;i}$ and using equations (8.10) and (8.2)

$$m_{x,j;} = \frac{GJ}{L}\psi_j = -m_{x,i;j}$$

Putting $\psi_i = 0$ in equations (8.12) we obtain

$$k_{jj;i} = \frac{GJ}{L} = -k_{ij;i}$$

The stiffness matrix is found to be (compare 5.10)

$$\mathbf{k_j} = \frac{GJ}{L}\begin{bmatrix} 1 & -1 \\ -1 & 1 \end{bmatrix} \qquad (8.13)$$

For an assembly of **n** bars with n nodes the nodal load and displacement vectors are

$$\mathbf{F} = \{m_{x,1} \quad m_{x,2} \quad \ldots \quad m_{x,i} \quad \ldots \quad m_{x,n}\} \qquad (8.14a)$$

$$\mathbf{\delta} = \{\psi_1 \quad \psi_2 \quad \ldots \quad \psi_i \quad \ldots \quad \psi_n\} \qquad (8.14b)$$

where $m_{x,i}$ and ψ_i are the twisting moment, and rotation, at the ith node of the assembly. The assembly stiffness matrix **K** is of size $(n \times n)$ and is formed in the usual way (see Section 5.3.2) by accumulating components from the element stiffness matrices.

One nodal displacement must always be known in order to eliminate rigid body displacement. If all other displacements are also known the assembly stiffness equations

$$\mathbf{F} = \mathbf{K\delta} \qquad (8.15)$$

can be used directly. Otherwise the reduced equations

$$\mathbf{F_R} = \mathbf{K_{RR}\delta_R} + \mathbf{K_{RP}\delta_P} \qquad (8.16)$$

(see equations 5.29) are solved for the unknown displacements (vector $\mathbf{\delta_R}$). In practice these are obtained from (8.15) by partitioning **δ** and striking out rows from **F** and **K** corresponding to nodes at which displacement is prescribed; $\mathbf{K_{RR}}$ is then obtained from **K** by removing columns corresponding to the deleted rows; the columns so removed constitute matrix $\mathbf{K_{RP}}$.

Example 8.6 Two rods are rigidly joined together as in Figure 8.6b; **1** is hollow with inner and outer diameters 90 and 100 mm and length 2 m; **2** is solid of diameter 80 mm and length 4 m. Taking $G = 80 \times 10^9$ N/m² find the angle of twist when the displacement of node 1 is -0.02 rad, node 2 is load free, and a twisting moment of 10 kNm acts at node 3.

Solution   Utilizing equations (8.8) and (8.7) the values of GJ/L for the hollow and solid rods are $134 \cdot 7 \times 10^3$ and $80 \cdot 5 \times 10^3$ Nm/rad. Then the complete stiffness equations are

$$\begin{bmatrix} m_{x,1} \\ 0 \\ 10^4 \end{bmatrix} = 10^3 \begin{bmatrix} 134 \cdot 7 & -134 \cdot 7 & 0 \\ -134 \cdot 7 & (134 \cdot 7 + 80 \cdot 5) & -80 \cdot 5 \\ 0 & -80 \cdot 5 & 80 \cdot 5 \end{bmatrix} \begin{bmatrix} -0 \cdot 02 \\ \psi_2 \\ \psi_3 \end{bmatrix}$$

The reduced equations are

$$\begin{bmatrix} 0 \\ 10^4 \end{bmatrix} = 10^3 \begin{bmatrix} 215 \cdot 2 & -80 \cdot 5 \\ -80 \cdot 5 & 80 \cdot 5 \end{bmatrix} \begin{bmatrix} \psi_2 \\ \psi_3 \end{bmatrix} + 10^3 \begin{bmatrix} -134 \cdot 7 \\ 0 \end{bmatrix} [-0 \cdot 02]$$

and so $\psi_2 = 0 \cdot 0543$, $\psi_3 = 0 \cdot 178$, and the angle of twist $(\psi_3 - \psi_1)$ is $0 \cdot 198$ rad.

Example 8.7   If the assembly of Example 8.6 is clamped at nodes 1 and 3 and a twisting moment of 10 kNm is exerted at node 2 find the greatest shear stress in each rod.

Solution   Here $\mathbf{F} = \{m_{x,1} \quad 10^4 \quad m_{x,3}\}$ and $\boldsymbol{\delta} = \{0 \quad \psi_2 \quad 0\}$. $\mathbf{K}$ is as in Example 8.6. Hence the reduced stiffness equations become

$$10^4 = 10^3[215 \cdot 2]\,[\psi_2]$$

Then $\psi_2 = 0 \cdot 0465$ and the angles of twist are $0 \cdot 0465$ for rods **1** and **2**. Using equation (8.3) with $d\theta/dx = \theta/L$ the greatest stress in rod **1** is 93 MN/m², and in rod **2** it is $37 \cdot 2$ MN/m².

Problems for Solution   8.12–8.14

8.4   Strain Energy in Torsion

Consider an infinitesimal length dx of rod subject to torsion moment T and experiencing an angle of twist $d\theta$. During load application moment and twist are proportional (equation 8.5) and so following the discussion in Article 4.1 the strain energy of the element is

$$dU = \tfrac{1}{2}T\,d\theta \tag{8.17}$$

We can use equations (8.5) to express the strain energy in terms of any of the parameters T, $d\theta$, $\tau_{x\theta}$, for a rod

e.g.

$$dU = \frac{T^2\,dx}{2GJ} \tag{8.18}$$

Example 8.8   A hollow shaft has outer and inner diameters of 100 mm and 80 mm and is 4 m long. Calculate the strain energy when the shaft is transmitting 400 kW at 600 r.p.m. $G = 80$ GN/m².

Solution Using (8.1) the torsion moment is found to be 6380 Nm and so the strain energy U of the whole shaft is from (8.18)

$$U = \int_0^L \frac{T^2\,\mathrm{d}x}{2GJ} = \frac{T^2L}{2GJ} = \frac{(6380)^2 \times 4 \times 2}{2 \times 80 \times 10^9 \times \pi \times (0 \cdot 05^4 - 0 \cdot 04^4)}$$

$$= 176\ \text{Nm}$$

Problems for Solution   8.15–8.17

8.5   Coil Springs

A coil spring consisting of a wire of outer radius r_0 wound into a helix of mean radius R_0 is shown in Figure 8.7a. When the helix is tightly wound each coil is nearly in a plane perpendicular to the axis and the spring is termed close-coiled helical. At the ends of the spring the wire is bent into

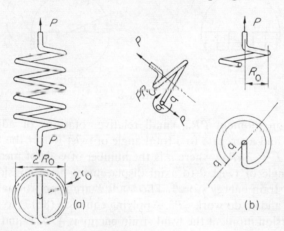

Figure 8.7

the axial direction as shown, and loaded by opposite axial forces P. To investigate the nature of the loading cut the wire at any position qq as in Figure 8.7b and apply loads to qq to maintain the remainder of the spring in the same condition as before cutting. The elevation of the plane of the cut shows that it must be subject to an axial force P and a moment PR_0 in its plane. The former will cause simple shearing of the wire with mean value

$$\tau_1 = \frac{P}{\pi r_0^2}$$

The moment PR_0 is a twisting moment and so the wire is subject to torsion moment $T = PR_0$. The greatest stress τ_2 due to this action is at the wire surface and is obtained by substituting $r = r_0$ in equation (8.5). Then

$$\tau_2 = \frac{Tr_0}{J} = \frac{2PR_0}{\pi r_0^3}$$

The distributions of τ_1 and τ_2 over the wire cross-section are shown in Figure 8.8. The critical condition is at the inside edge of the wire where both τ_1 and τ_2 have the same direction. Therefore

$$\tau_{\max} = \tau_1 + \tau_2 = \frac{P}{\pi r_0^2}\left(1 + \frac{2R_0}{r_0}\right) \tag{8.19}$$

The stiffness of the spring is the axial force to cause a unit axial relative displacement of the ends. In determining axial displacement δ it is usual to neglect any contribution due to the simple shearing. Then, due to the

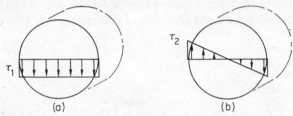

(a)   (b)

Figure 8.8

constant torsion moment PR_0, small relative rotations of adjacent cross-sections occur and aggregate to a total angle of twist θ over the length of the wire, which is $L = 2\pi R_0 n$ where n is the number of coils of mean radius R_0.

To relate angle of twist θ to axial displacement δ we consider the work done and the strain energy stored. The loads P are proportional to the axial displacements and so do work $\frac{1}{2}P\delta$. Applying equation (8.17) to a rod subject to uniform torsion moment the total strain energy is $\frac{1}{2}T\theta$. Equating the work done by the axial loads to the strain energy stored, we have

$$\tfrac{1}{2}P\delta = \tfrac{1}{2}T\theta$$

Substituting $T = PR_0$ we obtain $\delta = R_0\theta$. Using this result in (8.10) and putting $T = PR_0$, $L = 2\pi R_0 n$, $J = \pi r_0^4/2$ we obtain

$$P = \left(\frac{Gr_0^4}{4R_0^3 n}\right)\delta = k\delta \tag{8.20}$$

The quantity in brackets is the stiffness k of the spring.

Example 8.9 A helical spring is to be made of 25 mm diameter wire wound to a coil radius 75 mm. An extension of 100 mm is required without exceeding a shear stress of 100 MN/m². Find the minimum number of whole coils which can be used and the actual stiffness of the spring. Take $G = 80$ GN/m².

Solution Using equation (8.19) we have

$$\tau_{max} = \frac{P}{\pi \times 0.0125^2}\left(1 + \frac{2 \times 0.075}{0.0125}\right) \leqslant 100 \times 10^6$$

i.e.
$$P \leqslant 3780 \text{ N}.$$

Using the stiffness equation (8.20) $P = k\delta \leqslant 3780$ and after substitution we find $n \geqslant 30.5$. Therefore the required number of whole coils is 31 and the actual stiffness is 37.3 kN/m.

Problems for Solution 8.18–8.20

8.6 Non-circular Cross-sections

8.6.1 *Introduction*

The analysis of the torsion of rods of circular cross-section was based on the assumption that plane cross-sections remain undistorted but rotate with respect to other cross-sections (Section 8.2.2). This treatment requires that the only stress component is $\tau_{x\theta}$ which acts normal to a radius in the cross-section (Figure 8.3). If we consider a block of material intersecting the boundary, this stress component acts as in Figure 8.9a in a manner consistent with the boundary condition that there are no forces on the exterior surface of the rod.

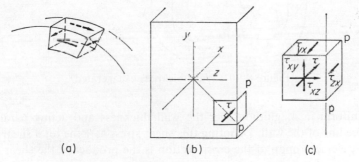

(a) (b) (c)

Figure 8.9 pp lic on the free surface of the bar

In early attempts to analyse the torsion of bars with non-circular cross-sections, it was also assumed that plane cross-sections remained undistorted. As we have seen this assumption requires that the resultant shear stress in the cross-section should act perpendicular to a radial line from the centre, e.g. the stress τ in Figure 8.9b. However if τ is resolved into components parallel to the y and z axes, as in Figure 8.9c, we see that there would be a stress in the free surface of the bar. Since there are no such stresses the assumption of undistorted cross-sections is evidently false. The correct solution of this problem was first given by St. Venant. The cross-sections warp as can easily be demonstrated by twisting a slab of rubber. The details of the resulting general analysis for solid-form cross-sections are beyond the scope of this text. (See for example references 1–3 for theoretical and numerical treatments.) However, simple yet satisfactory solutions can easily be obtained for many of the thin-walled sections widely used in practice.

8.6.2 *Thin-walled Tubes*

An infinitesimal length dx of tube of arbitrary cross-section subject to torsion moment T is shown in Figure 8.10. The wall thickness t is small compared with the overall dimensions of the cross-section. In these circumstances it is found satisfactory to assume that the stress in the cross-section is a shear

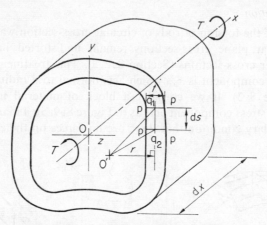

Figure 8.10   Wall thickness exaggerated

stress uniform in magnitude over the wall thickness and acting parallel to the centre line of the wall. Denoting this shear stress as τ the total shear force on a tiny element pppp of the cross-section is the product of the shear stress and the area of the element

i.e. $$\tau(t\,ds)$$

where ds is the length of pppp along the wall centre line. The moment of this force about any point O' is

$$(\tau t\,ds)\,r$$

where r is the perpendicular distance from O' to the tangent to the centre line of pppp. The total moment on the cross-section is the integral round the centre line of the wall and is equal to the torsion moment

i.e. $$T = \int_s \tau t r\,ds$$

where s is the length of the centre line of the wall, usually termed the mean perimeter. The further assumption is now made that the product of t and τ is a constant known as the shear flow q_x. Hence

$$T = t\tau \int_s r \, ds$$

Since $(r \, ds)$ is twice the area of the triangle $O'q_1q_2$ it follows that $\int_s r \, ds$ is just twice the area A_s enclosed within the mean perimeter s. Then

$$T = 2A_s t\tau = 2A_s q_x \qquad (8.21)$$

The angle of twist $d\theta$ over the length dx can be found by equating the work done by the torsion moment T to the strain energy stored in the material. Since τ is the only stress component for the chosen coordinates, we have, by analogy with equation (4.14) of Section 4.1.2, the strain energy density as

$$U_0 = \frac{\tau^2}{2G}$$

The strain energy of a prism of length dx and cross-section pppp is $U_0(t \, ds)dx$ and therefore that of the length dx of tube is

$$dU = \int_s \frac{\tau^2}{2G} t \, dx \, ds$$

Equating this to the work $\frac{1}{2}T \, d\theta$ of the torsion moment, substituting for τ from (8.21), and simplifying we obtain

$$T = \left(\frac{4GA_s^2}{\int_s \frac{ds}{t}}\right) \frac{d\theta}{dx}$$

In the particular case when the wall thickness is constant the integral in the denominator reduces to $\int_s ds = s$. Also if T is constant over a finite length L we can replace $(d\theta/dx)$ by θ/L to obtain finally

$$T = \left(\frac{4GA_s^2 t}{s}\right) \frac{\theta}{L} = \frac{k\theta}{L} \qquad (8.22)$$

Example 8.10 Figure 8.11a shows the centre line of a tube subject to constant torsion moment of 800 kNm. The wall thickness is 10 mm and $G = 80$ GN/m². Find the greatest shear stress and the stiffness of unit length.

Solution

$$A_s = (1 \times 1) + (\pi \times 0.5^2) = 1.786 \text{ m}^2$$

From (8.21)

$$\tau = \frac{800 \times 10^3}{2 \times 1.786 \times 0.01} = 22.4 \times 10^6 \text{ N/m}^2$$

$$s = 1 + 1 + (2\pi \times 0.5) = 5.14 \text{ m}$$

Then using (8.22) stiffness/unit length = 1.982×10^9 Nm/rad

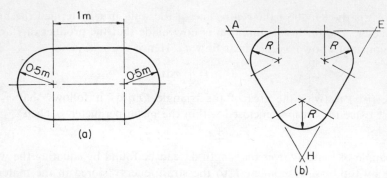

Figure 8.11   (b) $R = 0.2$ m, AE = EH = HA = 1 m

Problems for Solution   8.21, 8.22

8.6.3 *Membrane Analogy*

It can be shown that a bar of arbitrary cross-section, subject to uniform torsion moment, has certain properties analogous to some characteristics of the deflexion surface of a membrane clamped round the edges with tension $\bar{S}$ per unit length, subject to uniform lateral pressure p, and having the same shape as the bar cross-section. In a detailed discussion, as for example in reference 1, it is established that if

$$\frac{p}{\bar{S}} = \frac{2G\theta}{L}$$

then the following correspondences exist:

(a) Contour lines in the membrane correspond to directions of resultant shear stress in the torsion bar.

(b) The greatest slope at any point of the membrane is numerically equal to the resultant shear stress at the corresponding point in the cross-section of the torsion bar.

(c) Twice the volume contained between the deflected membrane and the plane of its outline is equal to the torsion moment on the bar.

For a rectangular membrane the deflected form is shown in Figure 8.12a. Since the central deflexion of the membrane is the same for sections along both of lines mm and nn it follows that the greatest slopes of the membrane and therefore the greatest stresses in a bar of rectangular cross-section are at points m which are the points on the periphery nearest the centre. In bending by contrast the greatest stress is at the greatest distance from the centre.

In the case of a hollow (tubular) section the corresponding membrane is a strip stretched between two plates (Figure 8.12b). The inner plate corresponds to the hollow interior of the bar and is forced out as a rigid body by

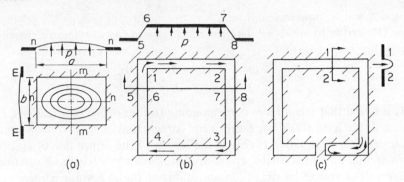

Figure 8.12

the fluid pressure, as illustrated. When the wall is thin the membrane has negligible curvature so that the greatest slope and therefore by analogy the greatest stress in the bar is uniform over the wall thickness. Also the contour lines in the membrane are parallel to the edges, corresponding to stress acting parallel to the edges. These interpretations give some confidence in the use of the assumptions made in Section 8.6.2. At the re-entrant corners 1, 2, 3, 4 of Figure 8.12b the membrane tends to kink giving greater slope and by analogy a stress concentration in the torsion bar. These effects can be reduced by rounding the corners.

8.6.4 *Open Cross-sections*

These are cross-sections such as channels, angles, etc. (Figure 8.13). As previously stated numerical analysis[3] may be carried out to obtain quantitative information, but the membrane analogy allows a qualitative appreciation by considering the torsional behaviour of a bar of narrow rectangular cross-section. Referring to Figure 8.12a it may be appreciated that if the rectangle is made relatively thinner by increasing side a the curvature parallel to the

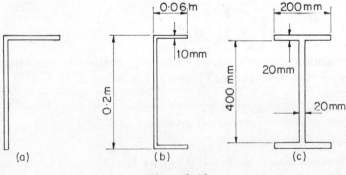

Figure 8.13

long side will become less, and as $a \to \infty$ the membrane assumes a prismatic form. Referring to tabulated data for rectangles in for example reference 1, it can then be shown that

$$T = \frac{ab^2}{3}\tau = \frac{Gab^3}{3}\frac{\theta}{L} = k\left(\frac{\theta}{L}\right) \tag{8.23}$$

It is found that membranes corresponding to thin open cross-sections, e.g. Figure 8.12c have this form except near corners and ends. It follows that if we interpret a in equation (8.23) as the length of the centre line of an open cross-section then a reasonable estimate of the torsional behaviour is obtained. There will of course be stress concentrations at the re-entrant corners.

Example 8.11   An angle section has sides of 400 mm and 200 mm, and wall thickness 5 mm (Figure 8.13a). Estimate the torsional stiffness factor. $G = 80 \times 10^9$ N/m².

Solution   The equivalent rectangle has $a = 595$ mm; $b = 5$ mm. Using (8.23)

$$k = \frac{80 \times 10^9 \times 0.595 \times 5^3 \times 10^{-9}}{3} = 1980 \text{ Nm/rad/m}$$

The angle section in the above example is exceedingly flexible in torsion, a characteristic feature of open sections, and one which can be further illustrated by comparing the membranes for a tube and a slit tube. Referring to Figures 8.12b and c it is clear that the volume enclosed is much greater for the tube and therefore by analogy so is the torsion moment for a given angle of twist. More precisely using equation (8.22) we find the stiffness of the tube of mean side c and wall thickness t as

$$k_1 = \frac{4Gc^4t}{4c} = Gc^3t$$

The slit tube corresponds to a rectangle of length $4c$ and width t so that from (8.23)

$$k_2 = \frac{G4ct^3}{3}$$

Hence $k_1/k_2 = 3c^2/4t^2$ and if $c/t = 10$ the stiffness ratio is 75.

The correlation between the areas enclosed by contour lines and the relative stiffness may also be noted (see Figures 8.12b and c).

Problems for Solution   8.23, 8.24

8.6.5 *Stiffness Equations*

In the preceding sections general results have been obtained for the torsional stiffness of thin tubular sections (equation 8.22) and of open sections (equation

8.23). These equations can both be written in the form

$$T = \frac{GJ'}{L}\theta \qquad (8.24)$$

where J' for a tube is $4A_s^2 t/s$ and for an open section is $ab^3/3$. Since equation (8.24) is similar to (8.10) for circular sections, all the matrix methods of Article 8.3 can be extended to assemblies of bars of non-circular form. However particularly in the case of open cross-sections some care is necessary. As discussed in Section 8.6.1 the analysis required the cross-section to have freedom to deform. If this is prevented as for example by clamps, bending effects occur. An introductory treatment may be found in reference 4.

Problem for Solution   8.25

8.7   Elastic/Plastic Analysis

We now apply to the case of a torsion bar the elastic/plastic method of analysis previously used in Articles 5.6 and 7.5 for direct loading and bending of bars. In the case of a rod of circular cross-section yielding begins at the outer surface (see equation 8.6). If the value of τ_0 in this equation is not known from direct experiment its value can be deduced from tensile test data; the Tresca and Von Mises Criteria (equations 1.19 and 5.52) predict values of $\sigma_0/2$ and $\sigma_0/\sqrt{3}$, respectively; experimental evidence favours the latter value but the former is conservative.

Using equations (8.5) and (8.6) the torsion moment to initiate yielding is

$$T_0 = \frac{J\tau_0}{r_0} \qquad (8.25)$$

If the torsion moment is further increased the additional moment is borne by the remaining material, more and more of which is brought to the yield stress. When yielding has penetrated to radius r_1 (Figure 8.14a) the torsion

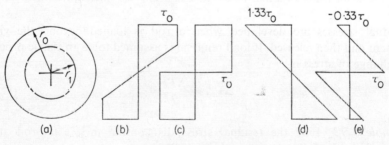

Figure 8.14

moment is the sum of contributions from the elastic and plastic zones of the cross-section (compare Section 7.5.1). Then for a solid rod

$$T = \int_0^{r_1} (2\pi r \mathrm{d}r) \, \tau_{x\theta} r + \int_{r_1}^{r_0} (2\pi r \mathrm{d}r) \, \tau_0 r$$

But from Figure 8.14b $\tau_{x\theta} = \tau_0 r / r_1$ and so

$$T = \frac{\pi \tau_0}{6} (4r_0^3 - r_1^3)$$

When $r_1 = r_0$ the torsion moment corresponds to equation (8.25). When $r_1 = 0$ there is a fully plastic stress distribution (Figure 8.14c) and the corresponding torsion moment for a solid rod is

$$\hat{T} = \frac{2\pi \tau_0 r_0^3}{3} \tag{8.26}$$

The shape factor in torsion is the ratio of the fully plastic and yield moments, i.e. $\hat{T}/T_0$, and for the solid rod has value 4/3. The shape factor is less for a hollow rod since in this case there is relatively less material under low stress when the outer surface is just yielding. (Compare bending of rectangular and I-sections.) Since yielding causes damage to the microscopic structure the additional strength implied by equation (8.26) is only available when fatigue of the material is not a hazard.

Example 8.12 Find the torsional shape factor for a hollow shaft having inner and outer radii in the ratio $r_i/r_0 = 0.9$.

Solution Using equations (8.25) and (8.8)

$$T_0 = 0.172 \pi \tau_0 r_0^3$$

Here

$$\hat{T} = \int_{r_i}^{r_0} (2\pi r \mathrm{d}r) \, \tau_0 r = 0.18 \pi \tau_0 r^3$$

and so

$$\hat{T}/T_0 = 1.05.$$

Thus the shape factor of a thin-walled tube is negligibly different from unity.

Residual stresses are developed when a rod is loaded beyond the yield moment and then released. If load removal is assumed to be an elastic process the change of stress is

$$\Delta \tau_{x\theta} = \frac{\Delta T r}{J} \tag{8.27}$$

Example 8.13 Find the residual stress distribution in a solid rod after release from the fully plastic condition.

Solution The initial stress distribution is uniform, value τ_0, as in Figure 8.14c. The required change of torsion moment is $\Delta T = -\hat{T}$ and so, using (8.26) in (8.27), we obtain (see Figure 8.14d)

$$\Delta\tau_{x\theta} = -\frac{4}{3}\tau_0\left(\frac{r}{r_0}\right)$$

Then the final stress distribution is (Figure 8.14e)

$$\tau_0\left(1 - \frac{4r}{3r_0}\right)$$

Since the shear stress due to torsion of thin-walled tubes of constant wall thickness is the same throughout the cross-section (Section 8.6.2), the yield and fully plastic moments are the same. Using equation (8.21)

$$T_0 = \hat{T} = 2A_s\tau_0 t \qquad (8.28)$$

However statically indeterminate assemblies of these tubes can have higher collapse than yield loads. Referring to Figure 8.5a total collapse requires yielding of both tubes since after only one has yielded the load can be supported by the other as a statically determinate structure. This process of development of a mechanism may be compared with the discussion for direct loading in Section 5.6.3.

Example 8.14 Determine the collapse load of the assembly in Figure 8.5a if the bars are rectangular hollow sections with mean wall dimensions 200×150 mm and 150×100 mm and the same wall thickness of 6 mm. Take $\tau_0 = 100$ MN/m², $a = 2$ m, $b = 1$ m.

Solution Applying equations (8.28) we find for tubes 1 and 2 fully plastic moments of 36 kNm and 18 kNm respectively. Both tubes must yield to develop a mechanism and since m (Figure 8.5b) is shared by the tubes it must have the value $(18+36) = 54$ kNm for collapse to occur. The value of m to initiate yielding can be shown to be only 43·6 kNm. It is interesting to note that the tube lengths were not needed to find the collapse load.

If the moment m were applied at C instead of a clamp the collapse load would be only 18 kNm.

Problems for Solution 8.26, 8.27

8.8 References

1. Timoshenko, S. and Goodier, J. N., loc. cit. Article 1.15.
2. Thom, A. and Apelt, C. J., *Field Computations in Engineering and Physics*, Van Nostrand, 1961.
3. Herrmann, L. R., 'Elastic torsional analysis of irregular shapes', *Proc. A.S.C.E.*, **91**, EM6, 1965.
4. Timoshenko, S., *Strength of Materials*, Vol. II, Van Nostrand, 1956.

8.9 Problems for Solution

(Take $G = 80$ GN/m² unless otherwise indicated.)

8.1. Draw the torsion moment diagram for the rod in Figure 8.1a when (a) the forces on the cross-bar at B are reversed, and (b) when both of the 4 kN forces act up.

8.2. A shaft ABC has operating speeds of 500 r.p.m. and 2000 r.p.m. Power of 100 kW is supplied at each of A and B, which are 1 m apart, and is delivered to load at C where BC = 2 m. Draw the torsion moment diagram for the more onerous condition.

8.3. A hollow shaft has inner and outer diameters of 80 mm and 100 mm respectively, and is 2 m long. A sample of the material yielded in a tensile test at 300 MN/m². Find the maximum power which can be transmitted at (a) 600 r.p.m., (b) 6000 r.p.m., if a safety factor of 2 is necessary. Find the angle of twist in each case.

8.4. Shaft ABC has AB = 1 m, BC = 2 m; power of 200 kW is supplied at B and 100 kW is delivered at each of A and C. Find the least diameter of solid shaft for a running speed of 300 r.p.m. if the allowable shear stress is 100 MN/m². Calculate the relative twist of A and C.

8.5. Shaft ABCD consists of a 1 m length AB and a 0·5 m length BC each of 50 mm diameter solid rod, and a 0·5 m length CD of hollow rod with inner and outer diameters 44 mm and 50 mm. Twisting moments of m, $-0·5m$, and $-0·5m$ are exerted at A, B and D respectively. Find the maximum value for m if $\sigma_0 = 240$ MN/m² and $f_s = 1·5$.

8.6. A line shaft ABC is driven at A and delivers 50 kW at each of stations B and C when running at 600 r.p.m. The shaft consists of a hollow portion AB of outer diameter 50 mm and inner diameter d, and a solid portion BC of diameter d. By considering conditions in the two parts of the shaft find the limits within which d must lie if the allowable shear stress is 100 MN/m². Choose that safe value of d which gives the lightest shaft assembly. Find the angle of twist for this value of d if AB = ½BC = 1 m.

8.7. Find the dimensions of a hollow shaft with outer and inner diameters in the ratio 1 : 0·9 for the duty described in Problem 8.4. Find the fractional saving of mass.

8.8. Repeat Problem 8.4 but with the speed doubled.

8.9. In a system like Figure 8.5a the rod AB is of steel ($G = 80$ GN/m²) and BC is of aluminium alloy ($G = 25$ GN/m²). The lengths of AB and BC are 1 m and 2 m, and the rod has diameter 50 mm. Find the angle of twist and stresses per unit twisting moment at B, the rod being completely constrained at A and C.

8.10. In Figure 8.5b let $m = 45$ kNm, $a = 1$ m, $b = 2$ m. Find the most suitable rod from those with the properties listed in Table 6.3 (Problem 6.25). (*Note* $J = 2I$ for a circular cross-section.) Allowable shear stress is 100 MN/m².

8.11. In Problem 8.9 let m be 6 kNm and the aluminium rod be hollow, inner and outer diameters, 40 mm and 60 mm. Find the greatest stress if the support at C allows a rotation of 0·02 radians there in the same sense as m.

8.12. Find the greatest stress in the system of Example 8.6 if $\psi_1 = 0$, and there are twisting moments of 5 and -10 kNm at nodes 2 and 3 respectively.

8.13. If $m = 2$ kNm find the angle of twist of the shaft described in Problem 8.5.

8.14. Rework Problems 8.9 and 8.11 using the stiffness method.

8.15. Find the diameter of a solid shaft which transmits the same power as that in Example 8.8 and has the same strain energy, ω and length.

8.16. Calculate the strain energy in a solid shaft which transmits the same power as in Example 8.8 and has the same maximum stress, ω and length.

8.17. Calculate the strain energy in the shaft of Problem 8.5 when $m = 2000$ Nm. (*Hint* The energy of the whole is the sum of the energies of the parts.)

8.18. A safety valve of 60 mm diameter which is required to open at a pressure of 1·5 MN/m² gauge is to be loaded by a spring having eight coils of mean diameter 150 mm. Wires of diameter 18, 24 and 32 mm are available in material having maximum permissible shear stress of 120 MN/m². Find the most suitable wire, and the required initial spring compression if the valve is to open correctly.

8.19. Two springs mounted coaxially are compressed between two rigid parallel surfaces. The springs have the same number of coils, but have coil diameters in the ratio 3 : 1. The wires have the same elastic modulus and diameters $2d$ and d. Find the proportion of the total load carried by each spring. (*Hint* Each spring experiences the same axial displacement.)

8.20. A rigid horizontal bar is supported by two springs A and B made from the same wire, and a distance l apart. Spring A has twice as many coils as B and its coil radius is half that of B. A load P is placed on the bar at q from spring A. Find an expression for the slope of the bar.

8.21. A thin-walled conduit has cross-section as in Figure 8.11b. The wall thickness is 4 mm. Find the greatest torsion moment which can be carried if the shear stress is not to exceed 20 MN/m² and calculate the unit twist at maximum load.

8.22. A cantilever structure consists of two lengths of square hollow section AB and BC rigidly connected together at B. Twisting moments of 20 kNm and 10 kNm act at B and C and may have opposite directions. AB has mean wall dimensions 150 × 150 mm and length 2 m. For BC the figures are 100 × 100 mm and 1 m. Both have 9 mm wall thickness. Find the greatest twist between A and C and the greatest stress.

8.23. Calculate the torsional stiffnesses of bars with the cross-sections in Figures 8.13b and c. Allowing for a stress concentration of 2 estimate the greatest torsion moment in each case if the allowable stress is 60 MN/m². (*Hint* Treat the stress concentration factor like a safety factor.)

8.24. Compare stiffness and maximum stress in a tube of circular cross-section with those in a similar tube having a single longitudinal slit of negligible width. The wall thickness is one-fifth of the mean radius of the tube wall.

8.25. Rework Problem 8.22 by matrix methods.

8.26. Determine the fully plastic moment for a rod of inner and outer diameter 0·1 and 0·2 m. Find the final stress distribution when the load is taken to the limit and then half removed. $\tau_0 = 100$ MN/m².

8.27. Find the ratio of collapse to yield load for the assembly of Example 8.7. Take $\tau_0 = 100$ MN/m² and remember that the shape factor is not unity.

Combined Loading Systems

9.1 Bending and Direct Loads

A straight uniform bar as described in Article 5.1 is shown in Figure 9.1 with its ends subject to axial forces X and to moments m_z. The axial forces

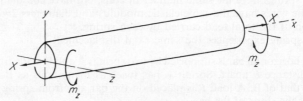

Figure 9.1

are equal and opposite and therefore constitute a direct loading which, acting alone, would cause a uniform state of stress of the form of array (A5.1) (see Article 5.2)

i.e.
$$\begin{pmatrix} \sigma_x & 0 & 0 \\ 0 & 0 & 0 \\ 0 & 0 & 0 \end{pmatrix} \qquad (A9.1)$$

Using equation (5.5) the normal stress due to the direct loading is obtained as

$$\sigma_x = \frac{X}{A} \qquad (9.1)$$

where A is the cross-sectional area of the bar and X is taken as positive if the forces have a tensile action and negative if they are compressive.

The moments m_z cause bending in the xy plane due to a uniform bending moment

$$M_z = m_z \qquad (9.2)$$

The state of stress (see array A6.1, Section 6.3.1) is of the same form as that for direct loading but the value of σ_x varies with position in the cross-section. If the xy plane is a plane of symmetry then we have from equation (6.18)

$$\sigma_x = \frac{M_z y}{I_z} \qquad (9.3)$$

Since the states of stress due to the direct and bending loads both have the form of array (A9.1) for the x, y, z axes, it follows that the state of stress for the combined load system also has this form and the total normal stress σ_x is the sum of the separate contributions given by equations (9.1) and (9.3), i.e.

$$\sigma_x = \frac{X}{A} + \frac{M_z y}{I_z} \qquad (9.4)$$

The greatest stress magnitudes occur at the extreme values of y within the cross-section, and are independent of x when M_z is uniform as in equation (9.2). When transverse forces in equilibrium are present, as in Figure 9.2, critical conditions are at the cross-section subject to greatest bending moment.

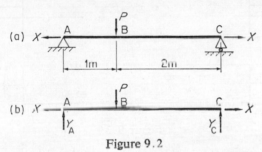

Figure 9.2

Example 9.1 The bar in Figure 9.2a is simply supported at A and C and is subject to a transverse force $P = 3$ kN and direct loads $X = 30$ kN. Find the extremes of normal stress, if the bar is a tube of internal and external diameters 100 mm and 110 mm.

Solution The supports exert reactive forces Y_A and Y_C (Figure 9.2b) which from equilibrium are 2 kN and 1 kN respectively.

The greatest bending moment is -2 kNm at B.

$$I_z = \frac{\pi}{4}(r_0^4 - r_i^4) = \frac{\pi}{4}(55^4 - 50^4) \times 10^{-12} = 228 \times 10^{-8} \text{ m}^4$$

$$A = \pi(r_0^2 - r_i^2) = 1650 \times 10^{-6} \text{ m}^2$$

Substituting in (9.4)

$$\sigma_x = \frac{30 \times 10^3}{1650 \times 10^{-6}} + \frac{-2000y}{228 \times 10^{-8}} = 18 \cdot 2 \times 10^6 - 878y \times 10^6$$

The extreme values of y are $\pm 0 \cdot 055$ m for which the stresses are -30 MN/m^2 and $66 \cdot 4$ MN/m^2.

A system of direct and bending loads also results when non-axial forces act parallel to the longitudinal axis of a bar. In Figure 9.3a the forces X act parallel to the x axis but displaced from it a distance y_e in the xy plane. The previous analysis of bars subject to direct loads required that these should act along the longitudinal axis (Article 5.2). However the forces X can be 'moved' to the axial position as follows. Pairs of equal and opposite axial forces X_1 and X_2 are applied at each end of the bar, as in Figure 9.3b. These forces are balanced and therefore do not disturb the equilibrium of the bar. If the forces X_1 and X_2 are made equal in magnitude to X then at each end the forces X and X_2 constitute a couple Xy_e. The original forces X are thus replaced by axial forces X and end moments Xy_e as shown in Figure 9.3c.

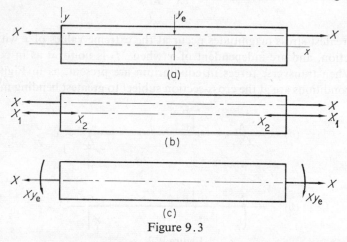

Figure 9.3

The original system has in this way been replaced by combined bending and direct loading as in Figure 9.1. The stresses can be calculated from equation (9.4) with $M_z = Xy_e$.

Close to the ends the actual stress distribution will be different from that given by equation (9.4). However by St. Venant's Principle (see Article 4.6) these differences become negligible at distances from the ends comparable to the transverse dimensions of the bar.

When slender bars are subject to compressive loads a special form of collapse known as buckling may take place. As this phenomenon is dealt with separately in Chapter 11 it will be understood that the examples in this chapter are not subject to this effect.

Example 9.2 A girder with cross-section as in Figure 9.4a is part of a testing frame and is required to withstand tensile forces of 1 MN acting parallel to its longitudinal axis. The forces may become offset y_e as in Figure 9.4b. Determine the greatest allowable offset if the permissible shear stress is 110 MN/m².

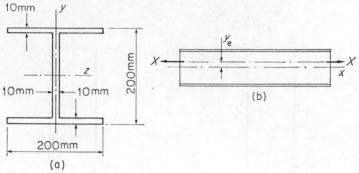

Figure 9.4

Solution The offset forces are replaced by axial tensile forces $X = 10^6$ N and opposed moments $Xy_e = 10^6 \, y_e$ Nm. These end moments exert bending moments

$$M_z = \pm 10^6 \, y_e \text{ Nm}$$

$$I_z = \left(\frac{200 \times 200^3}{12} - \frac{190 \times 180^3}{12} \right) \times 10^{-12} = 0{\cdot}410 \times 10^{-4} \text{ m}^4$$

$$A = ((400 \times 10) + (180 \times 10)) \times 10^{-6} = 5{\cdot}80 \times 10^{-3} \text{ m}^2$$

The extreme values of y are $\pm 0{\cdot}1$ m, and from (9.4) the extremes of normal stress are $172 \times 10^6 \pm 2{\cdot}44 \, y_e \times 10^9$ N/m^2.

Referring to array (A9.1) we have

$$\tau_{\max} = \tfrac{1}{2}\sigma_x \leqslant 110 \times 10^6 \text{ N/m}^2$$

and so

$$172 \times 10^6 + 2{\cdot}44 \, y_e \times 10^9 \leqslant 220 \times 10^6 \qquad \text{(i)}$$

i.e.

$$y_e \leqslant 19{\cdot}7 \times 10^{-3} \text{ m}$$

Example 9.3 A squat column is subject to a uniformly distributed lateral (transverse) load of 2 kN/m and a force of 10 kN parallel to the x axis and offset 0·2 m as in Figure 9·5. The column is a square tube of outside edge 0·2 m and wall thickness 0·01 m. Find the position and magnitude of the greatest normal stress.

Solution Taking the x axis along the column as shown and following the procedure of the previous example the offset force of 10 kN is replaced by an axial force of 10 kN and moment $10 \times 0{\cdot}2 = 2$ kNm. The reactive loads at the support are obtained from static equilibrium as

$$X_A = 10 \text{ kN}, \ Y_A = 4 \text{ kN} \quad \text{and} \quad m_A = 6 \text{ kNm}$$

The direct load is -10 kN (i.e. compressive) and the bending moment at x from A is

$$M_z = 6 - 4x + x^2$$

which has greatest value 6 kNm at A. Here

$$I_z = \frac{(0{\cdot}2^4 - 0{\cdot}18^4)}{12} = 4590 \times 10^{-8} \text{ m}^4$$

and so using (9.4) with $y = \pm 0{\cdot}1$ m the extreme stresses are 11·8 MN/m^2 and $-14{\cdot}4$ MN/m^2. The greatest stress magnitude is therefore at A'.

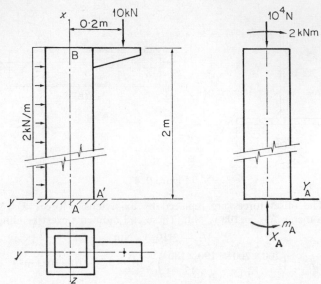

Figure 9.5

When there are bending moments in both xz and xy planes the total stress at any position on the bar is obtained by adding to (9.4) the additional contribution from bending in the xz plane derived in Section 6.3.7 (equation 6.19). Then

$$\sigma_x = \frac{X}{A} + \frac{M_z y}{I_z} + \frac{M_y z}{I_y} \qquad (9.5)$$

Bending moments in the xz plane may be caused by end moments, transverse loads, or by non-axial direct loads. For instance if the loads of 1 MN in Example 9.2 act parallel to the x axis but through the point B of the cross-section (Figure 9.6a) there are offsets y_e in the xy plane, and z_e in the xz plane. The equivalent load system is obtained as in the earlier discussion by adding equal and opposite axial forces as in Figure 9.6b. The couple has components

$$m_z = X y_e \quad \text{and} \quad m_y = X z_e$$

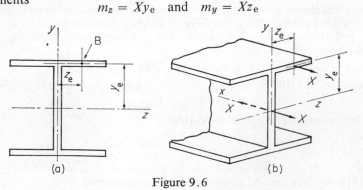

Figure 9.6

Example 9.4 The girder of Example 9.2 may be eccentrically loaded as in Figure 9.6. Determine the limiting value of z_e when $y_e = 10$ mm for the same maximum stress.

Solution
$$I_y = (20 \times 200^3 + 180 \times 10^3)/12 \text{ mm}^4 = 0 \cdot 1335 \times 10^{-4} \text{ m}^4$$

The end moments m_y cause uniform positive bending moment $M_y = 10^6 z_e$ (see Section 6.2.4). Using (9.5) with information from Example 9.2 we find

$$\sigma_x = (172 + (24,400\, y_e)y + (74,800\, z_e)z) \times 10^6 \text{ N/m}^2$$

The extreme values of y and z are $\pm 0 \cdot 1$ m, and $y_e = 0 \cdot 01$ m, so that the extremes of normal stress are $(172 \pm 24 \cdot 4 \pm 7480\, z_e)$ MN/m².

Proceeding as in (i) of Example 9.2 we obtain $z_e \leqslant 3 \cdot 15 \times 10^{-3}$ m.

Problems for Solution 9.1–9.7

9.2 Combined Torsion, Bending and Direct Load

The rod shown in Figure 9.7a is subject to direct load, bending in the xy plane, and also to end twisting moments m_x in equilibrium which impose a uniform torsion moment $T = m_x$. Therefore shear stresses $\tau_{x\theta}$ are developed in accordance with equation (8.5) with greatest magnitude at the periphery where

$$(\tau_{x\theta})_{r_0} = \frac{r_0 T}{J} \tag{9.6}$$

The state of stress has the form of array (A8.3) of Section 8.2.2 relative to polar coordinates x, θ, r. Now the radial and tangential directions r, θ coincide with the y and z directions at points B, C, D, E in the cross-section (Figure 9.7b). Therefore at these points the stress arrays (A8.3) and (A9.1) can be added to find the state of stress due to the combination of torsion with bending and direct load, relative to the polar coordinates, as

$$\begin{pmatrix} \sigma_x & \tau_{x\theta} & 0 \\ \tau_{\theta x} & 0 & 0 \\ 0 & 0 & 0 \end{pmatrix} \tag{A9.2}$$

where σ_x and $\tau_{x\theta}$ are given by equations (9.4) and (9.6).

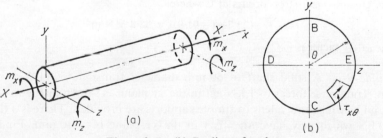

Figure 9.7 (b) Radius of rod r_0

As was noted in Article 9.1 the extremes of normal stress are at the cross-section subject to greatest bending m oment, and occur there at the extreme values of y, w hich in this case are $\pm r_0$. In Figure 9.7b these extremes are at points B and C of the cross-section and so, since $\tau_{x\theta}$ has the same value at all points on the periphery, the states of stress at B and/or C are the most severe. The principal stresses are found by Mohr Circle analysis and the greatest shear stress as half the greatest difference of principal stress.

Example 9.5   A solid shaft of diameter 0·1 m transmits power of 1 MW at 1000 r.p.m. over a length of 1 m. The ends of the shaft are also subject to moments $m_z = 4\,$kNm as in Figure 9.7a, and to an axial thrust (i.e. compressive direct load) of 200 kN. Find the maximum shear stress in the shaft material and the safety factor if $\sigma_0 = 200\,$MN/m^2.

Solution   The angular velocity is $\omega = 2\pi \times 1000/60\,$rad/sec and using equation (8.1) the end twisting moments are $10^6/\omega$, i.e. 9540 Nm; the torsion moment T is thus 9540 Nm. The polar moment of area (see equation 8.7) is $\pi(0{\cdot}05)^4/2\,$m^4 and so from equation (9.6)

$$(\tau_{x\theta})r_0 = 48{\cdot}6\,\text{MN/m}^2$$

The bending moment M_z is uniform, value 4 kNm, the second moment of area is $\pi(0{\cdot}05)^4/4\,$m^4 (equation 6.12), and the extreme values of y are $\pm 0{\cdot}05$ m. Then substituting in equation (9.4)

$$(\sigma_x)_{\perp r_0} = (-25{\cdot}5 \pm 40{\cdot}7)\,\text{MN/m}^2$$

The states of stress at B and C of Figure 9.7b are therefore

$$\begin{pmatrix} 15{\cdot}2 & 48{\cdot}6 & 0 \\ 48{\cdot}6 & 0 & 0 \\ 0 & 0 & 0 \end{pmatrix} \text{ and } \begin{pmatrix} -66{\cdot}2 & 48{\cdot}6 & 0 \\ 48{\cdot}6 & 0 & 0 \\ 0 & 0 & 0 \end{pmatrix} \text{units MN/m}^2$$

By drawing Mohr Circles (see Article 1.10) or by using equations (1.8) the corresponding principal stress arrays are found to be

$$\begin{pmatrix} 56{\cdot}8 & 0 & 0 \\ 0 & -41{\cdot}6 & 0 \\ 0 & 0 & 0 \end{pmatrix} \text{ and } \begin{pmatrix} 25{\cdot}7 & 0 & 0 \\ 0 & -91{\cdot}9 & 0 \\ 0 & 0 & 0 \end{pmatrix} \text{units MN/m}^2$$

The greatest shear stress occurs at C where

$$\tau_{\max} = \tfrac{1}{2}(25{\cdot}7 - (-91{\cdot}9)) = 58{\cdot}8\,\text{MN/m}^2$$

The safety factor is $\sigma_0/2\tau_{\max} = 1{\cdot}70$.

Example 9.6   A solid shaft of 50 mm diameter transmits 500 kW at 3000 r.p.m. Transverse forces of 1 kN act in the xy plane as shown in Figure 9.8a, and end bearings equivalent to simple supports are provided. There is a thrust of 20 kN which may have an offset in the xy plane of $+10$ mm. Find the greatest shear stress in the shaft.

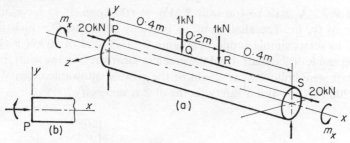

Figure 9.8

Solution   Here $\omega = 2\pi \times 50$ rad/sec and the end twisting moments are $500 \times 10^3/\omega$, i.e. 1590 Nm. Substituting $T = 1590$ Nm in equation (9.6) we have

$$(\tau_{x\theta})_{r_0} = 64 \cdot 7 \text{ MN/m}^2$$

From symmetry the transverse forces exerted by the end bearings are each 1 kN. The compressive forces are replaced by axial forces of 20 kN and moments of 200 Nm directed as in Figure 9.8b. The greatest bending moment is then $-0 \cdot 6$ kNm acting over QR. Hence from equation (9.4) we have

$$(\sigma_x)_{\pm r_0} = (-10 \cdot 2 \pm 49) \text{ MN/m}^2$$

Writing the extreme states of stress and proceeding as in the previous example the maximum shear stress is found to be $71 \cdot 1$ MN/m² at $y = -r_0$.

A formula which is often quoted for the stresses due to combined bending and twisting moments in a shaft will now be derived. The state of stress at the worst positions is of the form of array (A9.2) wherein $\tau_{x\theta}$ is given by (9.6) and σ_x by (9.4) with $X = 0$ and $y = \pm r_0$. The greatest shear stress for axes x, θ, r can be adapted from equation (1.10) and array (A8.1) as

$$\tau_{\max} = \sqrt{\left(\frac{\sigma_x - \sigma_\theta}{2}\right)^2 + \tau_{x\theta}^2}$$

From array (A9.2) we have $\sigma_\theta = 0$, and so using equations (9.4) and (9.6) we obtain

$$\tau_{\max} = \frac{1}{2}\sqrt{\left(\frac{M_z r_0}{I_z}\right)^2 + \left(\frac{2Tr_0}{J}\right)^2}$$

From equations (6.12) and (8.7) we have $J = 2I_z$ for a circular shaft. Hence

$$\tau_{\max} = \frac{r_0}{J}\sqrt{M_z^2 + T^2} \tag{9.7}$$

For a solid shaft $J = \pi r_0^4/2$ and so

$$\tau_{\max} = \frac{2}{\pi r_0^3}\sqrt{M_z^2 + T^2} \tag{9.8}$$

Example 9.7 A shaft to transmit 2 MW at 600 r.p.m. is to have an outer diameter of 0·2 m. The shaft runs in bearings at P and Q which may be assumed to act as simple supports (Figure 9.9). A load of 50 kN acts at the overhung end R of the shaft. Find the greatest shear stress in the shaft material. If the shaft were hollow what would be the greatest allowable internal radius if $\sigma_0 = 200$ MN/m² and a safety factor of 2 is necessary?

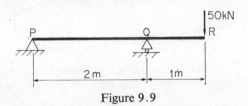

Figure 9.9

Solution The torsion moment is $2 \times 10^6/20\pi$, i.e. 31,800 Nm.

The supports at P and Q exert reactive forces which are found from static equilibrium to be -25 kN and 75 kN. Hence the greatest bending moment is 50 kNm at Q. There are no direct loads, and the shaft is solid. Hence substituting in equation (9.8) we have

$$\tau_{max} = \frac{2}{\pi \times 0·1^3} \sqrt{10^6 \times (50^2 + 31·8^2)} = 37·7 \times 10^6 \text{ N/m}^2$$

A hollow shaft with inner radius r_i and the same outer radius is required to have $\tau_{max} \leqslant \sigma_0/2f_s = 50$ MN/m². Then using equation (9.7) we have

$$\tau_{max} = \frac{r_o}{J} \sqrt{10^6 \times (50^2 + 31·8^2)} \leqslant 50 \times 10^6$$

where

$$J = \frac{\pi(0·1^4 - r_i^4)}{2}$$

Hence

$$r_i \leqslant 0·0705 \text{ m}$$

Problems for Solution 9.8–9.11

9.3 Statically Determinate Assemblies

If an assembly of bars is supported in such a manner that the reactive loads can be determined using only the equations of static equilibrium then it is said to be statically determinate. The example in Figure 9.10a is a simply supported beam constructed by joining together two bars **1** and **2** which have different cross-sections. The support at 1 can exert a reactive force with components X_1, Y_1 in the x and y directions. The roller support at 3 exerts only a reactive force, Y_3. From the three equations of static equilibrium for the xy plane viz.

$$\sum X = 0; \quad \sum Y = 0; \quad \sum m_z = 0 \tag{9.9}$$

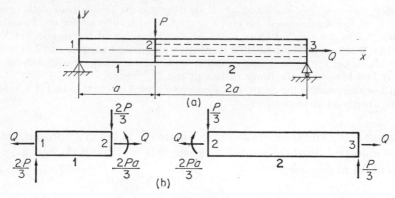

Figure 9.10

we can find $X_1 = -Q$, $Y_1 = 2P/3$, $Y_3 = P/3$. Using standard procedures as in Articles 6.2 and 6.3 we can draw bending moment and shearing force diagrams for the complete assembly, and, knowing the properties of the cross-sections, calculate the stresses. We can also, if we so desire, deduce the effective loading on each component bar of the assembly. For example referring to Figure 9.10b we have previously found that at 3 forces Q and $P/3$ act on bar **2** in the x and y directions. Since there are no intermediate loads this bar can only be maintained in equilibrium by loads exerted at 2. Using equations (9.9) these loads are found to consist of forces Q and $P/3$ and a moment $2Pa/3$ as shown in Figure 9.10b. We then have what is usually called the free body diagram of bar **2**. In a similar manner we can obtain the free body diagram for bar **1**, also shown in the figure.

It should be noted that the resultants of the end loads exerted on the two bars at their junction must be equal to the actual external loads at the junction. In this example the forces Q and moments $2Pa/3$ on the two bars are equal and opposite, corresponding to the absence of external force in the x direction, and of external moment. However there is a resultant force P in the y direction corresponding to the external force applied at 2. This discussion may be compared with that for nodal equilibrium used in developing stiffness-type equations for an assembly of bars (see Sections 5.3.2 and 7.4.5).

Once the free body diagram for a bar has been obtained we can deduce the stresses in the bar using the appropriate methods previously described in Articles 9.1 and 9.2.

Example 9.8 The assembly in Figure 9.10a consists of a solid rod **1** of 0·1 m diameter, and a hollow rod **2** of outer and inner diameters 0·1 m and 0·08 m. The rods have lengths of 1 m and 2 m respectively and the forces P and Q are 9 kN and 80 kN. Determine the maximum tensile stress.

Solution The reactive loads and free body diagrams are found from static equilibrium as previously described. Referring to the free body diagram for rod **1** shown in Figure 9.10b, we see that there is tensile direct loading of magnitude 80 kN; in addition there are negative bending moments with greatest value at end 2 of magnitude $2Pa/3$, i.e. -6 kNm. Using equation (9.4) the maximum tensile stress in this rod is 71·4 MN/m² at the lower surface at end 2.

In a similar manner the greatest tensile stress in rod **2** is found to be 131·6 MN/m² at the lower surface at end 2.

Example 9.9 The assembly shown in Figure 9.11a lies in a horizontal plane when unloaded. Find the greatest shear stress in each bar if the bars are solid rods of 0·1 m diameter.

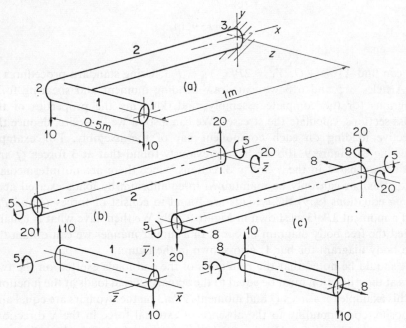

Figure 9.11 Units of force kN; units of moment kNm

Solution The reactive loads at the fixed support are found by using the general equilibrium equations (Appendix I). Thus from $\Sigma Y = 0$ we find $Y_3 = 20$ kN, and from $\Sigma m_z = 0$ and $\Sigma m_x = 0$ we obtain reactive moments of 20 kNm and 5 kNm respectively as shown in Figure 9.11b.

Since in this case the bars are not collinear or parallel it is convenient to introduce the idea of local coordinate axes (see also Section 14.3.1). Thus we take the longitudinal axis of any bar to be $\bar{x}$, and $\bar{y}$ and $\bar{z}$ are axes in the cross-sectional plane of the bar (see Figure 9.11b). Then applying equations (9.9) to the $\bar{x}\bar{y}$ plane of bar **1** we find that the loads to be exerted at end 2 to balance the known force at 1 are an upward force of 10 kN and a moment of 5 kNm in the $\bar{x}\bar{y}$ plane as in Figure 9.11b.

In the case of bar **2** the additional equilibrium equation $\Sigma m_{\bar{x}} = 0$ is required.

Then the loads at 2 to balance the known loads at 3 are as shown in Figure 9.11b. It may be noted that the resultant of the loads at 2 on bars **1** and **2** is a down force of 10 kN corresponding to the actual external load at 2 in Figure 9.11a.

The stresses are calculated from the free body diagrams as described in Example 9.8. Bar **1** is in simple bending (equation 9.4 with $X = 0$), and bar **2** is in combined bending and torsion (see Article 9.2). The stress computations are left as an exercise.

Example 9.10 If an axial outward force of 8 kN is imposed at 1 in Figure 9.11a draw the free body diagrams.

Solution The additional reactive loads at the support are 8 kN in the negative z direction and a moment of 8 kNm in the xz plane; the free body diagrams are as in Figure 9.11c. Bar **1** is in combined bending and direct loading and bar **2** is subject to torsion and bending in two planes.

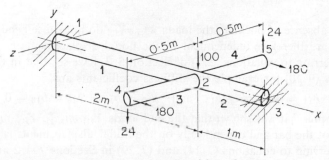

Figure 9.12 Units of force kN

If the reactive loads cannot be determined from static equilibrium as for example in Figure 9.12, the assembly is statically indeterminate. A general method of solving these systems is provided by the stiffness methods previously developed for direct loading, bending and torsion in Articles 5.3, 7.4 and 8.3. In Article 9.4 we consider stiffness methods for combined loading systems.

Problems for Solution 9.12, 9.13

9.4 Stiffness Equations for Combined Loading

9.4.1 *Bending and Direct Loads*

A finite bar element **j** is shown in Figure 9.13 subject to loads $X_{i;j}$, $Y_{i;j}$, $m_{z,i;j}$ at end i and $X_{j;j}$, $Y_{j;j}$, $m_{z,j;j}$ at end j. Corresponding to these are nodal displacements u_i, v_i, θ_i at i and u_j, v_j, θ_j at j, where u, v represent displacements in the x, y directions, and θ is rotation in the xy plane.

The stiffness equations expressing the end loads in terms of the end

9

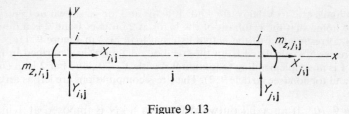

Figure 9.13

displacements will have the form

$$X_{i;j} = b_{11}u_i + b_{12}v_i + b_{13}\theta_i + b_{14}u_j + b_{15}v_j + b_{16}\theta_j$$
$$Y_{i;j} = b_{21}u_i + b_{22}v_i + b_{23}\theta_i + b_{24}u_j + b_{25}v_j + b_{26}\theta_j$$
$$m_{z,i;j} = b_{31}u_i + \ldots \qquad (9.10)$$
$$\cdot \qquad \cdot$$
$$\cdot \qquad \cdot$$
$$\cdot \qquad \cdot$$

In the absence of buckling the loads $X_{i;j}$, $X_{j;j}$ do not cause bending of the bar and are therefore independent of the nodal displacements v_i, θ_i, v_j, θ_j. Then referring to equations (5.7) of Section 5.3.1 we see that in the first of equations (9.10) for example the stiffness coefficients are

$$b_{11} = AE/L = -b_{14} \quad \text{and} \quad b_{12} = b_{13} = b_{15} = b_{16} = 0$$

In Chapter 7 it was shown that the end loads $Y_{i;j}$, $m_{z,i;j}$, $Y_{j;j}$, $m_{z,j;j}$ cause bending of the bar and depend only on the nodal displacements v_i, θ_i, v_j, θ_j. Then referring to equations (7.14) and (7.29) in Sections 7.4.2 and 7.4.3, we see that in the second of equations (9.10) the stiffness coefficients are

$$b_{22} = 12EI_z/L^3 = -b_{25}; \quad b_{23} = 6EI_z/L^2 = b_{26}; \quad b_{21} = b_{24} = 0$$

In a similar manner the other stiffness coefficients can be found, and we now write (9.10) in the matrix form

$$
\begin{bmatrix}
X_{i;j} \\[2mm]
Y_{i;j} \\[2mm]
m_{z,i;j} \\[2mm]
X_{j;j} \\[2mm]
Y_{j;j} \\[2mm]
m_{z,j;j}
\end{bmatrix}
=
\begin{bmatrix}
\dfrac{AE}{L} & 0 & 0 & -\dfrac{AE}{L} & 0 & 0 \\[3mm]
0 & \dfrac{12EI_z}{L^3} & \dfrac{6EI_z}{L^2} & 0 & -\dfrac{12EI_z}{L^3} & \dfrac{6EI_z}{L^2} \\[3mm]
0 & \dfrac{6EI_z}{L^2} & \dfrac{4EI_z}{L} & 0 & -\dfrac{6EI_z}{L^2} & \dfrac{2EI_z}{L} \\[3mm]
-\dfrac{AE}{L} & 0 & 0 & \dfrac{AE}{L} & 0 & 0 \\[3mm]
0 & -\dfrac{12EI_z}{L^3} & -\dfrac{6EI_z}{L^2} & 0 & \dfrac{12EI_z}{L^3} & -\dfrac{6EI_z}{L^2} \\[3mm]
0 & \dfrac{6EI_z}{L^2} & \dfrac{2EI_z}{L} & 0 & -\dfrac{6EI_z}{L^2} & \dfrac{4EI_z}{L}
\end{bmatrix}
\begin{bmatrix}
u_i \\[2mm]
v_i \\[2mm]
\theta_i \\[2mm]
u_j \\[2mm]
v_j \\[2mm]
\theta_j
\end{bmatrix}
\qquad (9.11)
$$

or, in the usual symbolic form
$$\mathbf{F_j} = \mathbf{k_j}\boldsymbol{\delta_j} \tag{9.11a}$$

Example 9.11 Determine the end displacements for the girder of Example 9.2 when $y_e = 10$ mm. Take $L = 2$ m, $E = 210$ GN/m^2, and the ends to be simply supported.

Solution We call the beam 1 and its ends, nodes 1 and 2.

The non-axial loads are equivalent to direct loads of 1 MN and opposed end moments of $(10^6 \times 0.01)$, i.e. 10^4 Nm, positive at 1, negative at 2 (Figure 9.4). Hence
$$\mathbf{F_j} = \{X_{1;1} \quad Y_{1;1} \quad 10^4 \quad 10^6 \quad Y_{2;1} \quad -10^4\}$$

The supports prevent deflexion at the ends, i.e. $v_1 = v_2 = 0$, and to locate the system in the x direction we let $u_1 = 0$ so that
$$\boldsymbol{\delta_j} = \{0 \quad 0 \quad \theta_1 \quad u_2 \quad 0 \quad \theta_2\}$$

The complete equations for the element are obtained by substituting for $\mathbf{F_j}$, $\boldsymbol{\delta_j}$ in (9.11). Rows 1, 2, 5 and columns 1, 2, 5 of $\mathbf{k_j}$ can then be eliminated to obtain the reduced equations

$$
\begin{bmatrix} 10^4 \\ \\ 10^6 \\ \\ -10^4 \end{bmatrix}
=
\begin{bmatrix} \dfrac{4EI_z}{L} & 0 & \dfrac{2EI_z}{L} \\ \\ 0 & \dfrac{AE}{L} & 0 \\ \\ \dfrac{2EI_z}{L} & 0 & \dfrac{4EI_z}{L} \end{bmatrix}
\begin{bmatrix} \theta_1 \\ \\ u_2 \\ \\ \theta_2 \end{bmatrix}
$$

Substituting for E, I_z, A, L and solving we find
$$u_2 = 1.64 \times 10^{-3} \text{ m}; \qquad \theta_1 = -\theta_2 = 1.16 \times 10^{-2} \text{ rad.}$$

The stiffness equations (9.11) treat only loads in the xy plane. If bending moments act in the xz plane additional end loads $Z_{i;j}$, $m_{y,i;j}$, $Z_{j;j}$, $m_{y,j;j}$ are required; the stiffness matrix of the bar is of size 10×10 and is obtained in a similar way (to 9.11) using I_y in equation (7.29).

9.4.2 *Combined Direct, Bending and Torsion Loads*

In Figure 9.14 positive twisting moments $m_{x,i;j}$, $m_{x,j;j}$ have been added to the ends of the finite bar element of Figure 9.13. The end displacements corresponding to these twisting moments are rotations ψ_i, ψ_j in yz planes. Following the same procedure as in Section 9.4.1 stiffness equations can be written expressing each of the eight end loads

$$\mathbf{F_j} = \{X_{i;j} \quad Y_{i;j} \quad m_{x,i;j} \quad m_{z,i;j} \quad X_{j;j} \quad Y_{j;j} \quad m_{x,j;j} \quad m_{z,j;j}\} \tag{9.12}$$

in terms of the eight nodal displacements.

$$\boldsymbol{\delta_j} = \{u_i \quad v_i \quad \psi_i \quad \theta_i \quad u_j \quad v_j \quad \psi_j \quad \theta_j\} \tag{9.13}$$

and stiffness coefficients b_{12}, etc. In Chapter 8 it was shown that in the

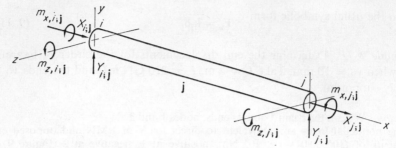

Figure 9.14

particular case of a rod (i.e. bar of circular cross-section) the twisting moments depend only on the rotations ψ_i, ψ_j in yz planes. Then referring to equations (8.12), (8.13) and (9.11) the stiffness matrix of the rod is obtained as

$$
\mathbf{k_j} =
\begin{bmatrix}
\dfrac{AE}{L} & 0 & 0 & 0 & -\dfrac{AE}{L} & 0 & 0 & 0 \\[2mm]
0 & \dfrac{12EI_z}{L^3} & 0 & \dfrac{6EI_z}{L^2} & 0 & -\dfrac{12EI_z}{L^3} & 0 & \dfrac{6EI_z}{L^2} \\[2mm]
0 & 0 & \dfrac{GJ}{L} & 0 & 0 & 0 & -\dfrac{GJ}{L} & 0 \\[2mm]
0 & \dfrac{6EI_z}{L^2} & 0 & \dfrac{4EI_z}{L} & 0 & -\dfrac{6EI_z}{L^2} & 0 & \dfrac{2EI_z}{L} \\[2mm]
-\dfrac{AE}{L} & 0 & 0 & 0 & \dfrac{AE}{L} & 0 & 0 & 0 \\[2mm]
0 & -\dfrac{12EI_z}{L^3} & 0 & -\dfrac{6EI_z}{L^2} & 0 & \dfrac{12EI_z}{L^3} & 0 & -\dfrac{6EI_z}{L^2} \\[2mm]
0 & 0 & -\dfrac{GJ}{L} & 0 & 0 & 0 & \dfrac{GJ}{L} & 0 \\[2mm]
0 & \dfrac{6EI_z}{L^2} & 0 & \dfrac{2EI_z}{L} & 0 & -\dfrac{6EI_z}{L^2} & 0 & \dfrac{4EI_z}{L}
\end{bmatrix}
\qquad (9.14)
$$

If the bar is of non-circular cross-section we could replace J in (9.14) by the modified quantity J' as discussed in Section 8.6.5. The stiffness matrix contains a large population of zeros due to the absence of any coupling of the direct, bending and torsional effects. For example the direct loads have no contribution from the displacements v, θ associated with bending, or from the torsional displacement ψ.

9.4.3 *Assemblies of Elements*

When a collinear assembly of **n** bar elements is formed joining n nodes, stiffness equations may be written relating the nodal loads to the nodal displacements in the general form

$$F = K\delta$$

where

$$F = \{F_1 \quad F_2 \quad F_3 \quad \ldots \quad F_n\}$$

and the F_i are the nodal load vectors. Similarly

$$\delta = \{\delta_1 \quad \delta_2 \quad \delta_3 \quad \ldots \quad \delta_n\}$$

where the δ_i are nodal displacement vectors.

K is a matrix of stiffness coefficients.

The number of components in the vectors F_i, δ_i is equal to the number of degrees of freedom of displacement at the nodes. Denoting this number D_f then the number of components in each of F and δ is

$$\eta = (n \times D_f) \qquad (9.15)$$

For example for combined direct load and bending in the xy plane we saw in Section 9.4.1 that there were three degrees of freedom of nodal displacement. Therefore for an assembly having 6 nodes the matrices F and δ have $\eta = (6 \times 3) = 18$ components arranged in (18×1) column matrices.

The assembly stiffness matrix K is therefore of size $(\eta \times \eta)$ and is formed from the separate stiffness matrices of the n bar elements as has already been described in considerable detail in Chapters 5, 7 and 8. Their components are added into locations in the assembly stiffness matrix according to the following rules. Let p and q be the nodal numbers of the ends of an element j. Then

Row in k_j	Row address in K	Column in k	Column address in K
1	$D_f \times p - (D_f - 1)$	1	$D_f \times p - (D_f - 1)$
2	$D_f \times p - (D_f - 2)$	2	$D_f \times p - (D_f - 2)$
.	.	.	.
.	.	.	.
.	.	.	.
D_f	$D_f \times p$	D_f	$D_f \times p$
$D_f + 1$	$D_f \times q - (D_f - 1)$	$D_f + 1$	$D_f \times q - (D_f - 1)$
$D_f + 2$	$D_f \times q - (D_f - 2)$	$D_f + 2$	$D_f \times q - (D_f - 2)$
.	.	.	.
.	.	.	.
$2D_f$	$D_f \times q$	$2D_f$	$D_f \times q$

$$(9.16)$$

As an example of the use of (9.16) consider an element joining nodes 2 and 3 of an assembly of bars with 3 nodal degrees of freedom (e.g. combined direct and bending loads), i.e. $D_f = 3$. The element stiffness matrix k_j is of size 6×6 since the bar links two nodes.

The component in row 4 column 3 of k_j is allocated as follows, using $p = 2$, $q = 3$. Row 4 is $D_f + 1$. Therefore the row address in **K** is $D_f \times q - (D_f - 1) = 3 \times 3 - (3 - 1) = 7$. Column 3 equals D_f. Therefore column address in **K** is $D_f \times p = 3 \times 2 = 6$.

These equations are applied as discussed in Section 7.4.6. Sufficient nodal displacements must be introduced to prevent rigid body displacement, so giving reduced stiffness equations of the form of (7.49) which can be inverted, as in (7.50), to obtain the unknown displacements. The reactive loads can then be found, free body diagrams drawn, and the stresses calculated as in Articles 9.1–9.3. It should be noted that special methods are necessary with non-parallel bars (see Article 14.3).

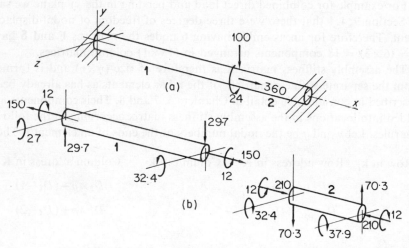

Figure 9.15   Units of force kN; units of moment kNm

Example 9.12   Find the nodal displacements, and the free body diagrams for bars **1** and **2** of the assembly shown in Figure 9.12. The properties *AE*, *EI*, *GJ* for these bars are: Bar **1** 6000 MN, 30, 24, MNm²; Bar **2** 4200 MN, 15, 12 MNm².

Solution   Bars 3 and 4 are statically determinate and by drawing their free body diagrams the effective loads on node 2 of the collinear assembly 1, 2, 3 are found to be as in Figure 9.15a.

After substituting the properties of bars 1 and 2 in (9.14) and then applying the rules (9.16) the assembly stiffness matrix is found. However all displacements at nodes 1 and 3 are zero and so we obtain finally reduced stiffness equations

$$10^3 \begin{bmatrix} 360 \\ -100 \\ 24 \\ 0 \end{bmatrix} = 10^6 \begin{bmatrix} 7200 & 0 & 0 & 0 \\ 0 & 225 & 0 & 45 \\ 0 & 0 & 24 & 0 \\ 0 & 45 & 0 & 120 \end{bmatrix} \begin{bmatrix} u_2 \\ v_2 \\ \psi_2 \\ \theta_2 \end{bmatrix}$$

Solving we find

$$u_2 = 5 \times 10^{-5} \text{ m}; \, v_2 = -0 \cdot 48 \times 10^{-3} \text{ m}; \, \psi_2 = 10^{-3} \text{ rad}; \, \theta_2 = 0 \cdot 18 \times 10^{-3} \text{ rad}.$$

Then using the element stiffness equations element end loads are found giving free body diagrams as shown in Figure 9.15b.

Problems for Solution 9.14, 9.15

9.5 Problems for Solution

9.1. A squat tubular column of length 3 m and inside and outside diameters 0·16 m and 0·2 m is subject to an axial load of 100 kN, and a parallel load of 50 kN offset 0·2 m (compare Figure 9.5). Find the stress distribution in any cross-section of the tube. If the axial load were removed find the reduction in the offset load which would be necessary for the same maximum tensile stress in the tube.

9.2. A sample of sheet steel 50 mm wide, 0·5 m long and 1 mm thick is held in the grips of a tensile testing machine. The axis of the specimen is parallel to that of the testing machine but may be offset. Investigate the effect of such misalignment on the stress measurements.

9.3. A bar of 4 m length has box-type cross-section (cf. Figure 6.14a) with outside breadth and depth 0·1 m and 0·2 m and wall thickness 10 mm. The bar is loaded as in Figure 6.7a, and may also be subject to an axial force of P of 200 kN. Using a safety factor of 1·5 find the maximum allowable value of w with and without the axial force. $\sigma_0 = 240 \text{ MN/m}^2$.

9.4. The offset load in Problem 9.1 is replaced by a distributed load of uniformly varying intensity with maximum value w_0 at the base and zero value at the top of the column. Find the greatest w_0 for the same maximum tensile stress as in Problem 9.1.

9.5. The bar of cross-section described in Problem 9.3 is subject only to offset tensile forces of 200 kN. If the offsets in each plane are always equal find the greatest offset if the normal stress is not to exceed $\pm 160 \text{ MN/m}^2$.

9.6. Repeat Problem 9.3 if a distributed load of intensity w also acts on the bar in the xz plane.

9.7. A squat column (cf. Figure 9.5) has solid rectangular cross-section with sides 30 mm and 60 mm, and is subject to a vertical compressive force P which may be offset. Delineate the region of the cross-section within which the load must act if no tensile stresses are to develop in the material.

9.8. The shaft of Example 9.5 may be subject to overload due to power surges. Find the greatest value which can be tolerated within a safety factor of 1.25.

9.9. Repeat Example 9.6 but with the load of 1 kN at R reversed in direction.

9.10. An overhung shaft supported as in Figure 9.9 is subject also to an anti-clockwise moment $m_z = 80$ kNm at P. Find the minimum diameter of solid shaft for this duty if $\sigma_0 = 200$ MN/m² and $f_s = 2$.

9.11. A 25 mm diameter shaft transmits a torque of 250 Nm and a thrust of 25 kN which may be offset. Find the greatest allowable offset if the maximum shear stress is not to exceed 100 MN/m².

9.12. Draw free body diagrams for the rods in Figure 8.1a.

9.13. Draw the free body diagrams for the system of Figure 9.11a with a clockwise twisting moment of 12 kNm (looking along 2, 1) added at 1 and a force of 10 kN in the negative z direction added at 2.

9.14. Repeat Example 9.12 if the force of 24 kN at node 5 is reversed.

9.15. Obtain free body diagrams for the system of Example 9.8 (Figure 9.10a), but with a fixed support at node 1. Take $E = 200$ GN/m².

Axisymmetric Loading
(Cylinders and Disks)

10.1 Thin Cylinders Under Internal Pressure

10.1.1 *Stresses in Thin Cylinders*

In this chapter we consider the behaviour of bodies of cylindrical form subject to loading which is symmetric with respect to the longitudinal axis. Such loading may be due to fluid pressure or the effects of rotation. We begin by studying the simple case of a cylindrical vessel with comparatively thin walls and containing fluid under uniform pressure p (Figure 10.1). As a working rule we shall consider the wall to be 'thin' when its thickness t is

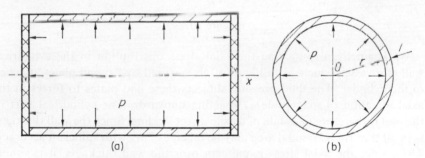

Figure 10.1   Sectional views in planes (a) containing, (b) normal
to, the longitudinal axis

less than 10 % of the internal radius r of the cylinder. It can be shown that under these conditions certain simplifying assumptions about the stress distribution can reasonably be made.

Choosing a polar coordinate system x, θ, r as in Figure 10.1, the state of stress at a point is expressed in general by the stress components shown in Figure 8.3. However, because of the symmetry of the fluid pressure loading the coordinate directions at any point can be considered to be the principal axes for stress and so the state of stress has the form shown below.

$$\begin{pmatrix} \sigma_x & 0 & 0 \\ 0 & \sigma_\theta & 0 \\ 0 & 0 & \sigma_r \end{pmatrix} \tag{A10.1}$$

The normal stress components σ_x, σ_θ, σ_r are known respectively as the axial, hoop and radial stresses and are the principal stresses at the point.

For a thin-walled cylinder it is now assumed that

(a) The state of stress is uniform over the wall thickness.
(b) The radial stress component σ_r is negligible.

These assumptions allow the cylinder to be treated as if it were a membrane, and make it possible to find the stresses in the wall (see Figure 10.2a) by considering only equilibrium of forces.

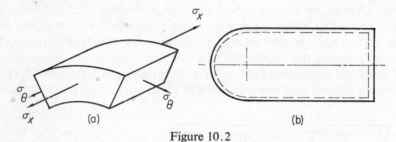

Figure 10.2

We begin by investigating the axial stress distribution in the cylindrical wall of a vessel, which, as in Figure 10.1, is closed by flat end plates attached to the cylinder. The fluid pressure subjects these end plates to forces in the axial direction of magnitude $\pi r^2 p$ acting outwards. The cylindrical part of the wall is therefore in a state of tensile direct loading. Since the wall thickness is small the cross-sectional area may be taken as $2\pi rt$ with sufficient accuracy. Then, since the axial stress is uniform over the wall thickness, it is simply the ratio of these quantities

i.e. $$\sigma_x = \frac{pr}{2t} \tag{10.1}$$

Although flat ends were specified above the same result would be obtained for any other shape of end closure since the resultant axial force $p\pi r^2$ is not affected. This could be proved mathematically but is easily justified as follows by considering a system such as in Figure 10.2b. It is clear that there can be no difference between the axial forces on the ends or the cylinder could not be in axial equilibrium under pressure.

Let us now consider the forces exerted on the section of the cylindrical wall shown in Figure 10.1a. This section has total area $2Lt$ where L is the internal length of the cylinder, and the normal stress component acting on these surfaces is the hoop stress σ_θ (see Figure 10.2a). The forces on these surfaces are found by observing that the plane of the section in Figure 10.1a divides the cylinder into two semicircular shells (Figure 10.3a). The total force on say the upper shell in the direction n normal to the section plane is

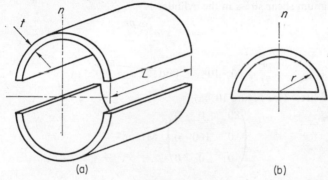

(a)   (b)

Figure 10.3

$p2rL$. (This result can be established by similar reasoning to that used in connexion with the axial forces on the ends by considering for example a vessel of semicylindrical form with cross-section as in Figure 10.3b.) Since there is clearly an equal and opposite force on the lower shell of Figure 10.3a it follows that the material in the section plane is subject to normal tensile loading of magnitude $p2rL$. As the normal stress σ_θ is uniformly distributed we obtain

$$\sigma_\theta = \frac{p2rL}{2Lt} = \frac{pr}{t} \qquad (10.2)$$

Thus both σ_θ and σ_x are tensile for a cylinder with closed ends, and σ_θ is the greatest normal stress. In array (A10.1) σ_r is zero and so the maximum shear stress is

$$\tau_{\max} = \tfrac{1}{2}\sigma_\theta$$

According to the maximum shear stress theory (see Article 1.12) yielding will therefore occur when

$$\sigma_\theta = \sigma_0$$

It should be noted that the expressions (10.1) and (10.2) for axial and hoop stress are valid only at some distance from the ends of the cylinder, as will be explained in Section 10.1.3. The pressure p is of course the gauge pressure, i.e. the internal pressure measured relative to the external pressure (usually atmospheric). Because of hydrostatic effects the pressure cannot strictly be uniform. However, this effect is negligible for gases and may also be ignored for liquids except when the pressure is mainly or entirely hydrostatic as in the case of storage tanks.

Example 10.1 A steel cylinder of 2 m internal diameter is to retain fluid under pressure of 4 MN/m². Find the required wall thickness if $\sigma_0 = 240$ MN/m² and a safety factor of 1·5 is necessary.

Solution The allowable shear stress is $\sigma_0/2f_s = 80$ MN/m².
The maximum shear stress in the cylinder is

$$\tau_{max} = \tfrac{1}{2}\sigma_\theta = \frac{pr}{2t}$$

Hence

$$\frac{pr}{2t} \leqslant 80 \times 10^6 \quad \text{and} \quad t \geqslant 0.025 \text{ m}$$

The state of stress (see Figure 10.4a) is

$$\begin{pmatrix} 80 & 0 & 0 \\ 0 & 160 & 0 \\ 0 & 0 & 0 \end{pmatrix} \text{MN/m}^2$$

Example 10.2 A closed tube of 4 mm wall thickness and 200 mm internal diameter transmits a maximum torsion moment of 8 kNm and contains fluid under a pressure of 4 MN/m². Find the maximum shear stress at the inner surface of the cylindrical wall.

Solution The torsion moment causes a shear stress $\tau_{x\theta}$ (Figure 10.4b) found by using equation (8.5). Hence

$$\tau_{x\theta} = \frac{Tr}{J} = \frac{1}{J}(8 \times 10^3 \times 0.1)$$

Here

$$J = \frac{\pi}{2}(0.104^4 - 0.1^4) = 0.26 \times 10^{-4} \text{ m}^4$$

and so

$$\tau_{x\theta} = 30.8 \text{ MN/m}^2$$

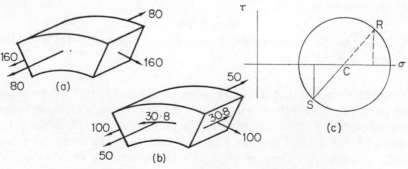

Figure 10.4   Units of stress MN/m²

The fluid pressure causes axial and hoop stresses found by substituting in equations (10.1) and (10.2). The state of stress (see Figure 10.4b) is

$$\begin{pmatrix} 50 & 30 \cdot 8 & 0 \\ 30 \cdot 8 & 100 & 0 \\ 0 & 0 & 0 \end{pmatrix} \text{MN/m}^2$$

By drawing a Mohr Circle (see Figure 10.4c) we find the principal stress array as

$$\begin{pmatrix} 35 \cdot 3 & 0 & 0 \\ 0 & 114 \cdot 7 & 0 \\ 0 & 0 & 0 \end{pmatrix} \text{MN/m}^2$$

and the greatest shear stress is $114 \cdot 7/2 = 57 \cdot 4$ MN/m^2.

Certain other end conditions are possible for a cylinder. For example if pressure is generated by a piston (Figure 10.5a) the axial forces on the ends due to the fluid pressure are resisted by external forces and the axial stress in the cylindrical wall is zero. Alternatively end plates with sealing rings might be held on by external forces P as in Figure 10.5b. Clearly $P \geqslant p\pi r^2$ and in practice there must be net forces $(P - \pi p r^2)$ causing axial compression of the cylinder wall with axial stress

$$\sigma_x = -\frac{(P - p\pi r^2)}{2\pi r t} \tag{10.3}$$

Since this axial stress is always negative while the hoop stress is always positive for internal pressure it follows from the stress array (A10.1) that the greatest difference of principal stress is now $(\sigma_\theta - \sigma_x)$ and the greatest shear stress is

$$\tau_{max} = \tfrac{1}{2}(\sigma_\theta - \sigma_x)$$

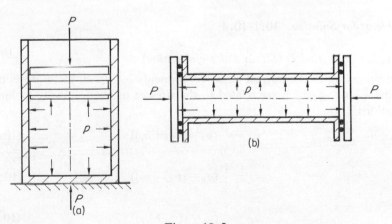

Figure 10.5

Example 10.3 A cylinder of 0·5 m diameter is to be sealed by end plates retained by external forces exerted by a testing machine (Figure 10.5b). These forces are to be maintained at 1·5 times the outward fluid pressure forces on the ends. Find the minimum wall thickness for a pressure of 2 MN/m² if the allowable shear stress is 40 MN/m².

Solution Since the external forces P are maintained equal to $1 \cdot 5p\pi r^2$ we have, substituting in (10.3)

$$\sigma_x = -\frac{pr}{4t}$$

Taking the hoop stress from (10.2) we obtain

$$\tau_{\max} = \tfrac{1}{2}(\sigma_\theta - \sigma_x) = \tfrac{1}{2}\left(\frac{pr}{t} + \frac{pr}{4t}\right) = \frac{5pr}{8t}$$

The shear stress must not exceed 40 MN/m² and so we find

$$t \geqslant 7 \cdot 81 \text{ mm}$$

In the above discussion only internal pressure has been considered. Equations (10.1) and (10.2) are also valid for external pressure if $-p$ is substituted for p. However, care is necessary in the use of these formulae since the cylindrical wall may buckle under this compressive loading. It can be shown (see for example reference 1) that the critical external pressure to cause this form of collapse is

$$p_{\text{crit}} = \frac{Et^3}{4(1 - \nu^2)\,r^3}$$

For example if the cylinder discussed in Example 10.2 were subject to external pressure the critical value (taking $E = 200 \text{ GN/m}^2$, $\nu = 0 \cdot 3$) is found to be 3·52 MN/m².

Problems for Solution 10.1–10.4

10.1.2 *Thin Cylinders* (*Strain and Deformation*)

Corresponding to the normal stress components σ_x, σ_θ, σ_r there are normal strains ϵ_x, ϵ_θ, ϵ_r which are related to the stresses by the following (compare equations 3.1)

$$\epsilon_x = \frac{1}{E}\left(\sigma_x - \nu(\sigma_\theta + \sigma_r)\right) \tag{10.4a}$$

$$\epsilon_\theta = \frac{1}{E}\left(\sigma_\theta - \nu(\sigma_x + \sigma_r)\right) \tag{10.4b}$$

$$\epsilon_r = \frac{1}{E}\left(\sigma_r - \nu(\sigma_x + \sigma_\theta)\right) \tag{10.4c}$$

Using these equations the normal strains can be deduced for any of the states of stress previously derived. It is also of interest to know the change of the enclosed volume of the cylinder. Let Δr and ΔL be the increases of internal radius r and axial length L of a cylinder due to internal fluid pressure p. The change of internal volume is

$$\Delta V = \pi(r + \Delta r)^2(L + \Delta L) - \pi r^2 L$$

Neglecting products of small quantities and dividing by the original volume, we obtain the fractional volume change as

$$\frac{\Delta V}{V} = \frac{2\Delta r}{r} + \frac{\Delta L}{L}$$

The second term on the right-hand side may at once be recognized as the axial strain ϵ_x. To interpret the first term we note that at zero pressure the internal circumference is $2\pi r$ and at pressure p it is $2\pi(r + \Delta r)$. Therefore the strain of this circumference, or hoop strain as it is called, is

$$\epsilon_\theta = \frac{2\pi(r + \Delta r) - 2\pi r}{2\pi r} = \frac{\Delta r}{r} \qquad (10.5)$$

This hoop strain, which equals the fractional change of radius, is not to be confused with radial strain ϵ_r which refers to changes in wall thickness (usually negligible). Then

$$\frac{\Delta V}{V} = 2\epsilon_\theta + \epsilon_x \qquad (10.6)$$

Example 10.4 A cylinder of 0.4 m diameter and 2 m length, internally, has 4 mm wall thickness and retains a pressure of 1 MN/m². Find the change of volume if $E = 200$ GN/m², $\nu = 0.3$.

Solution Using equations (10.1) and (10.2) we find

$$\sigma_x = 25 \text{ MN/m}^2; \qquad \sigma_\theta = 50 \text{ MN/m}^2$$

Substituting these quantities in (10.4a and b) we have, noting that $\sigma_r = 0$,

$$\epsilon_x = 0.05 \times 10^{-3} \qquad \text{and} \qquad \epsilon_\theta = 0.2125 \times 10^{-3}$$

From (10.6)

$$\Delta V/V = 0.475 \times 10^{-3}$$

and so

$$\Delta V = 0.475 \times 10^{-3} \times \pi \times 0.2^2 \times 2 = 0.1192 \times 10^{-3} \text{ m}^3$$

It should be noted that to raise the pressure from atmospheric to the working pressure (at constant temperature) it will be necessary to introduce more fluid. The amount required is partly due to the increased cylinder volume at the higher pressure and partly to compression of the fluid. For a liquid we can write (compare equation 3.6)

$$p = K_f \frac{\Delta V_f}{V_f} \tag{10.7}$$

where K_f is the bulk modulus of the liquid and V_f is its volume at atmospheric pressure. The volume of liquid to be supplied is thus $(\Delta V + \Delta V_f)$. If the fluid is a gas or vapour the mass to be introduced will be evaluated using the appropriate thermodynamic equation of state.

Example 10.5   An air receiver in the form of a thin-walled cylinder has to have internal diameter of 0·5 m and length 4 m. The working pressure is to be 2 MN/m² and axial displacement will be prevented. Find the minimum wall thickness required if $\sigma_0 = 160$ MN/m², $E = 200$ GN/m², $\nu = 0·3$ and a safety factor of 2 is advisable.

Solution   The axial strain $\epsilon_x = 0$. Substituting in equation (10.4a) and recalling that $\sigma_r = 0$ for a thin-walled vessel, we find

$$\sigma_x = \nu\sigma_\theta = 0·3\sigma_\theta$$

From (10.2) $\sigma_\theta = pr/t = 2 \times 10^6 \times 0·25/t$ N/m² and since $\sigma_\theta > \sigma_x > 0$ we have $\tau_{max} = \frac{1}{2}\sigma_\theta$. But

$$\tau_{max} \leqslant \frac{\sigma_0}{2_{/s}} = 40 \times 10^6 \text{ N/m}^2$$

and so

$$t \geqslant 6·25 \text{ mm}$$

Problems for Solution   10.5–10.8

10.1.3 *Thin Spherical Vessel*

Although strictly outside the pronounced scope of this chapter, it is useful to consider the behaviour of a thin-walled spherical vessel under uniform internal pressure both for its intrinsic interest and because it can help in the understanding of conditions near the ends of a thin cylinder. Following the membrane assumptions listed in Section 10.1.1 we take the normal stress in the radial direction to be zero and the two tangential, or hoop, stress components (Figure 10.6a) to be uniform over the wall thickness. From the symmetry of the system the hoop stresses are clearly equal. Denoting these as σ the state of stress is

$$\begin{pmatrix} \sigma & 0 & 0 \\ 0 & \sigma & 0 \\ 0 & 0 & 0 \end{pmatrix} \tag{A10.2}$$

and the maximum shear stress is $0·5\sigma$.

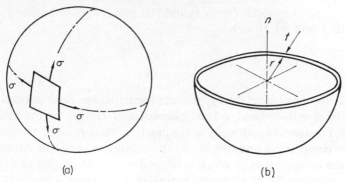

Figure 10.6

To find the stress σ we first note that the cross-sectional area of the vessel in a diametral plane (Figure 10.6b) may be taken as $2\pi rt$ where r is the internal radius and t is the wall thickness. The normal force exerted on this area by the upper half of the sphere is $p\pi r^2$ and so the hoop stress is

$$\sigma = \frac{pr}{2t} \tag{10.8}$$

Thus we see that the hoop stress in a thin spherical vessel is half that in a thin cylinder for the same values of p, r and t.

We can find the hoop strains in the cylinder wall using a stress/strain equation of the usual form. Taking values from (A10.2) we have

$$\epsilon = \frac{1}{E}(\sigma - \nu(\sigma + 0)) = \frac{\sigma(1 - \nu)}{E} \tag{10.9}$$

The enclosed volume strain of the vessel is $4\pi r^3/3$ and so (using 10.5 and 10.9)

$$\frac{\Delta V}{V} = \frac{3\Delta r}{r} = 3\epsilon = \frac{3\sigma(1 - \nu)}{E}$$

In a thin cylinder with hemispherical end closures (e.g. the left-hand end in Figure 10.2b) the hoop stress in the end closure (equation 10.8) is half that in the cylinder wall (equation 10.2) if both have the same thickness. We can now evaluate the radial expansions of the end closure and of the cylinder in the plane of their junction using equation (10.5).

For the hemisphere the hoop strain is given by equation (10.9) and substituting for σ from equation (10.8) we have

$$(\Delta r)_{\text{end}} = \frac{pr^2(1 - \nu)}{2Et}$$

The corresponding result for the cylindrical wall is obtained using equations (10.4b, 10.1 and 10.2) with $\sigma_r = 0$.

$$(\Delta r)_{\text{cyl}} = \frac{pr^2(2 - \nu)}{2Et}$$

Clearly the membrane theory predicts incompatible deformations at the junction of a cylinder and a hemispherical end closure of the same wall thickness. To maintain continuity at the junction, there must be some bending of the two components resulting in additional stresses known as discontinuity stresses, the computation of which is beyond the scope of this text (see for example reference 2). It is worth noting that to make $(\Delta r)_{\text{end}} = (\Delta r)_{\text{cyl}}$ the thickness of the end closure would be $(1 - \nu)/(2 - \nu)$ of that for the cylinder. However, since ν for steel is $0 \cdot 3$, this ratio would be only $0 \cdot 412$ and therefore the membrane stress in the end given by equation (10.8) would exceed that given by (10.2) for the cylinder.

Discontinuity stresses also occur with other forms of end closure for a cylinder; for a treatment of these shell theory must be studied (see for example references 2–4).

Problem for Solution   10.9

10.2   Thick-walled Cylinders

10.2.1   *Stresses*

We now drop the requirement that the wall thickness of a cylinder be small in relation to its internal radius, and consider the general case of a cylinder of inner and outer radii R_1 and R_2 (Figures 10.7a and b) subject to uniform internal and external pressures p_1 and p_2. We again use polar coordinates x, θ, r (where x is the longitudinal axis) for which the general state of stress at a point would be as illustrated in Figure 8.3. However, because of the symmetry of both the loading and the geometrical form it can be taken that the coordinate directions are the directions of the principal axes for stress so that the state of stress is as in array (A10.1). The principal stresses σ_x, σ_θ, σ_r are (as in Section 10.1.1) termed the axial, hoop and radial stress components respectively.

The stresses acting on an infinitesimal element at radius r are as shown in Figure 10.7c. It will be seen that the radial stress at the outer surface, radius $(r + dr)$, differs from that at the inner surface by an amount $d\sigma_r$. This increment of radial stress is necessary because, as shown in Figure 10.7d, the hoop stresses are not normal to the radial direction at the centre of the element and so exert a net radial force on the element.

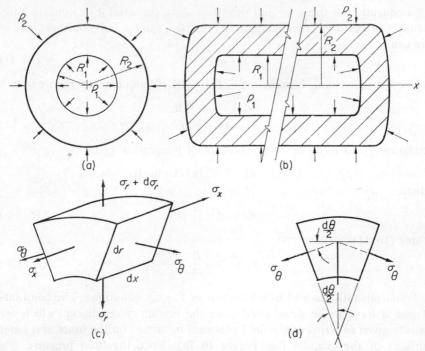

Figure 10.7   Sectional views in planes (a) normal to (b) containing, the longitudinal axis

We begin by writing the equation for equilibrium of forces on the element in the radial direction. The outward radial force on the element is

$$(\sigma_r + d\sigma_r)\,((r + dr)\,d\theta\,dx) \tag{i}$$

The inward radial force is

$$\sigma_r(r\,d\theta\,dx) + 2(\sigma_\theta\,dr\,dx)\sin\tfrac{1}{2}d\theta \tag{ii}$$

where the first term is due to σ_r and the second term is the sum of the radial components of the forces ($\sigma_\theta\,dr\,dx$). Since $d\theta$ is infinitesimal $\sin\tfrac{1}{2}d\theta = \tfrac{1}{2}d\theta$. Therefore, equating (i) and (ii) for radial equilibrium of the element, dividing throughout by $d\theta\,dx$, expanding, and neglecting products of infinitesimals, we obtain

$$r\,d\sigma_r + \sigma_r\,dr = \sigma_\theta\,dr$$

i.e.

$$\frac{d\sigma_r}{dr} + \frac{\sigma_r - \sigma_\theta}{r} = 0 \tag{10.10}$$

A second relation between σ_r and σ_θ is found by assuming that away from the ends of the cylinder original plane cross-sections remain plane under load.

Consequently the strain ϵ_x and the stress σ_x in the axial direction are independent of r and θ. Then in equation (10.4a) ϵ_x and σ_x are constants and so we can write

$$\sigma_r + \sigma_\theta = 2A \qquad (10.11)$$

where A is a constant. Eliminating σ_θ from (10.10) and rearranging we obtain

$$\frac{\mathrm{d}\sigma_r}{\sigma_r - A} = -\frac{2\mathrm{d}r}{r}$$

Integrating and using ln (B) as constant of integration gives

$$\ln(\sigma_r - A) = -2\ln(r) + \ln(B)$$

Hence

$$\sigma_r = A + \frac{B}{r^2} \qquad (10.12a)$$

Using (10.11) we have

$$\sigma_\theta = A - \frac{B}{r^2} \qquad (10.12b)$$

Equations (10.12a and b) are known as Lamé's equations. The constants A and B have to be determined from the boundary conditions which are usually given in terms of the fluid pressures p_1 and p_2 on the inner and outer surfaces of the cylinder (see Figure 10.7a). Since the fluid pressure acts normal to the surface there is a compressive radial stress of this magnitude in the material at the cylinder surface.

i.e.   at $r = R_1$ $\sigma_r = -p_1$;   at $r = R_2$ $\sigma_r = -p_2$ (10.13)

Substituting these conditions in (10.12a) we obtain

$$(\sigma_r)_{R_1} = -p_1 = A + \frac{B}{R_1^2} \quad \text{and} \quad (\sigma_r)_{R_2} = -p_2 = A + \frac{B}{R_2^2} \quad (10.14)$$

These equations can be solved for the constants A and B which can then be substituted in equations (10.12) to obtain explicit expressions for the radial and hoop stresses (see Section 10.2.3). In the examples given later in this section we shall solve for A and B in numerical terms.

Since, as previously stated, the axial stress σ_x is uniform over the cross-section it can be determined by considering only the equilibrium of axial forces on the cylinder as in the case of a thin-walled vessel. Usually the end closures are directly attached to the cylindrical wall, as in Figure 10.7b, and there are then net outward forces $(p_1 \pi R_1^2 - p_2 \pi R_2^2)$ which cause axial tension distributed uniformly over the cross-sectional area $\pi(R_2^2 - R_1^2)$ of the cylinder.

i.e. $\sigma_x = \dfrac{(p_1 R_1^2 - p_2 R_2^2)}{(R_2^2 - R_1^2)}$

Substituting for p_1 and p_2 from equations (10.14) and simplifying we find

$$\sigma_x = A \qquad (10.15)$$

Thus once the constants A and B in the Lamé equations are known the state of stress at any point in the cylinder can be determined. If internal pressure is retained by end sealing plates as in Figure 10.5b net inward forces $(P - \pi p_1 R_1^2)$ cause axial compression of the wall and, with p_2 zero

$$\sigma_x = -\frac{P - \pi p_1 R_1^2}{\pi(R_2^2 - R_1^2)} \qquad (10.16)$$

Example 10.6 A closed cylinder with internal and external diameters 0·1 m and 0·2 m respectively is subject to an internal pressure of 120 MN/m². Find the states of stress at the inner and outer surfaces.

Solution Here $p_1 = 120$ MN/m² so that at the inner surface

$$(\sigma_r)_{R_1} = -120 \times 10^6 = A + \frac{B}{0.05^2} \qquad (i)$$

Similarly, since $p_2 = 0$, we have

$$(\sigma_r)_{R_2} = 0 = A + \frac{B}{0.1^2} \qquad (ii)$$

Solving (i) and (ii) $A = 40 \times 10^6$ N/m² and $B = -0.4 \times 10^6$ N. Substituting these in (10.12b) and putting $r = R_1 = 0.05$ m and $r = R_2 = 0.1$ m we find $(\sigma_\theta)_{R_1} = 200$ MN/m² and $(\sigma_\theta)_{R_2} = 80$ MN/m². From (10.15) $\sigma_x = 40$ MN/m². Then the states of stress at the inner and outer surfaces are respectively (units MN/m²)

$$\begin{pmatrix} 40 & 0 & 0 \\ 0 & 200 & 0 \\ 0 & 0 & -120 \end{pmatrix} \quad \text{and} \quad \begin{pmatrix} 40 & 0 & 0 \\ 0 & 80 & 0 \\ 0 & 0 & 0 \end{pmatrix}$$

The distribution of hoop and radial stress over the wall thickness is illustrated in Figure 10.8 where we see that the greatest stress magnitudes are at the inner surface.

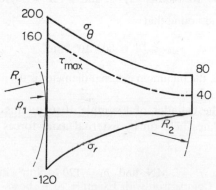

Figure 10.8 Units of stress MN/m²

For both states of stress shown in Example 10.6 the axial stress is (in the algebraic sense) intermediate between σ_θ and σ_r, and reference to equations (10.12) and (10.15) shows that this is a general result for attached ends. Therefore at radius r the greatest shear stress is, using (1.12) and (10.12),

$$(\tau_{max})_r = \tfrac{1}{2}\,|\sigma_r - \sigma_\theta| = \left|\frac{B}{r^2}\right| \qquad (10.17)$$

from which we see that $(\tau_{max})_r$ is greatest at the smallest radius

i.e. $$\tau_{max} = \left|\frac{B}{R_1^2}\right| \qquad (10.18)$$

The distribution of $(\tau_{max})_r$ for Example 10.6 is shown in Figure 10.8. Note that if the ends are held on by external forces the axial stress is not necessarily the intermediate principal stress.

Example 10.7 If the material of the cylinder of Example 10.6 has yield stress of 400 MN/m² find the safety factor.

Solution Critical conditions are at the inner surface and so using the stress array in Example 10.6 $\tau_{max} = \tfrac{1}{2}|200-(-120)|$ MN/m². Then

$$f_s = \frac{\sigma_0}{2\tau_{max}} = 1 \cdot 25$$

Example 10.8 A closed cylinder is to have an internal diameter of 0·1 m and retain a pressure of 75 MN/m². Find the minimum external diameter required if the material has $\sigma_0 = 240$ MN/m² and a safety factor of 1·5 is necessary.

Solution The inner radius is $R_1 = 0 \cdot 05$ m; let the outer radius be R_2. Substituting $p_1 = 75$ MN/m², $p_2 = 0$ in (10.14) we obtain

$$A = 75 \times 10^6/(400R_2^2 - 1) \qquad \text{and} \qquad B = -AR_2^2$$

Then using (10.18) and noting that

$$\tau_{max} \leqslant \frac{\sigma_0}{2f_s} = 80 \times 10^6 \text{ N/m}^2$$

we find $R_2 \geqslant 0 \cdot 2$ m, i.e. the minimum external diameter is 0·4 m.

Example 10.9 If the cylinder of Example 10.6 retains the same pressure using end sealing plates held on by external axial forces of 2 MN find the axial stress.

Solution Substituting $P = 2$ MN and $p_1 = 120$ MN/m² in (10.16) we have $\sigma_x = -44 \cdot 9$ MN/m². Comparison with Example 10.6 shows that σ_x remains the intermediate principal stress.

Problems for Solution 10.10–10.13

10.2.2 *Strain and Deformation*

The normal strains ϵ_x, ϵ_θ, ϵ_r at any point can be calculated from the stress components using equations (10.4). The hoop and radial strains clearly depend on r whereas ϵ_x is constant by assumption. Denoting by u the radial displacement of a point originally at radius r, expression (10.5) for the hoop strain can be rewritten as

$$\epsilon_\theta = \frac{u}{r} \qquad (10.19)$$

If at radius $(r+dr)$ the radial displacement is $(u+du)$ then du is the net increase of the original distance dr and the radial strain is

$$\epsilon_r = \frac{du}{dr} \qquad (10.20)$$

The change of enclosed volume is as expressed by equation (10.6) with the hoop strain evaluated at R_1, i.e. $(\epsilon_\theta)_{R_1}$.

Example 10.10 Find the axial forces required to prevent axial displacement of the cylinder of Example 10.6. If the internal length is 2 m find the volume of water ($K_f = 2100$ MN/m²) to be supplied to attain the working pressure of 120 MN/m². Take $E = 210$ GN/m², $\nu = 0.3$.

Solution Since axial displacement is prevented ϵ_x is zero and from (10.4a)

$$\sigma_x = \nu(\sigma_\theta + \sigma_r)$$

Using (10.11) and taking A from Example 10.6 we have

$$\sigma_x = 2\nu A = 24 \text{ MN/m}^2$$

Referring to Figure 10.7b the outward forces on the ends are $p_1 \pi R_1^2$; let P be the inward forces preventing axial displacement. Then, since

$$\sigma_x = \frac{(p_1 \pi R_1^2 - P)}{\pi(R_2^2 - R_1^2)}$$

we obtain $P = 377$ kN.

Taking σ_θ and σ_r at the inner surface from Example 10.6 we obtain from (10.4b)

$$(\epsilon_\theta)_{R_1} = \frac{1}{E}(200 - 0.3(24 - 120)) \times 10^6 = 1.09 \times 10^{-3}$$

Then using (10.6) with ϵ_x zero $\Delta V/V$ is 2.18×10^{-3}. From (10.7) we obtain $\Delta V_f/V_f$ as 0.0571 and so the total volume of fluid required is

$$(57.1 + 2.18) \times 10^{-3} \times (\text{cylinder volume}) = 0.931 \times 10^{-3} \text{ m}^3$$

Problems for Solution 10.14, 10.15

10.2.3 *Some General and Particular Results*

Introducing the boundary conditions $(\sigma_r)_{R_1} = -p_1$ and $(\sigma_r)_{R_2} = -p_2$ as in equation (10.14) general expressions can be deduced for the constants A and B in equations (10.12) which are then

$$\sigma_r = \frac{p_1 R_1^2 - p_2 R_2^2}{R_2^2 - R_1^2} - \frac{R_1^2 R_2^2 (p_1 - p_2)}{(R_2^2 - R_1^2) r^2} \tag{10.21}$$

$$\sigma_\theta = \frac{p_1 R_1^2 - p_2 R_2^2}{R_2^2 - R_1^2} + \frac{R_1^2 R_2^2 (p_1 - p_2)}{(R_2^2 - R_1^2) r^2} \tag{10.22}$$

Provided σ_x is intermediate between σ_r and σ_θ, as is usual,

$$(\tau_{max})_r = \tfrac{1}{2}(\sigma_\theta - \sigma_r) = \frac{R_1^2 R_2^2 (p_1 - p_2)}{(R_2^2 - R_1^2) r^2} \tag{10.23}$$

Clearly the greatest shear stress is always at the inner surface ($r = R_1$).

(a) *Internal Pressure Only*   Setting $p_2 = 0$ in (10.21–10.23) we find

$$\sigma_r = \frac{p_1 R_1^2}{R_2^2 - R_1^2}\left(1 - \frac{R_2^2}{r^2}\right); \quad \sigma_\theta = \frac{p_1 R_1^2}{R_2^2 - R_1^2}\left(1 + \frac{R_2^2}{r^2}\right); \quad (\tau_{max})_r = \frac{p_1 R_1^2 R_2^2}{(R_2^2 - R_1^2) r^2}$$

$$\tag{10.24a, b, c}$$

Since $r \leqslant R_2$ the radial and hoop stresses are always compressive and tensile respectively (Figure 10.8). The magnitude of the hoop stress always exceeds that of the radial stress and both have their greatest magnitudes at the inner surface. The greatest shear stress is at $r = R_1$.

When the wall thickness $t = R_2 - R_1$ is small compared with R_1 equations (10.24a and b) can be compared with

$$\sigma_r = 0; \qquad \sigma_\theta = \frac{p_1 R_1}{t}$$

which are the stresses obtained for a thin cylinder in Section 10.1.1 using the membrane assumptions. Using (10.24a and b) we can calculate the actual stresses in a cylinder with radii $R_2 = 1\cdot1 R_1$ which corresponds to the limiting value proposed in Section 10.1.1 for the validity of these assumptions. Writing $\mu = p_1 R_1/t$ the radial stress varies from $-0\cdot1\mu$ to zero, and the hoop stress from $1\cdot05\mu$ to $0\cdot95\mu$, over the wall thickness. Thus the membrane assumptions are reasonable within the stated limit of wall thickness.

In order to consider the case of very large wall thickness it is convenient to express the maximum shear stress (putting $r = R_1$ in 10.24c) in the form

$$\tau_{max} = \frac{p_1}{\left(1 - \left(\dfrac{1}{\lambda}\right)^2\right)} \tag{10.25a}$$

where $\lambda = R_2/R_1$. Then since $\tau_{max} \leqslant \tau_A$ where τ_A is the allowable shear stress, we find that

$$\frac{p_1}{\tau_A} \leqslant 1 - \frac{1}{\lambda^2} \qquad (10.25b)$$

As is clear from Figure 10.9b $(p_1/\tau_A) \rightarrow 1$ when the wall thickness is large. Thus even for an infinitely thick wall the internal pressure cannot exceed the allowable shear stress if yielding is not to occur.

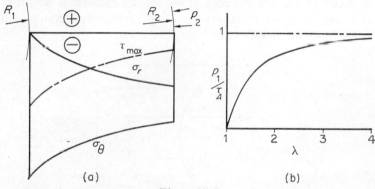

(a)   (b)

Figure 10.9

(b) *External Pressure Only*   Setting $p_1 = 0$ in (10.21 and 10.22) we find

$$\sigma_r = -\frac{p_2 R_2^2}{R_2^2 - R_1^2}\left(1 - \frac{R_1^2}{r^2}\right); \quad \sigma_\theta = -\frac{p_2 R_2^2}{R_2^2 - R_1^2}\left(1 + \frac{R_1^2}{r^2}\right) \qquad (10.26a, b)$$

Since $r \geqslant R_1$ both σ_θ and σ_r are compressive at all radii (Figure 10.9a), σ_θ being greater in magnitude than σ_r. The greatest shear stress in the material is at the inner surface $(r = R_1, p_1 = 0$ in equation 10.23). For a cylinder of very small internal radius, i.e. $R_1 \ll R_2$, we have, from (10.26b), at the inner radius

$$(\sigma_\theta)_{R_1} = -\frac{2p_2 R_2^2}{R_2^2 - R_1^2} \rightarrow -2p_2 \qquad (10.27a)$$

For a solid cylinder $(R_1 = 0)$, subject to pressure p_2, equations (10.26a and b) reduce to

$$\sigma_r = \sigma_\theta = -p_2 \qquad (10.27b)$$

Comparing (10.27a and b) we see that the presence of a tiny circular hole (e.g. crack, flaw) at the centre of a solid cylinder may cause doubling of the maximum hoop stress.

10.2.4 *Compound Cylinders*

In the previous section it has been shown that for either internal or external pressure critical conditions occur at the inner surface of a cylinder and that the hoop stress is always numerically greatest. In the common case of internal pressure only, $(\sigma_\theta)_{R_1}$ is tensile; whereas with external pressure only, $(\sigma_\theta)_{R_1}$ is compressive. These results suggest the use of superimposed external pressure to reduce the tensile stress due to internal pressure, and thereby circumvent the limit expressed by (10.25b).

External pressure can be applied mechanically by winding on wire under tension or by shrinking on reinforcing rings or cylinders. In the latter case a second cylinder is manufactured with internal diameter slightly less than the outer diameter of the first cylinder. If the second cylinder is heated it may be passed over the inner, forming on cooling a compound cylinder with an interference fit at the interface.

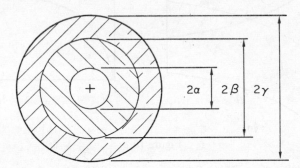

Figure 10.10

Let α, γ be the inner and outer radii of a compound cylinder (Figure 10.10), and β the nominal radius of the interface. Let b be the total radial interference, i.e. the difference between the actual outer radius of the inner component cylinder and the actual inner radius of the outer cylinder. Then since the two component cylinders are forced to adopt a common interface:

(Decrease of outer radius of inner cylinder) + (Increase of inner radius

of outer cylinder) = Radial interference (b)

(10.28)

i.e. the sum of the radial displacements of the component cylinders at radius β is equal to the radial interference. The inner cylinder behaves as if subject to an external pressure and the outer cylinder as if subject to an internal pressure. Since their surfaces are in contact these apparent pressures are equal in magnitude and will be denoted by p. This pressure has no axial component and therefore $\sigma_x = 0$. Using (10.19) and (10.4b)

$$\epsilon_\theta = \frac{u}{r} = \frac{1}{E}(\sigma_\theta - \nu\sigma_r) \qquad (10.29)$$

For the inner component cylinder $R_1 = \alpha$; $R_2 = \beta$; $p_1 = 0$; $p_2 = p$ and so σ_θ and σ_r can be derived from equations (10.26a and b). Substituting in (10.29) we find at $r = \beta$ after some simplification

$$(u_{\text{inner}})_{r\,=\,\beta} = -\frac{p\beta}{E}\left(\frac{(\beta^2+\alpha^2)}{(\beta^2-\alpha^2)}-\nu\right) \tag{10.30a}$$

which is an inward displacement.

For the outer component cylinder $R_1 = \beta$; $R_2 = \gamma$; $p_1 = p$; $p_2 = 0$. Using equations (10.24a and b) and proceeding in a similar manner we find (outward)

$$(u_{\text{outer}})_{r\,=\,\beta} = \frac{p\beta}{E}\left(\frac{(\gamma^2+\beta^2)}{(\gamma^2-\beta^2)}+\nu\right) \tag{10.30b}$$

Substituting from (10.30) in (10.28) and simplifying we obtain

$$p = \frac{Eb(\beta^2-\alpha^2)\,(\gamma^2-\beta^2)}{2\beta^3(\gamma^2-\alpha^2)} \tag{10.31}$$

Using this result the mechanical pressure due to a given interference b can be calculated. The methods of Section 10.2.1 can then be used to obtain the initial stress distribution in the cylinder. Figure 10.11a shows a typical

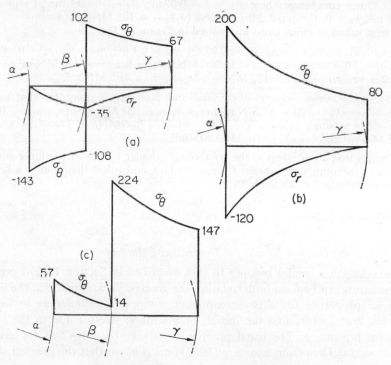

Figure 10.11   Units of stress MN/m²

distribution of these stresses due to assembly of a compound cylinder. Figure 10.11b shows a typical stress distribution in a homogeneous cylinder of the same inner and outer radii as the compound cylinder and subject to an internal fluid pressure p_1. The stresses in the compound cylinder wall when subject to fluid pressure p_1 are the algebraic sum of those in Figures 10.11a and b. The total hoop stress magnitude (see Figure 10.11c) is thus reduced at $r = \alpha$, and increased at the inner surface of the outer component cylinder. The critical condition is at one of these surfaces.

Example 10.11 Let us suppose that the cylinder of Example 10.6 was made as a two-component compound cylinder with interface at radius 0·07 m, and radial interference 10^{-3} mm per mm of radius. Find the maximum shear stress for the same working pressure. Take $E = 210$ GN/m^2.

Solution Putting $\alpha = 0\cdot05$ m, $\beta = 0\cdot07$ m, $\gamma = 0\cdot1$ m, $b = 10^{-3} \times 0\cdot07$ m in equation (10.31) we find $p = 35 \times 10^6$ N/m^2. The initial stresses are now found for each component cylinder.

(a) Inner component: here $R_1 = \alpha = 0\cdot05$ m; $R_2 = \beta = 0\cdot07$ m; $p_1 = 0$; $p_2 = p = 35$ MN/m^2. Using (10.26b) we have $(\sigma_\theta)_{0\cdot05} = -143$ MN/m^2.

(b) Outer component: here $R_1 = \beta = 0\cdot07$ m; $R_2 = \gamma = 0\cdot1$ m; $p_1 = p = 35$ MN/m^2; $p_2 = 0$. From (10.24b) we have $(\sigma_\theta)_{0\cdot07} = 102$ MN/m^2. These values of initial stress are plotted in Figure 10.11a.

(c) Fluid pressure: now $R_1 = 0\cdot05$ m; $R_2 = 0\cdot1$ m; $p_1 = p_1$; $p_2 = 0$. From Example 10.6 we have $(\sigma_\theta)_{0\cdot05} = 200$ MN/m^2; $(\sigma_r)_{0\cdot05} = -120$ MN/m^2. Using (10.24) we find $(\sigma_\theta)_{0\cdot07} = 122$ MN/m^2; $(\sigma_r)_{0\cdot07} = -41\cdot6$ MN/m^2.

(d) Total stresses: superposing initial and fluid distributions (in that order) $(\sigma_\theta)_{0\cdot05} = -143 + 200 = 57$ MN/m^2; $(\sigma_r)_{0\cdot05} = -120$ MN/m^2; $(\sigma_\theta)_{0\cdot07_+} = 102 + 122 = 224$ MN/m^2; $(\sigma_r)_{0\cdot07_+} = -35\cdot0 - 41\cdot6 = -76\cdot6$ MN/m^2. Then $(\tau_{max})_{0\cdot05} = 88\cdot5$ MN/m^2 and $(\tau_{max})_{0\cdot07_+} = 150\cdot3$ MN/m^2.

The greatest shear stress in the compound cylinder is thus at the inner surface of the outer component cylinder (Figure 10.11c), and is less than that in a homogeneous cylinder (see Example 10.7).

Problems for Solution 10.16–10.18

10.2.5 *Some General Results for Compound Cylinders*

Proceeding in a similar manner to that described in Section 10.2.4 general expressions can be found interrelating the greatest shear stress and the interference properties for a two-component compound cylinder as in Figure 10.10. We consider first the initial, or assembly, stresses due to the interference pressure p. The usual practice is to take $(\sigma_\theta - \sigma_r)/2$ as the critical shear stress. Then from equation (10.23) it is known that the greatest shear stress in each component is at its inner surface.

The inner tube has $R_1 = \alpha$, $R_2 = \beta$, and is subject to external pressure, i.e. $p_1 = 0$, $p_2 = p$. Using equation (10.23) and introducing $\lambda_1 = \beta/\alpha$ we obtain at the inner surface ($r = \alpha$) the greatest shear stress as

$$\frac{-p\lambda_1^2}{(\lambda_1^2 - 1)} \tag{i}$$

Similarly for the outer tube we put $R_1 = \beta$, $R_2 = \gamma$, $p_1 = p$, $p_2 = 0$ in equation (10.23) and obtain at the inner surface ($r = \beta$), the greatest shear stress as

$$\frac{p\lambda_2^2}{(\lambda_2^2 - 1)} \tag{ii}$$

where $\lambda_2 = \gamma/\beta$.

We now consider the shear stress due to an internal fluid pressure p_1 acting on a homogeneous cylinder with inner and outer radii $R_1 = \alpha$, $R_2 = \gamma$. Again using equation (10.23), with $p_2 = 0$ and $\lambda = \gamma/\alpha$, we find the greatest shear stress at radius $r = \alpha$ to be

$$\frac{p_1\lambda^2}{(\lambda^2 - 1)} \tag{iii}$$

and at radius $r = \beta$ it is

$$\frac{p_1\lambda_2^2}{(\lambda^2 - 1)} \tag{iv}$$

The total shear stress at the inner surface of each tube is now obtained by superposition. Using (i) and (iii) we have for the inner tube

$$(\tau_{\max})_{r=\alpha} = \frac{-p\lambda_1^2}{\lambda_1^2 - 1} + \frac{p_1\lambda^2}{\lambda^2 - 1} \tag{10.32a}$$

and using (ii) and (iv)

$$(\tau_{\max})_{r=\beta} = \frac{p\lambda_2^2}{\lambda_2^2 - 1} + \frac{p_1\lambda_2^2}{\lambda^2 - 1} \tag{10.32b}$$

For given fluid pressure and dimensions the relative magnitude of these stresses depends on the interference pressure p (equation 10.31). The most efficient choice of interference pressure is clearly that for which the greatest shear stress is the same in each tube. Equating (10.32a) and (10.32b) and simplifying we obtain the relation

$$p\left(\frac{\lambda_1^2}{\lambda_1^2 - 1} + \frac{\lambda_2^2}{\lambda_2^2 - 1}\right) = p_1\left(\frac{\lambda^2 - \lambda_2^2}{\lambda^2 - 1}\right) \tag{10.33}$$

From this equation the interference pressure p is found for the design fluid pressure p_1; as p is fixed it follows that if the cylinder is used at some other fluid pressure the shear stresses given by equations (10.32) are not equal. Substituting for p from (10.33) in either of (10.32) we obtain, after simplification, the maximum shear stress in the compound cylinder, as

$$\tau_{max} = \frac{\lambda^2 p_1}{2\lambda^2 - \lambda_1^2 - \lambda_2^2} \qquad (10.34)$$

for any particular geometry expressed by the ratios λ, λ_1, λ_2. Noting that $\lambda = \lambda_1\lambda_2$, it is easily shown, by differentiation with respect to either λ_1 or λ_2, that the shear stress is least for a given λ and p_1 when

$$\lambda_1 = \lambda_2 = \sqrt{\lambda} \qquad (10.35)$$

Using this result equations (10.33), (10.31) and (10.34) become

$$p = \tfrac{1}{2}p_1\left(\frac{\lambda-1}{\lambda+1}\right) \qquad (10.36)$$

$$\frac{b}{\beta} = \frac{p_1}{E} \qquad (10.37)$$

$$\tau_{max} = \frac{p_1\lambda}{2(\lambda-1)} \qquad (10.38)$$

The maximum shear stress must not exceed the allowable shear stress τ_A and so

$$p_1 \leqslant 2\tau_A\left(1-\frac{1}{\lambda}\right) \qquad (10.39)$$

From this expression the value of λ is obtained for the required pressure p_1. Then using (10.37) the fractional interference can be found. It should be noted that when the wall thickness becomes very great, i.e. as $\lambda \to \infty$ in equation (10.39), the pressure which can be retained tends to the limiting value $2\tau_A$. Referring to equation (10.25b) and Figure 10.9b we see that this is double that for a homogenous cylinder of the same dimensions.

Example 10.12 Reconsider the compound cylinder of Example 10.11 to determine the optimum interference pressure for the previous working pressure, and the greatest pressure the cylinder could then withstand.

Solution Here $\lambda_1 = 0.07/0.05 = 1.4$, $\lambda_2 = 0.1/0.07 = 1.429$, $\lambda = 2$. Substituting these quantities in equation (10.33) we find the optimum interference pressure to be

$$p = 0.1633 p_1$$

or 19.6 MN/m² for a fluid pressure of 120 MN/m². Thus the value used for the system of Example 10.11 was excessive.

Substituting in (10.34) we find $\tau_{max} = p_1$ and so the greatest pressure which can be used is equal to the allowable shear stress. It is interesting to note that a single cylinder to retain this pressure would require to be infinitely thick (see equation 10.25b).

Example 10.13 Find optimum dimensions for a cylinder of 0·05 m inner diameter to retain a fluid pressure of 120 MN/m², if the allowable shear stress is 160 MN/m². $E = 210$ GN/m².

Solution Using (10.38) with $\tau_{max} = 160$ MN/m², $p_1 = 120$ MN/m² we find $\lambda = 1·6$. Then $\lambda_1 = \lambda_2 = \sqrt{\lambda} = 1·264$ and so $\alpha = 0·05$ m, $\beta = 0·0632$ m, $\gamma = 0·08$ m. Finally from (10.37) the fractional interference is $p_1/E = 0·573 \times 10^{-3}$.

The single cylinder for this duty was found in Examples 10.6 and 10.7 and had outer radius of 0·1 m requiring 1·92 times as much material.

For further discussion see for example references 5 and 6.

Problems for Solution 10.19–10.21

10.2.6 *Elastic/Plastic Analysis*

Since σ_θ and σ_r are non-uniform under purely elastic conditions (Figure 10.8) we would expect, as in bending and torsion, to require a greater internal pressure to produce fully plastic conditions than to initiate yielding. As discussed in Section 10.2.1 σ_x is intermediate between σ_θ and σ_r so that the Tresca yield criterion is

$$\sigma_0 = 2 |\tau_{max}| = |\sigma_\theta - \sigma_r|_{max} \qquad (10.40)$$

Only internal pressure will be considered, and from equation (10.24c) the greatest shear stress is at the inner surface $(r - R_1)$. Introducing this result in the yield condition (10.40) the internal pressure p_0 to initiate yielding is

$$p_0 = \frac{\sigma_0(R_2^2 - R_1^2)}{2R_2^2} \qquad (10.41)$$

If the pressure is further increased more material is brought to the yield point so that a plastic zone is formed in the wall extending from the inner surface to a cylindrical surface of radius R_s. The elastic part of the wall lying between $r = R_s$ and $r = R_2$ still has its greatest shear stress at its inner radius R_s where the yield condition is just satisfied. Let p_s be the effective pressure on the inner surface of the elastic zone. Then putting $p_1 = p_s$ and $R_1 = R_s$ in (10.24c) and combining with (10.40)

$$p_s = \frac{\sigma_0(R_2^2 - R_s^2)}{2R_2^2} \qquad (10.42)$$

From this result p_s can be calculated for any known position of the yield surface and is of course zero when the wall is fully plastic $(R_s = R_2)$. Using equations (10.24a and b) the radial and hoop stress distributions in the elastic zone can be determined. However, it is not known how p_s is related to the actual fluid pressure p_1^P at the inner surface of the cylinder wall. Nor is the stress distribution in the plastic zone known except that it must satisfy the yield condition, i.e. for $R_1 \leqslant r \leqslant R_s$

$$\sigma_\theta - \sigma_r = \sigma_0 \qquad (10.43)$$

However, the stress distribution must also obey the stress equilibrium equation (10.10) for the radial direction. Combining these gives

$$r \frac{d\sigma_r}{dr} = \sigma_0$$

Rearranging and integrating this becomes

$$\sigma_r = \sigma_0 \ln(r) + C$$

where C is a constant of integration to be determined from the boundary condition $(\sigma_r)_{R_1} = -p_1^P$. Hence $C = -p_1^P - \sigma_0 \ln(R_1)$ and we have

$$\sigma_r = \sigma_0 \ln\left(\frac{r}{R_1}\right) - p_1^P \qquad (10.44)$$

At the other boundary of the plastic zone $(r = R_s)$ the radial stress is compressive and equal in magnitude to the effective pressure p_s on the elastic zone, i.e. $(\sigma_r)_{R_s} = -p_s$. Then (10.44) can be written

$$p_1^P = p_s + \sigma_0 \ln\left(\frac{R_s}{R_1}\right)$$

Substituting for p_s from (10.42) we have

$$p_1^P = \frac{\sigma_0}{2}\left(1 - \frac{R_s^2}{R_2^2}\right) + \sigma_0 \ln\left(\frac{R_s}{R_1}\right) \qquad (10.45)$$

This equation determines the pressure needed to attain a penetration R_s of the plastic zone. When $R_s = R_1$ equation (10.45) corresponds to (10.41) for $p_1^P = p_0$. When $R_s = R_2$ the wall is fully plastic and we have the fully plastic pressure

$$\hat{p} = \sigma_0 \ln\left(\frac{R_2}{R_1}\right) \qquad (10.46)$$

The ratio of this to the yield pressure is

$$\frac{\hat{p}}{p_0} = \frac{2 \ln\left(\dfrac{R_2}{R_1}\right)}{\left(1 - \left(\dfrac{R_1}{R_2}\right)^2\right)} \qquad (10.47)$$

For example when $R_2 = 2R_1$ the ratio is 1·84.

The radial stress in the plastic zone is given by (10.44) and hence using (10.43)

$$\sigma_\theta = \sigma_0 \left(1 + \ln\left(\frac{r}{R_1}\right)\right) - p_1^P \qquad (10.48)$$

For a particular plastic zone penetration the stress distribution is obtained from (10.44), (10.45) and (10.48). Figures 10.12a and b show the stress distribution for fully and partly plastic conditions respectively. In Figure 10.12a the curves are parallel since they differ by the constant σ_0 (equation 10.43).

During yielding the material of the plastic zone increases its free radial dimensions. Consequently when the pressure is released the elastic zone is prevented from recovering its original dimensions and is left in hoop tension

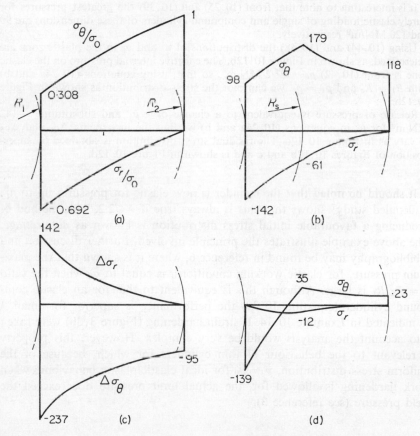

Figure 10.12   Units of stress MN/m²; $R_2 = 2R_1$

while the plastic zone is forced into a condition of hoop compression. This situation is analogous to that caused by the interference fit in a compound cylinder (see Section 10.2.4). The residual stress distribution can be obtained by a superposition procedure similar to that used in Section 7.5.2 for bending. The residual stress distribution is the sum of those due to p_1^P and $\Delta p = 0 - p_1^P$ as is described in Example 10.14 and illustrated by Figures 10.12b, c and d. This method is of course based on the assumption that pressure removal is an elastic process.

Example 10.14 Calculate the internal pressure to cause yielding to a depth of 20 mm in the wall of a cylinder with inner and outer radii 50 mm and 100 mm. Find the residual stress distribution. $\sigma_0 = 240$ MN/m².

Solution Here $R_s = 0.07$ m and from (10.45) $p_1^P = 142$ MN/m².

It is interesting to note that from (10.25) and (10.39) the greatest pressures for purely elastic loading of single and compound cylinders of these dimensions are 90 and 120 MN/m² respectively.

Using (10.44) and (10.48) the distributions of σ_r and σ_θ in the plastic zone are determined, as shown in Figure 10.12b. The effective internal pressure on the elastic zone is from (10.42) $p_s = 61.2$ MN/m², so that, using equations (10.24a and b) with $R_1 = R_s$ and $p_1 = p_s$, we can plot the stress distribution as shown in Figure 10.12b.

Release of pressure is equivalent to a change of $-p_1^P$ and substituting -142 MN/m² for p_1 in equations (10.24a and b) we find the increments $\Delta\sigma_r$ and $\Delta\sigma_\theta$ to vary as in Figure 10.12c. The residual stress distribution is obtained by superposition of Figures 10.12b and c and is shown in Figure 10.12d.

It should be noted that the cylinder is now elastic for pressures up to p_1^P. A detailed study[7] shows that this is always true if $\lambda \leqslant 2.2$. This method of producing a favourable initial stress distribution is known as *autofrettage*. The above example illustrates the principle involved. Further discussion and a bibliography may be found in reference 6, where it is shown that the maximum pressure, for elastic working conditions, is equal to σ_0 when the ratio $\lambda = R_2/R_1$ is large. Although this is equivalent to that for an elastic compound cylinder (equation 10.39) the performance is superior for small λ as indicated in Example 10.14. If strain hardening (Figure 3.3b) were taken into account the analysis would be very complex. However, this property is relevant to the behaviour of thin cylinders for which, because of the uniform stress distribution, $\hat{p} = p_0$ for ideal elastic/plastic behaviour; when work hardening is allowed for, the actual limit pressure does exceed the yield pressure (see reference 3).

Problems for Solution   10.22, 10.23

10.3 Stresses Due to Rotation

10.3.1 *Thin Rings and Rods*

The thin ring in Figure 10.13a has mean radius r, cross-sectional area A, density ρ, and rotates with angular velocity ω. An element of the ring (Figure 10.13b) is maintained in a circular path by a centripetal force equal to the

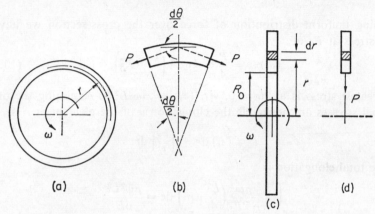

Figure 10.13

product of its mass $\rho A r\, d\theta$ and the centripetal acceleration $\omega^2 r$. This centrally directed force can only be supplied by tensile forces P exerted by the rest of the ring (Figure 10.13b). The inward radial component of these forces is $P\, d\theta$ and so, equating the expressions for this force,

$$P\, d\theta = \rho \omega^2 r^2 A\, d\theta$$

Assuming that the forces P are uniformly distributed over the cross-section of the element, then we have a tensile hoop stress

$$\sigma_\theta = \frac{P}{A} = \rho \omega^2 r^2 \qquad (10.49)$$

There are no radial or axial stresses.

Example 10.15 A thin-walled closed steel cylinder of internal diameter 1·2 m and wall thickness 12 mm rotates at 10 r.p.s. If the maximum shear stress is not to exceed 50 MN/m² find the greatest internal pressure allowable. $\rho = 7900$ kg/m³.

Solution Here $\omega = 2\pi n = 20\pi$ rad/sec. From (10.2), (10.49) and (10.1) we obtain $\sigma_\theta = 50p + 11\cdot2 \times 10^6$, $\sigma_x = 25p$ N/m². Since $\sigma_r = 0$

$$\tau_{max} = \tfrac{1}{2}|\sigma_\theta - \sigma_r| = 25p + 5\cdot6 \times 10^6$$

But $\tau_{max} \leqslant 50 \times 10^6$ so that $p \leqslant 1\cdot78 \times 10^6$ N/m².

Figure 10.13c shows a slender rod of cross-section A, length L rotating with angular velocity ω about an axis through its midpoint. The element at radius r has mass $\rho A\, dr$ and experiences a centripetal force $\rho\omega^2 A r\, dr$. Hence the total force P on all the material between radius R_0 and the end is (Figure 10.13d)

$$P = \int_{R_0}^{\frac{1}{2}L} \rho\omega^2 A r\, dr = \tfrac{1}{2}\rho\omega^2 A\left(\frac{L^2}{4} - R_0^2\right)$$

Assuming uniform distribution of force over the cross-section we have a radial stress at R_0 of

$$(\sigma_r)_{R_0} = \frac{P}{A} = \tfrac{1}{2}\rho\omega^2\left(\frac{L^2}{4} - R_0^2\right) \tag{10.50}$$

The greatest stress is at $R_0 = 0$, viz. $(\sigma_r)_0 = \rho\omega^2 L^2/8$. Assuming the other principal stresses are negligible the elongation u of the element at r is

$$u = (\epsilon_r)\, dr = \left(\frac{\sigma_r}{E}\right) dr$$

and the total elongation is

$$\int_{-\frac{1}{2}L}^{\frac{1}{2}L} \frac{\rho\omega^2}{2E}\left(\frac{L^2}{4} - r^2\right) dr = \frac{\rho\omega^2 L^3}{6E}$$

10.3.2 *Thin Disks*

In the case of a thin disk rotating with angular velocity ω about an axis through its centre it is reasonable to assume that there are no axial stresses σ_x. Proceeding as in Section 10.2.1 the radial forces on an element (Figure 10.7c) due to the stresses σ_r, $\sigma_r + d\sigma_r$, σ_θ are:

$$\text{Inward force} = (r\sigma_r + \sigma_\theta\, dr)\, d\theta\, dx \tag{i}$$

$$\text{Outward force} = (\sigma_r + d\sigma_r)\,(r + dr)\, d\theta\, dx \tag{ii}$$

The element is subject to a centripetal force $((\rho\, dr\, dx\, r\, d\theta)\,\omega^2\, r)$. Equating this force to the difference of (i) and (ii) we find

$$\frac{\sigma_\theta}{r} - \frac{d\sigma_r}{dr} - \frac{\sigma_r}{r} = \rho\omega^2 r \tag{10.51}$$

Putting $\sigma_x = 0$ in equations (10.4b and c) and solving for σ_θ and σ_r we have

$$\sigma_\theta = \frac{E}{1 - \nu^2}(\epsilon_\theta + \nu\epsilon_r) \qquad \text{and} \qquad \sigma_r = \frac{E}{1 - \nu^2}(\epsilon_r + \nu\epsilon_\theta) \tag{10.52a, b}$$

Substituting from equations (10.52), (10.19) and (10.20) the radial force equation (10.51) may be written as

$$\frac{d^2u}{dr^2} + \frac{1}{r}\frac{du}{dr} - \frac{u}{r^2} + \frac{(1 - \nu^2)\,\rho\omega^2 r}{E} = 0 \tag{10.53}$$

Rewriting the first three terms as

$$\frac{d}{dr}\left(\frac{1}{r}\frac{d}{dr}(ur)\right)$$

the solution is easily shown to be of the form

$$u = Ar + \frac{B}{r} - \frac{(1-\nu^2)}{E}\frac{\rho\omega^2 r^3}{8} \qquad (10.54)$$

where A and B are constants to be determined from the boundary conditions. Using equations (10.54), (10.52), (10.19) and (10.20) we obtain

$$\sigma_\theta = \frac{E}{1-\nu^2}\left(A(1+\nu) + \frac{B}{r^2}(1-\nu) - \frac{(1-\nu^2)}{E}\frac{\rho\omega^2 r^2}{8}(1+3\nu)\right) \qquad (10.55a)$$

$$\sigma_r = \frac{E}{1-\nu^2}\left(A(1+\nu) - \frac{B}{r^2}(1-\nu) - \frac{(1-\nu^2)}{E}\frac{\rho\omega^2 r^2}{8}(3+\nu)\right) \qquad (10.55b)$$

(a) *Solid Disk of Radius* R   If the stresses are not to be infinite when $r = 0$ it follows from (10.55) that B must be zero. At the outer surface $r = R$ there is no radial force and A is determined by setting $(\sigma_r)_R = 0$ in (10.55b). Then equations (10.55) become

$$\sigma_\theta = \frac{(3+\nu)\,\rho\omega^2 R^2}{8}\left(1 - \frac{r^2}{R^2}\left(\frac{1+3\nu}{3+\nu}\right)\right)$$

and   (10.56)

$$\sigma_r = \frac{(3+\nu)\,\rho\omega^2 R^2}{8}\left(1 - \frac{r^2}{R^2}\right)$$

Since $\nu \leqslant 0.5$ (equation 3.7) $(1+3\nu)/(3+\nu) < 1$. Therefore both σ_θ and σ_r are always tensile (Figure 10.14a) and decrease with increasing radius. Also

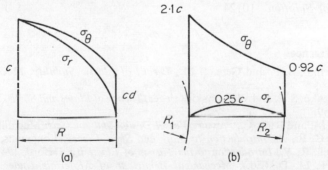

Figure 10.14   Stresses in (a) solid disk of radius R, (b) disk with radii $R_2 = 2R_1$. Values are for $\nu = 0.3$. $c = (3+\nu)\,\rho\omega^2 R^2/8$; $d = 2(1-\nu)/(3+\nu)$; in (b) R in c is R_2

$\sigma_r < \sigma_\theta$ except at $r = 0$ where they are equal with their maximum values

$$(\sigma_\theta)_0 = (\sigma_r)_0 = \frac{(3+\nu)\,\rho\,\omega^2 R^2}{8} \qquad (10.57)$$

(b) *Disk with Central Hole* At the inner and outer free surfaces (radii R_1 and R_2) the radial stress must be zero. Substituting $(\sigma_r)_{R_1} = (\sigma_r)_{R_2} = 0$ in (10.55b) to find A and B we obtain finally

$$\sigma_\theta = \left(\frac{3+\nu}{8}\right)\rho\,\omega^2 R_2^2 \left(\frac{R_1^2}{R_2^2}+1+\frac{R_1^2}{r^2}-\left(\frac{1+3\nu}{3+\nu}\right)\frac{r^2}{R_2^2}\right)$$

$$\sigma_r = \left(\frac{3+\nu}{8}\right)\rho\,\omega^2 R_2^2 \left(\frac{R_1^2}{R_2^2}+1-\frac{R_1^2}{r^2}-\frac{r^2}{R_2^2}\right)$$

σ_θ and σ_r are tensile at all radii and $\sigma_\theta > \sigma_r$ (Figure 10.14b). The greatest hoop stress is at the inner surface

i.e. $$(\sigma_\theta)_{R_1} = \left(\frac{3+\nu}{4}\right)\rho\,\omega^2 R_2^2 \left(1+\frac{R_1^2}{R_2^2}\left(\frac{1-\nu}{3+\nu}\right)\right)$$

Note that as $R_1 \to 0$

$$(\sigma_\theta)_{R_1} \to \left(\frac{3+\nu}{4}\right)\rho\,\omega^2 R_2^2$$

Comparison with equation (10.57) shows that for a central hole of negligible radius the stress is twice that in a solid disk of the same external radius. The radial stress is zero at R_1 and R_2 and is easily shown to have a turning value at $r = \sqrt{R_1 R_2}$ where

$$(\sigma_r)_{\max} = \left(\frac{3+\nu}{8}\right)\rho\,\omega^2 (R_2 - R_1)^2$$

Problem for Solution   10.24

10.4   References

1. Timoshenko, S. P. and Gere, J. M., *Theory of Elastic Stability*, McGraw-Hill, 1961.
2. Timoshenko, S. P. and Woinowski-Krieger, *Theory of Plates and Shells*, McGraw-Hill, 1959.
3. Bickell, M. and Ruiz, C., *Pressure Vessel Design and Analysis*, Macmillan, 1967.
4. Turner, C. E., *An Introduction to Plate and Shell Theory*, Longmans, 1965.
5. Southwell, R. V., *Introduction to the Theory of Elasticity*, Oxford, 1941.
6. Pugh, H. Ll. D. (Ed.), *Mechanical Behaviour of Materials Under Pressure*, Elsevier, 1970, Chap. 2.
7. Turner, L. B., 'The stresses in a thick hollow cylinder subjected to internal pressure', *Camb. Phil. Soc. Trans.*, **21**, 377 (1913).

10.5 Problems for Solution

(For steel take $E = 210 \text{ GN/m}^2$; $\nu = 0.3$.)

10.1. A thin-walled closed cylindrical pressure vessel is required to have internal diameter of 0·5 m and retain a pressure of 4 MN/m². Find the minimum wall thickness if a sample of material yielded in a tensile test at 300 MN/m² and a factor of safety of 3 is necessary.

10.2. Referring to the tube of Example 10.2 find the value of torsion moment for which two of the principal stresses are zero, if the pressure is unchanged.

10.3. If the ends of the tube of Example 10.2 are subjected to axial compressive forces P of magnitude $\pi p r^2$ find the torsion moment for which two of the principal stresses can be zero.

10.4. A thin-walled cylinder has wall thickness of 20 mm and internal diameter of 1 m. The ends are sealed by plates held on by external forces P. Find the limiting values of P if the greatest shear stress in the cylinder wall is not to exceed 50 MN/m². The fluid pressure is 3 MN/m².

10.5. A thin-walled closed steel cylinder of 400 mm internal diameter and length 2 m has to contain a pressure of 10 MN/m². If $\sigma_0 = 200 \text{ MN/m}^2$ find the minimum wall thickness. Find $\Delta V/V$ and the volume of fluid ($K_f = 2400 \text{ MN/m}^2$) required to attain working pressure.

10.6. Find the fractional change of the cylinder volume for each of the limiting values of P found in Problem 10.4. The material is steel.

10.7. A cylinder of 0·4 m internal diameter and 10 mm wall thickness is designed to retain a pressure of 4 MN/m². To provide a signal for a warning system a strain gauge is mounted on the cylinder in the circumferential direction. Calculate the reading on this gauge when the steel cylinder is at 105% of design pressure.

10.8. A closed cylinder of 200 mm internal diameter, 10 mm wall thickness, and 2 m length is subject to an internal pressure of 10 MN/m². Find the axial extension. Find the force to prevent further axial extension if p is raised to 12 MN/m². Take $E = 200 \text{ GN/m}^2$, $\nu = 0.25$.

10.9. A thin-walled, steel, spherical vessel is to have internal diameter of 1 m, and retain a pressure of 4 MN/m². If the allowable shear stress is 50 MN/m² find the necessary wall thickness. Find the change of enclosed volume for a pressure change of 1 MN/m².

10.10. Find the minimum wall thickness for a cylinder of 0·3 m internal diameter if it is to withstand an internal pressure of 35 MN/m² without exceeding a shear stress of 84 MN/m² in the steel.

10.11. A submersible instrument container has inner and outer diameters of 0·24 and 0·3 m. Find the greatest depth at which it can be used if a sample of the material yielded in a tensile test at 240 MN/m², and a safety factor of 1·2 is required. $\rho(\text{water}) = 1000 \text{ kg/m}^3$.

10.12. A closed cylinder has $R_1 = 0.06 \text{ m}$, $R_2 = 0.12 \text{ m}$. Find the maximum internal pressure if the shear stress is not to exceed 100 MN/m². Find the external pressure which would be needed to allow a 50% increase of the working internal pressure. Assume critical conditions occur at the inner surface.

10.13. A cylinder with $R_1 = 0.04$ m, $R_2 = 0.06$ m is subject to internal pressure of 40 MN/m². End sealing plates are held on by axial forces. Find the limiting values of these forces if $\sigma_0 = 200$ MN/m² and $f_s = 1.25$.

10.14. Determine the fractional change of the enclosed volume in Problem 10.10 and estimate the volume of water of bulk modulus 2100 MN/m² required to raise the gauge pressure from 0 to 35 MN/m². The internal length of the cylinder is 2 m.

10.15. A closed cylinder has $R_1 = 0.075$ m, $R_2 = 0.1$ m and a working internal pressure of 30 MN/m², If axial displacement is prevented by rigid attachments find the forces they must exert.

10.16. A compound cylinder is made up of a 25 mm thick tube shrunk on to one of inner and outer diameters 100 mm and 150 mm, with interference pressure of 21 MN/m². Sketch the distribution of initial stress and find the greatest internal pressure which can be applied if a sample of material yielded in a tensile test at 300 MN/m², and a safety factor of 1.25 is necessary.

10.17. A compound tube has inner and outer diameters of 0.24 m and 0.44 m, and an interference surface with diameter 0.32 m. If the diametral interference is 0.32 mm determine the state of stress at the inner surface when the internal pressure is 150 MN/m². The material is steel.

10.18. A sleeve of 100 mm length and 30 mm thickness is to be driven on to a solid shaft of 100 mm diameter. If the fractional interference is 0.0005 calculate the greatest shear stress in the sleeve, and in the shaft. Estimate the maximum force required to complete the assembly assuming a friction coefficient of 0.2 for steel.

10.19. A tube of inner and outer diameters 0.2 and 0.24 m respectively, is to be fitted with a sleeve 0.03 m thick of the same material in order to raise the safe internal pressure. Find the maximum possible relative pressure increase.

10.20. If the material of the cylinder of Problem 10.19 has allowable shear stress of 100 MN/m² find the greatest shear stress when the pressure is 80% of the design value.

10.21. A cylinder of 0.1 m internal diameter is to retain a pressure of 150 MN/m². Find the optimum dimensions for a two-component cylinder for this duty. The allowable shear stress is 100 MN/m².

10.22. A cylinder has internal and external diameters 0.2 and 0.4 m; find the pressure to initiate yielding. If a 20% overload occurs estimate the depth of yielding. Find the fully plastic pressure. Take $\sigma_0 = 200$ MN/m².

10.23. A cylinder has internal and external diameters 0.4 m and 0.8 m; find the pressure to induce yielding to a radius of 0.28 m and sketch the radial and hoop stress distributions. If the pressure is then released find the residual hoop stress at the inner surface. Take $\sigma_0 = 200$ MN/m².

10.24. Compare the rotation speeds to cause yielding in steel disks of 0.4 m diameter (a) solid, (b) with central hole of diameter 0.04 m. Take $\sigma_0 = 240$ MN/m², $\rho = 7900$ kg/m³.

Instability of Bars

11.1 Systems with Localized Elasticity

11.1.1 *Introduction*

To introduce the concepts of instability we can consider systems, such as in Figures 11.1a and 11.2a, consisting of pin-jointed rigid bars and linear elastic spring(s). Clearly in these systems unlike those considered in earlier chapters, the elasticity is localized, or lumped. The restoring force P exerted by a linear elastic spring is related to its deformation by

$$P = k\delta \tag{11.1}$$

where k is the stiffness, i.e. force/unit displacement. The close-coiled helical springs discussed in Article 8.5 obey equation (11.1). Following Figure 4.1b we note that the strain energy of the spring is

$$U = \tfrac{1}{2}P\delta = \tfrac{1}{2}k\delta^2 \tag{11.2}$$

Pinned connexions as between AB and BC in Figure 11.1a offer no resistance to relative rotation of the bars, which therefore exert only forces on each other as in Figure 11.1d. (If this cannot be accepted intuitively further discussion will be found in Section 14.1.2.)

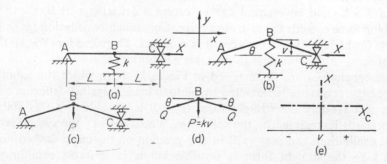

Figure 11.1

Considering now the system in Figure 11.1a we see that if the spring were not present, the straight configuration, although theoretically possible, would in practice prove unstable and the system would jack-knife. However, with the spring present a deflexion from the straight form, as in Figure 11.1b, results in deformation of the spring, which therefore in accordance with equation (11.1) exerts on the joint a force $P = kv$ where v is the displacement of B. Since this force is always directed towards the original position of B (Figure 11.1c) it tends to restore the straight configuration and is termed the restoring force.

Since the bars are in compressive loading they must exert forces towards joint B (as in Figure 11.1d) and from symmetry these are equal. Denoting these forces by Q we see that each has a component $Q \sin \theta$ acting away from the original position of B. We therefore say that there is a disturbing force of $2Q \sin \theta$. Now by considering equilibrium of joints A or C we obtain $X = Q \cos \theta$ and therefore the disturbing force is $2X \tan \theta$ which for small angles θ can be written as $2Xv/L$.

For the deflected form of the system to be in equilibrium we must have equality of the restoring and disturbing forces on joint B,

i.e.

$$kv = \frac{2Xv}{L} \tag{11.3}$$

Introducing X_C to designate the special value of X for which the deflected form is in equilibrium, we can write

$$X_C = \tfrac{1}{2}kL \tag{11.3a}$$

Since X_C is independent of deflexion v it follows that the equilibrium is neutral. Now suppose that X is different from X_C and consider what happens if some external agency causes the system to adopt a deflected form. If $X < X_C$ the disturbing force $2Xv/L$ is less than the restoring force kv so that on removal of the external agency the system returns to the straight configuration. Therefore if $X < X_C$ the straight form is in stable equilibrium. On the other hand if $X > X_C$ and an external agency causes a deflexion v at B, then the disturbing force exceeds the restoring force. Consequently deflexion tends to increase further even when the agency is removed. Therefore for $X > X_C$ the straight form of the system is in unstable equilibrium.

These conclusions are summarized in Figure 11.1e in which the equilibrium configurations are represented by plotting load X against deflexion v at B. For $X < X_C$ only the straight form ($v = 0$) is possible and this stable equilibrium is represented by the solid line. When $X = X_C$ the system is in neutral equilibrium and may exist in any position on the chain-dashed line. For $X > X_C$ the straight form is possible but is in unstable equilibrium indicated by the dashed line.

It is important to note that in the above discussion we determined the stability of the straight form of the system by considering the forces acting on it after introducing a small departure or perturbation from this configuration. In summary we found the system to be stable only when the load is less than a critical value which depends on the elastic properties of the system. This critical load can also be determined by comparing the external work done and the strain energy developed when a perturbation is introduced.

For a displacement v at B (Figure 11.1b) the strain energy is, from (11.2), $\frac{1}{2}kv^2$. The external force X moves through a distance $2(L - \sqrt{L^2 - v^2})$ which, by expansion in series, is found to be v^2/L for small values of v. Then for equilibrium $\frac{1}{2}kv^2 = Xv^2/L$ giving $X_C = \frac{1}{2}kL$ as before. If $X < X_C$ the strain energy exceeds the work done by the force X. Consequently when the perturbing agency is removed the spring is able to restore the straight form which is therefore stable. In contrast when $X > X_C$ the strain energy of the spring is insufficient to restore the straight form even after the agency is removed.

11.1.2 *Two Degrees of Freedom*

A system of three bars can deflect in either of the forms of Figures 11.2a and b and is said to have two degrees of freedom. Consider now whether there can be equilibrium in these deflected forms. As in the previous example it can be

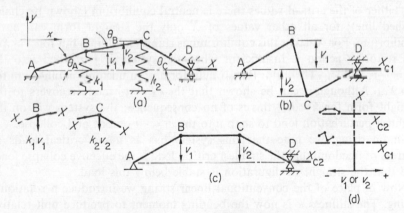

Figure 11.2 AB = BC = CD = L (d) not to scale

shown that for small displacements v_1, v_2 each bar is in compressive direct loading of magnitude X. Therefore each bar exerts a force X towards the joints and for equilibrium of B and C in the y direction ($k_1 = k_2 = k$):

$$kv_1 = X \sin \theta_A - X \sin \theta_B; \qquad kv_2 = X \sin \theta_B + X \sin \theta_C$$

Substituting $\sin \theta_A = v_1/L$ etc. and rearranging

$$\left(k - \frac{2X}{L}\right)v_1 + \frac{X}{L}v_2 = 0$$

$$\frac{X}{L}v_1 + \left(k - \frac{2X}{L}\right)v_2 = 0 \qquad (11.4)$$

These are homogeneous equations which are satisfied by $v_1 = v_2 = 0$ corresponding to the straight form and also when the determinant of coefficients is zero (Appendix II).

$$\begin{vmatrix} \left(k - \dfrac{2X}{L}\right) & \dfrac{X}{L} \\ \dfrac{X}{L} & \left(k - \dfrac{2X}{L}\right) \end{vmatrix} = 0 \text{ is equivalent to } k^2 - \frac{4kX}{L} + \frac{3X^2}{L^2} = 0$$

Solving we find

$$X_{C1} = \frac{kL}{3} \quad \text{or} \quad X_{C2} = kL \qquad (11.5)$$

X_{C1} and X_{C2} are characteristic values of X for the system. Substituting X_{C1} in equations (11.4) we find $v_1 = -v_2$ corresponding to the deflected form in Figure 11.2b. Similarly for X_{C2} we find $v_1 = v_2$ corresponding to Figure 11.2c.

Figure 11.2d charts the stability of equilibrium configurations. When X has either of the critical values there is neutral equilibrium (shown by chain-dashed line); for all other values of X only the straight form can be in equilibrium. For $X < X_{C1}$ this equilibrium is stable (solid line) but for $X > X_{C1}$ the restoring action is insufficient when the system is perturbed into the $v_1 = -v_2$ mode. Therefore the straight form is an unstable equilibrium for $X > X_{C1}$. Although it can be shown that the $v_1 = v_2$ mode recovers to the straight form for $X < X_{C2}$ this is of no consequence; the system will on the slightest perturbation tend to snap into the $v_1 = -v_2$ mode and collapse.

Summarizing, we note that this system has as many critical loads as degrees of freedom, that the smallest critical load is the effective collapse load, and that the straight configuration is stable below this load.

Now in place of the conventional linear spring we introduce a rotational spring. The stiffness k is now the bending moment to produce unit relative rotation of the bars joined by the spring. The governing equation is

$$M = k\theta \qquad (11.6)$$

and the strain energy is

$$U = \tfrac{1}{2}k\theta^2 \qquad (11.7)$$

Replacing the linear springs of Figure 11.2 with rotational springs (Figure 11.3) the system continues to have two degrees of freedom (v_1, v_2 having the

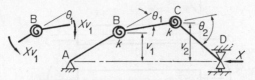

Figure 11.3   AB = BC = CD = L

same or opposite signs). Since moments can be exerted at joints B and C the bars are in a state of bending and direct loading, except when in the straight configuration. By considering equilibrium of the external forces we find that these consist of equal and opposite forces X at A and D. Then at B there are bending moments Xv_1, as shown, which tend to disturb the system.

The relative rotation of the bars at B is θ_1 so that from equation (11.6) the restoring moment exerted by the spring is $k\theta_1$. Then for equilibrium at B we require $k\theta_1 = Xv_1$. Similarly, for equilibrium at C we need $k\theta_2 = Xv_2$. Substituting (for small angles)

$$\theta_1 = \frac{v_1}{L} - \frac{v_2 - v_1}{L} \quad \text{and} \quad \theta_2 = \frac{v_2 - v_1}{L} + \frac{v_2}{L}$$

and rearranging we have

$$\left(\frac{2k}{L} - X\right)v_1 - \frac{k}{L}v_2 = 0$$

$$-\frac{k}{L}v_1 + \left(\frac{2k}{L} - X\right)v_2 = 0$$

(11.8)

Setting the determinant of coefficients to zero the characteristic values are found to be

$$X_{C1} = \frac{k}{L}; \quad X_{C2} = \frac{3k}{L}$$

(11.9)

Substituting in equations (11.8) it is found that X_{C1} corresponds to the mode $v_1 = v_2$ and X_{C2} to the mode $v_1 = -v_2$. Comparing with the linear spring system we note the reversal of mode corresponding to the lower critical load, i.e. the collapse load. The stability of equilibrium chart is otherwise the same as Figure 11.2d.

Problems for Solution   11.1–11.3

11.2   Stability of a Bar in Compression

11.2.1   *Offset Compression*

We shall now consider the case of bending and direct compressive loading of a *slender* bar which was specifically excluded in Article 9.1. As shown in that article offset axial loading (Figure 11.4a) is equivalent to combined bending

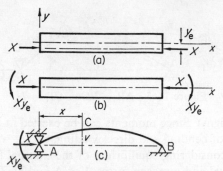

Figure 11.4

and direct loading (Figure 11.4b). A bar supported effectively in this condition is shown in deflected form in Figure 11.4c. The bending moment at a point C on the bar distant x from A is

$$M = Xy_e + Xv \qquad (11.10)$$

where v is the deflexion at C. The term Xv due to the axial force X makes the bending moment a function of the deflexion of the bar. This is different from any situation previously encountered and has important consequences.

Introducing (11.10) in the differential equation of bending (7.1)

$$EI \frac{d^2v}{dx^2} = -M = -Xy_e - Xv \qquad (11.11)$$

Let

$$p^2 = \frac{X}{EI} \qquad (11.12)$$

Then

$$\frac{d^2v}{dx^2} + p^2v = -p^2y_e \qquad (11.13)$$

As is easily verified by back substitution the solution has the form

$$v = C \cos px + D \sin px - y_e \qquad (11.14)$$

where C and D are constants of integration to be determined from the support conditions $v = 0$ at $x = 0$ and $x = L$. Hence

$$C = y_e; \qquad D = \frac{y_e(1 - \cos pL)}{\sin pL} \qquad (11.15)$$

From symmetry the greatest deflexion will occur at $x = L/2$. Substituting this value of x in equation (11.14) and taking C and D from (11.15) we obtain after simplification

$$v_{max} = (v)_{x=\frac{1}{2}L} = \frac{y_e\left(1 - \cos\dfrac{pL}{2}\right)}{\cos\dfrac{pL}{2}} \qquad (11.16)$$

Since from equation (11.12) $p \propto \sqrt{X}$ it follows that, in general, this deflexion is not proportional to the load. The superposition principle cannot be used for a non-linear structure of this type.

If in equation (11.16) the quantity $pL/2$ is small, we can write, using series expansion of the cosine,

$$v_{\max} = y_e \left(1 - \left(1 - \tfrac{1}{2}\left(\frac{pL}{2}\right)^2\right)\right) = \frac{(Xy_e)L^2}{8EI}$$

which is the deflexion due to the end moments only (see Figure 7.5). Thus when X is small the effect of the axial force on the bending is negligible. If instead $p \rightarrow \pi/L$ the central deflexion tends to infinity.

The bending moment at any position can be found by substituting from equations (11.14) and (11.15) in (11.10). However, the greatest value corresponds to the greatest deflexion and so using equation (11.16) we find the greatest bending moment as

$$M_{\max} = Xy_e \sec \frac{pL}{2} \qquad (11.17)$$

As p tends to π/L the bending moment also tends to infinity, and therefore the maximum stress calculated from equation (9.4) will also become infinite. The value $p = \pi/L$ would appear to be of special importance, and using equation (11.12) we find

$$\frac{\pi^2}{L^2} = p^2 = \frac{X_{CE}}{EI}$$

i.e. $$X_{CE} = \frac{\pi^2 EI}{L^2} \qquad (11.18)$$

where X_{CE} is called the Euler critical or buckling load after the famous mathematician.

Equation (11.16) is illustrated in Figure 11.5a for various values of offset y_e. At large offsets substantial deflexions may occur at small loads, but for small offsets there is very little deflexion until the load is nearly equal to the

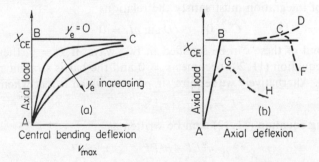

Figure 11.5

load X_{CE} at which the deflexion becomes very large. This sudden collapse is called buckling and may have disastrous consequences.

In the particular case of zero offset there is no bending until the critical load is attained, though as shown in Figure 11.5b there is axial compression of the bar. At the critical load the Figures show an indeterminate amount of deflexion due to the use in equation (11.11) of the approximate expression (d^2v/dx^2) for small bending curvatures (see Section 7.1.1). An exact analysis (e.g. reference 1, p. 76) shows that while the load gradually increases (line ABCD in Figure 11.5b) the critical load X_{CE} is still an effective limit to the usefulness of the bar.

These findings for a system in which the elasticity is distributed may be compared with those for cases of lumped elasticity discussed in Article 11.1 We see that in each case there is a characteristic or critical value of axial load at which large lateral deflexions occur. A straight or near straight form is only stable when the load is less than this value. Finally it should be noted that the above discussion assumes the maintenance of perfectly elastic conditions. As will be discussed later the onset of yielding results in a reduction of the load at which equilibrium can be maintained (see Figure 11.5b lines AGH or ABCF).

11.2.2 *Axial Compression*

The particular case of zero offset $y_e = 0$ corresponds to the absence of end moments in Figure 11.4c. The bar has to be perfectly straight and uniform and has to have perfectly axial loads. Although those requirements may seem a little unreal, nevertheless it is worth considering this case further. Since $y_e = 0$ equation (11.13) reduces to

$$\frac{d^2v}{dx^2} + p^2v = 0 \tag{11.19}$$

with solution

$$v = C\cos px + D\sin px \tag{11.20}$$

Introducing the support conditions $v = 0$ for $x = 0$ and $x = L$, the constants of integration must satisfy the relations

$$C = 0; \qquad D\sin pL = 0 \tag{11.21}$$

The second of these can be satisfied in two ways. If $D = 0$, then, with C also zero, equation (11.20) becomes $v = 0$ and the bar is straight regardless of the load. Alternatively we require $\sin pL = 0$ which is met when

$$pL = 0, \pi, 2\pi, \ldots, n\pi, \ldots \tag{11.22}$$

which, using equation (11.12), can be written as

$$X = 0, \frac{\pi^2 EI}{L^2}, \frac{4\pi^2 EI}{L^2}, \ldots, \frac{n^2\pi^2 EI}{L^2}, \ldots \tag{11.23}$$

Apart from the trivial case of zero load we see that there are an infinite number of characteristic loads for which a deflected form can be in equilibrium as an alternative to the straight configuration. The lowest of these $\pi^2 EI/L^2$ corresponds to the Euler critical load X_{CE} deduced from analysis of offset loading (equation 11.18). By substitution in equation (11.20) with D an arbitrary constant, it is easily confirmed that the deflected form for this load corresponds to that in Figure 11.4c. If higher characteristic loads are substituted in (11.20) it is found that the deflected forms (Figure 11.6a) consist of sine curves with the number of half waves equal to the value of n in the series (11.23).

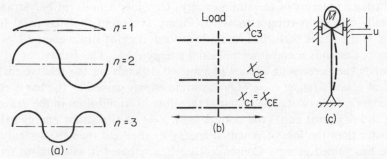

Figure 11.6

In the stability chart of Figure 11.6b the straight form is always a possible equilibrium configuration. For $X < X_{C1}$ the straight form is stable since as shown by Figure 11.5a a deflected form can then only be maintained in the presence of a perturbing moment. We therefore conclude that when $X < X_{C1}$ the restoring forces developed when the bar is given a small deflexion v must exceed the disturbing effects due to the bending moments Xv. At $X = X_{C1}$ these restoring and disturbing effects are evidently equal since, as has been shown, the bar can be in equilibrium in a deflected form. The straight form is therefore in neutral equilibrium when $X = X_{C1} = X_{CE}$. It is reasonable to suppose that if $X > X_{C1}$ the disturbing effects exceed the restoring effects so that the straight form is not recovered if a small deflexion is introduced. This can be proved, but requires the use of the large deflexion theory previously mentioned and is beyond the scope of this work. Accepting this result it follows that the straight form is an unstable equilibrium for $X > X_{C1}$. In Figure 11.6b (as in Figure 11.2d) stable, neutral and unstable equilibrium of the straight bar is indicated by full, chained and dashed lines; the abscissa is some measure of the transverse deflexion v of the bar. The figure also shows the condition of neutral equilibrium at the higher critical loads X_{C2} and X_{C3} which correspond to the deflected forms for $n = 2$ and $n = 3$ in equation (11.23). However, since the straight bar is only in stable equilibrium for $X < X_{C1}$ and may undergo very large bending deflexions when

$X = X_{C1}$, we conclude that $X_{C1} = X_{CE}$ is the effective buckling load and may in normal circumstances ignore the existence of X_{C2}, etc.

Since the bar has distributed elasticity it can deform at all points and therefore may be said to have an infinite number of degrees of freedom compared with the lumped elasticity systems discussed in Article 11.1. As is shown by (11.23) the bar has an infinite number of characteristic, or critical, loads and, as shown by Figure 11.6b, is stable in the straight form only for loads less than the lowest critical load. The correspondence of these properties with the discussion in Section 11.1.2 may be noted.

As in the case of lumped elasticity (see Section 11.1.1) the stability can also be discussed in terms of the strain energy. Consider a uniform bar arranged vertically and supporting a mass $\bar{M}$ (Figure 11.6c) so that the axial force is $\bar{M}g$. If the bar is perturbed into a bent form it gains strain energy U while the mass descends u and loses potential energy $\bar{M}gu$. The behaviour of the bar when the perturbing agency is removed depends on the relative magnitudes of U and $\bar{M}gu$. If $U = \bar{M}gu$ the strain energy gained by the bar is equal to the work done on it. The system is therefore in equilibrium in the deflected form and $\bar{M}g$ must equal the critical load X_{CE}. If the strain energy gained is greater than the loss of potential energy by the load then the system as a whole has gained energy. Consequently when released it will tend to return to the straight form which has less energy. Lastly if $U < \bar{M}gu$ the energy of the system has decreased by bending and whether or not the perturbing agency is removed further bending, corresponding to even lower energy, will take place. Thus according as U is greater than, equal to, or less than $\bar{M}gu$, the equilibrium of the straight form is stable, neutral or unstable.

Example 11.1 A girder with cross-section as in Figure 11.7a is formed by welding together two beams having cross-sections as in Figure 7.13b. The

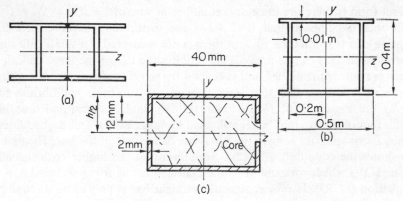

Figure 11.7

latter section has area 5800 mm² and second moments of area about its axes of symmetry of $0·410 \times 10^{-4}$ m⁴ and $0·133 \times 10^{-4}$ m⁴. Find the Euler buckling load for a 12 m length of the girder if the support conditions in both the xy and xz planes correspond to those in Figure 11.4c. Take $E = 210$ GN/m².

Solution Clearly $I_z = 2 \times 0·410 \times 10^{-4}$ m⁴ and using the parallel axes theorem (see Section 6.3.2) we find

$$I_y = 2(0·133 \times 10^{-4} + 5800 \times 10^{-6} \times 0·1^2) = 1·426 \times 10^{-4} \text{ m}^4.$$

Referring to equation (11.18) we see that the Euler load for buckling in the xy plane is less than that for the xz plane since $I_z < I_y$. Therefore the effective buckling load is

$$X_{\text{CE}} = \pi^2 \times 210 \times 10^9 \times 0·82 \times 10^{-4}/12^2 = 1·18 \times 10^6 \text{ N}.$$

Problems for Solution 11.4, 11.5

11.2.3 *Other End Conditions and Inelastic Effects*

To determine critical loads for bars subject to other end conditions we proceed as before by writing the differential equation for bending of the bar, obtaining a general solution and introducing the support conditions. For example the Euler load for the bar in Figure 11.8a is $\pi^2 EI/4L^2$ which is a

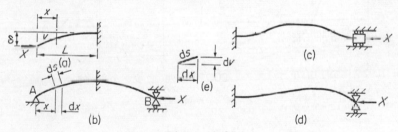

Figure 11.8

quarter of that found for a pinned-end bar of the same length. Referring to Figure 11.8b we see that the pinned bar has zero slope at its midpoint and so is equivalent to two bars of half the length supported as in Figure 11.8a. Thus we might obtain the Euler load for pinned ends by putting $L = L/2$ in the expression for the bar with a free end. Alternatively we can cover both cases using the expression

$$X_{\text{CE}} = \frac{\pi^2 EI}{(L/\mu)^2} = \frac{\pi^2 EI}{L_e^2} \qquad (11.24)$$

In the first of these expressions μ is the number of half sine waves in the deflected form, i.e. $\mu = \frac{1}{2}$ for Figure 11.8a, $\mu = 1$ for Figure 11.8b. We might then extrapolate and predict that $\mu = 2$ for the fixed-end bar in Figure 11.8c; this is verified by exact analysis. For the case of one fixed and one pinned end in Figure 11.8d the number of half waves is evidently slightly less than 1·5 and exact analysis gives $\mu = 1·43$. The quantity $L_e = L/\mu$ is known as the effective length, and is the length of a pinned-end column which has the same critical load as the given column, e.g. for a fixed-end column of length L the effective length is $L_e = L/2$.

Putting $I = Ar^2$ where A is the cross-sectional area and r is called the radius of gyration, equation (11.24) becomes

$$\frac{X_{CE}}{A} = \frac{\pi^2 E}{(L/r\mu)^2} \tag{11.25}$$

The quantity $(L/r\mu) = L_e/r$ is called the effective slenderness ratio of the bar, and (X_{CE}/A) is the mean normal stress at the critical load. Thus by plotting equation (11.25) as in Figure 11.9a a single curve is obtained for

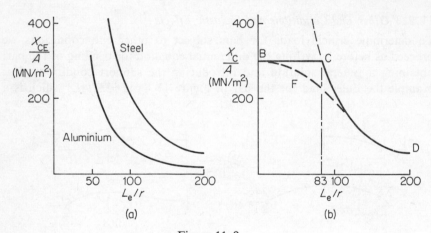

Figure 11.9

all bars having the same value of E. At large values of L_e/r the Euler load is clearly small whereas for squat bars it becomes very large. Obviously a practical limit is imposed at small L_e/r by the yield strength in simple compression, which is $A\sigma_0$ where σ_0 is the yield stress of the material. The yield stress of an ideal elastic/plastic material therefore provides an upper limit to the validity of equation (11.25) at a slenderness ratio obtained by setting $\sigma_0 = X_{CE}/A$. For a steel with $\sigma_0 = 300$ MN/m² this value is 83 and the buckling loads for perfect axial loading are obtained from the graph BCD (Figure 11.9b). Due to lack of straightness, small offsets of the load, and variations from ideal elastic/plastic behaviour, experimental results depart from this graph to some extent, e.g. graph BFD. A full discussion of these effects is beyond the scope of this text (see for example reference 2). In addition the problem is complicated, in the case of structural columns, by the existence of many empirically based formulae stipulated in Codes of Practice and government regulations. Therefore it should be noted that the example given below is intended only to illustrate in an elementary manner how we might allow for both elastic and inelastic buckling effects.

Example 11.2 Show that under the conditions of Example 11.1 the buckling is elastic. If the end conditions are fixed what is the buckling load. $\sigma_0 = 300$ MN/m².

Solution Substituting $(X_{CE}/A) = 300$ MN/m² in (11.25) the limiting slenderness ratio for elastic buckling is found to be 83.

The area of the section (Figure 11.7a) is 11,600 mm² and the least I is (from Example 11.1) 0.82×10^{-4} m⁴. Hence the least radius of gyration is $\sqrt{I/A} = 0.084$ m and the slenderness ratio for a 12 m pinned-end column is 12/0.084, i.e. 143. The column therefore experiences elastic buckling.

If the ends are fixed $\mu = 2$, the effective length becomes 6 m and the effective slenderness ratio is 71.5 which (see Figure 11.9b) corresponds to inelastic buckling at a load of $A\sigma_0$, i.e. 3.48 MN.

Problems for Solution 11.6-11.10

11.3 Approximate Methods

In Section 11.2.2 the relation between the strain energy of bending and the potential energy lost by a gravitational load was discussed; at the critical load these quantities were equal. This principle can be extended to other arrangements by equating the strain energy of bending ΔU to the work done by the axial forces during bending ΔW.

In developing the bent form from the straight configuration the axial force X (Figure 11.8b) moves through a distance λ equal to the difference between the curve AB and the chord AB. An element ds of the curve contributes dx to the chord. Referring to Figure 11.8e we can write

$$ds = \sqrt{dx^2 + dv^2} = dx\sqrt{1 + \left(\frac{dv}{dx}\right)^2}$$

$$\doteq dx\left(1 + \tfrac{1}{2}\left(\frac{dv}{dx}\right)^2\right)$$

Then

$$d\lambda = ds - dx = \tfrac{1}{2}\left(\frac{dv}{dx}\right)^2 dx$$

and for the whole bar

$$\lambda = \int_0^L \tfrac{1}{2}\left(\frac{dv}{dx}\right)^2 dx \qquad (11.26)$$

The procedure is to assume a suitable form for the bending deflexion equation, e.g. $v = a \sin(\pi x/L)$ for Figure 11.8b, where a is some constant. Then using (11.26)

$$\lambda = \frac{a^2\pi^2}{2L^2} \int_0^L \cos^2 \frac{\pi x}{L} \, dx = \frac{a^2\pi^2}{4L}$$

The bending moment is $M = Xv = Xa \sin(\pi x/L)$ and so using equation (7.7) the strain energy of bending is $(X^2 a^2 L/4EI)$.

At the critical load $X = X_{CE}$ and $\Delta U = \Delta W$ where $\Delta W = \lambda X_{CE}$. Hence $X_{CE} = \pi^2 EI/L^2$ which is the result obtained in Section 11.2.2. This exact correspondence is due to the fact that the assumed deflexion equation has the correct form (equation (11.20) with $C = 0$ from (11.21)). Using the approximate form $v = ax(L-x)$ which satisfies the support conditions $v = 0$ at $x = 0$ and $x = L$, we obtain a value $10 \, EI/L^2$ which differs from the exact solution by 1·3%. However, a poor choice of deflexion equation can lead to very poor estimates, e.g. an error of 42% using $v = ax^2(L-x)$. It is found that the estimate of X_{CE} obtained in this way always exceeds the correct value. This is because the assumed approximate deflexion curve could only be maintained with the aid of additional constraints and is thus analogous to a higher-order mode of buckling.

Lastly we note that some of the results of Article 11.1 for lumped elasticity provide approximate values for the critical load of an elastic bar. Thus we might crudely approximate a pinned-end bar by two rigid bars of half the original length and connected together by a rotational spring with stiffness equal to the average flexural rigidity of the elastic bar, i.e. EI/L. A better approximation is obtained if we imagine the bar as composed of two elements of length $L/2$ and lump the stiffness of each at its midpoint (compare Figure 11.10b). With one, two and three sub-divisions the critical loads are 4α, 8α, 9α where the exact solution for the pinned elastic bar is $\pi^2\alpha$.

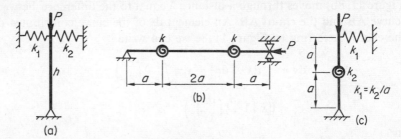

Figure 11.10

11.4 References

1. Timoshenko, S. P. and Gere, J. M. *Theory of Elastic Instability*, 2nd ed., McGraw-Hill, New York, 1961.
2. Horne, M. R. and Merchant, W., *The Stability of Frames*, Pergamon, Oxford, 1965.

11.5 Problems for Solution

(Take $E = 210$ GN/m^2 unless otherwise stated.)

11.1. Solve the systems of Figures 11.2 and 11.3 using energy methods.

11.2. Find the critical loads for the systems in Figure 11.10.

11.3. Find the critical load in Figure 11.2 if $k_1 = \frac{1}{2}k_2 = k$.

11.4. Assuming pinned-end conditions in both xy and xz planes find the Euler load and plane of buckling for (a) a 12 m length with cross-section as in Figure 7.13b, (b) a 20 m length with section as in Figure 11.7b.

11.5. A bar subject to axial compression has cross-section as in Figure 11.7c. The channels are aluminium and the core material has negligible EI value. If the end conditions are the same in both principal planes determine the most efficient value of h. Find the Euler load for a 2 m length of this section, taking $E = 70$ GN/m^2.

11.6. Find the buckling load of a 20 m length with cross-section as in Figure 11.7b if the ends are fixed. $\sigma_0 = 210$ MN/m^2. If $L = 10$ m, and the ends are pinned find the buckling load.

11.7. Find the maximum buckling load for the column cross-section found in Problem 11.5. $\sigma_0 = 120$ MN/m^2.

11.8. A stock list has five tubular sections with I and A as follows (units of cm^4, cm^2); 'A', 28·1, 8·30; 'B', 53·1, 9·22; 'C', 80·5, 9·22; 'D', 49·4, 7·41; 'E', 75·1, 12·0. Find which of these are suitable for a 5 m long column with pinned ends which is to support a compressive axial load of 20 kN with a safety factor of 3. $\sigma_0 = 210$ MN/m^2.

11.9. Repeat Problem 11.8 with a fixed-end column.

11.10. If the bar of Figure 11.7a has fixed ends for bending in the xy plane and pinned ends for the xz plane, find the critical load of a 12 m length. $\sigma_0 = 300$ MN/m^2.

12

Plane Elasticity

12.1 Basic Relationships

12.1.1 *Plane Stress and Plane Strain*

In this chapter we make a formal study of bodies which are loaded in such a manner that stress and strain can be considered to vary in only one plane, which we shall take to be the xy plane. This condition can occur in a prismatic body, which has end faces normal to the longitudinal axis z (Figure 12.1a), when it is loaded by forces parallel to the xy plane and uniform over z.

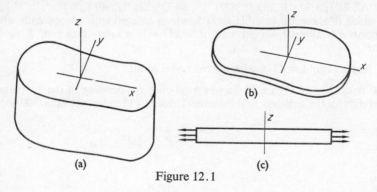

Figure 12.1

When the prism is short it has the form of a thin plate loaded only on its lateral surfaces (Figures 12.1b and c). Since the plate is thin and the stress components σ_z, τ_{xz}, τ_{yz} (see Article 1.3, array A1.1) are clearly zero at the free surfaces, the assumption is made that they are zero throughout the plate. We therefore have a system in which the state of stress at any point depends only on the components σ_x, σ_y, τ_{xy} which are further assumed not to vary with z. This condition is called plane stress (see Section 3.3.4 where the appropriate forms of the stress/strain relationships are quoted).

If the prism is long and displacements of the end surfaces are completely prevented then it is reasonable to suppose that the displacements in the lateral directions are independent of the longitudinal direction z. We may then say that the strain components ϵ_z, γ_{xz}, γ_{yz} (see Article 2.5, array A2.1) are zero throughout the prism, giving a state of plane strain. From the stress/strain relationships (3.11) quoted in Section 3.3.4 it is seen that although σ_z is non-zero its value depends on the strains in the xy plane. The stresses σ_z of course require the existence of normal forces on the end surfaces, and these are the forces which prevent longitudinal displacements.

Thus we may investigate either situation by considering only the xy plane. Many practical situations can be interpreted in this way including the thick-walled cylinders and rotating disks studied by elementary methods in Chapter 10.

12.1.2 *Equilibrium Equations*

Consider an infinitesimal rectangular block of dimensions dx, dy, dz as in Figure 12.2. In the general case the state of stress will vary from point to point in the material. Therefore differences in the magnitudes of stress

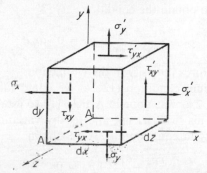

Figure 12.2 $\sigma_x' = \sigma_x + d\sigma_x;\ \tau_{xy}' = \tau_{xy} + d\tau_{xy}$, etc.

components may be expected between opposite faces of this block. For example if σ_x is the normal stress on the negative x face then the normal stress on the positive x face is taken to be $(\sigma_x + d\sigma_x)$.

The total force on the block due to a given stress component is evaluated by the simple product of that stress component and the area of the face on which it acts, e.g. force due to τ_{xy} is $\tau_{xy}\,dy\,dz$ acting in the negative direction of the y axis. The variations of stress components over the areas of the faces on which they act (e.g. variation of σ_x with respect to y) are neglected, since detailed analysis shows them to involve small quantities of higher order.[1]

For equilibrium of the block we require (see Appendix I)

$$\Sigma X = 0; \quad \Sigma Y = 0; \quad \Sigma m_z = 0 \qquad \text{(12.1a, b, c)}$$

The forces in the x direction due to the stress components are $-\sigma_x\,dy\,dz + (\sigma_x + d\sigma_x)\,dy\,dz$ on the x faces, and $-\tau_{yx}\,dx\,dz + (\tau_{yx} + d\tau_{yx})\,dx\,dz$ on the y faces.

In addition a body force component Π_x may act in the x direction. Substituting in equation (12.1a) and simplifying we find

$$d\sigma_x \, dy \, dz + d\tau_{yx} \, dx \, dz + \Pi_x = 0 \tag{12.2}$$

Now $d\sigma_x$ and $d\tau_{yx}$ are the changes of σ_x and τ_{yx} over dx and dy respectively and can be written as

$$d\sigma_x = \left(\frac{\partial \sigma_x}{\partial x}\right) dx \quad \text{and} \quad d\tau_{yx} = \left(\frac{\partial \tau_{yx}}{\partial y}\right) dy$$

Introducing these expressions and dividing by $(dx \, dy \, dz)$, the volume of the block, (12.2) becomes

$$\frac{\partial \sigma_x}{\partial x} + \frac{\partial \tau_{yx}}{\partial y} + p_x = 0 \tag{12.3}$$

where $p_x = \Pi_x/(dx \, dy \, dz)$ is the body force per unit volume, i.e. body force intensity, in the x direction.

For equilibrium of forces in the y direction (equation 12.1b) we proceed in a similar manner to obtain the condition

$$\frac{\partial \tau_{xy}}{\partial x} + \frac{\partial \sigma_y}{\partial y} + p_y = 0 \tag{12.4a}$$

where p_y is the body force intensity in the y direction.

To satisfy equilibrium condition (12.1c) we could take moments about line AA of Figure 12.2 as was done in Article 1.4 to establish the symmetry of the stress array. This result is not affected by variation of the stresses,[1] and putting $\tau_{xy} = \tau_{yx}$ in equation (12.3) we obtain

$$\frac{\partial \sigma_x}{\partial x} + \frac{\partial \tau_{xy}}{\partial y} + p_x = 0 \tag{12.4b}$$

Equations (12.4) are the differential equations of equilibrium for plane elasticity. The only admissible variations of σ_x, σ_y, τ_{xy}, with respect to x and y are those which satisfy these equations.

12.1.3 *Strain/Displacement Equations*

The displacement of a body due to external loads is completely described when the displacements of all parts of it are defined. To specify the displaced position of any one point three components of displacement are required parallel to the three coordinate axes. Let these displacements be u, v and w parallel to the x, y and z axes. Then the point originally at (x, y, z) is displaced to $(x+u, y+v, z+w)$. In general the displacement will vary from point to point within a body so that u, v and w are functions of x, y and z. In the case of a plane area, as in Figure 12.3, the displacements u and v are considered to be expressed by functions $u(x, y)$ and $v(x, y)$ which, for convenience, are selected so that rigid body displacements are excluded. The point O with original coordinates (x, y) is displaced to O' $(x+u, y+v)$. For a neighbouring point, e.g. C, with coordinates $(x+dx, y+dy)$ the displacements will be slightly different from those at O because of the changed values $x+dx$, $y+dy$ to be substituted in the functions u, v. Let these displacements be $(u+du)$ and $(v+dv)$.

The differentials of $u(x, y)$ and $v(x, y)$ are

$$du = \frac{\partial u}{\partial x}\,dx + \frac{\partial u}{\partial y}\,dy \quad \text{and} \quad dv = \frac{\partial v}{\partial x}\,dx + \frac{\partial v}{\partial y}\,dy$$

The displacements at $(x+dx, y+dy)$ are therefore

$$u(x+dx, y+dy) = u+du = u + \frac{\partial u}{\partial x}\,dx + \frac{\partial u}{\partial y}\,dy \qquad (12.5a)$$

$$v(x+dx, y+dy) = v+dv = v + \frac{\partial v}{\partial x}\,dx + \frac{\partial v}{\partial y}\,dy \qquad (12.5b)$$

In particular the point A originally at $(x+dx, y)$ will have displacements

$$u + \frac{\partial u}{\partial x}\,dx \quad \text{and} \quad v + \frac{\partial v}{\partial x}\,dx$$

obtained by substituting $dy = 0$ in equations (12.5). The displaced position A' of A then has coordinates (see Fig. 12.3) of

$$\left((x+dx)+\left(u+\frac{\partial u}{\partial x}\,dx\right)\right) \quad \text{and} \quad \left(y+\left(v+\frac{\partial v}{\partial x}\,dx\right)\right)$$

The x component of the new length of OA is the difference between the x coordinates of O' and A'

i.e.
$$\left((x+dx)+\left(u+\frac{\partial u}{\partial x}\,dx\right)\right) - (x+u) \quad \text{or} \quad dx\left(1+\frac{\partial u}{\partial x}\right)$$

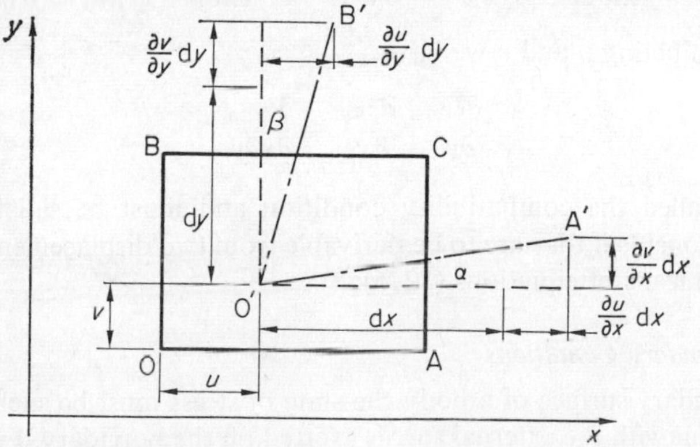

Figure 12.3   O is at (x, y); OA = dx; OB = dy

By definition, normal strain of OA is

$$\epsilon_x = \frac{\text{New length} - \text{Old length}}{\text{Old length}} = \frac{dx\left(1 + \frac{\partial u}{\partial x}\right) - dx}{dx} = \frac{\partial u}{\partial x} \quad (12.6a)$$

i.e. the normal strain in the x direction is the rate of change of x displacement with respect to x.

In a similar fashion we find

$$\epsilon_y = \frac{\partial v}{\partial y} \quad (12.6b)$$

The shear strain γ_{xy} associated with the xy plane was defined in Article 2.3 as the decrease of an original right angle of the undeformed body. In Figure 12.3 $A\hat{O}B$ is an original right angle and $A'\hat{O}'B'$ its deformed condition so that $\gamma_{xy} = A\hat{O}B - A'\hat{O}'B' = \alpha + \beta$. However, for the small angles of infinitesimal strain α is just $(\partial v/\partial x)$ and β is $(\partial u/\partial y)$, i.e.

$$\gamma_{xy} = \frac{\partial u}{\partial y} + \frac{\partial v}{\partial x} \quad (12.6c)$$

Equations (12.6) express the three strain components associated with the xy plane in terms of the displacement functions u and v. Repetition of this procedure for yz and xz planes would result in analogous strain/displacement equations, e.g.

$$\epsilon_z = \frac{\partial w}{\partial z} \quad (12.7)$$

Equations (12.6) express three strain components ϵ_x, ϵ_y, γ_{xy} in terms of two displacement functions u, v; by differentiating these equations we can obtain the following expressions

$$\frac{\partial^2 \epsilon_x}{\partial y^2} = \frac{\partial^3 u}{\partial x \partial y^2}; \qquad \frac{\partial^2 \epsilon_y}{\partial x^2} = \frac{\partial^3 v}{\partial x^2 \partial y}; \qquad \frac{\partial^2 \gamma_{xy}}{\partial x \partial y} = \frac{\partial^3 u}{\partial x \partial y^2} + \frac{\partial^3 v}{\partial x^2 \partial y}$$

Then eliminating u and v we have

$$\frac{\partial^2 \epsilon_x}{\partial y^2} + \frac{\partial^2 \epsilon_y}{\partial x^2} = \frac{\partial^2 \gamma_{xy}}{\partial x \partial y} \quad (12.8)$$

This is called the compatibility condition and must be satisfied by the strain components if they are to be derivable from two displacement functions u and v by means of equations (12.6).

12.1.4 *Boundary Conditions*

At the boundary surface of a body the state of stress must be such that there is equilibrium with the external forces exerted on the boundary. Figure 12.4a shows an elemental prism isolated from a body, with plane AABB part of

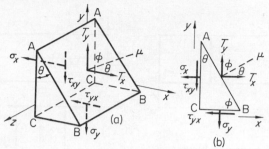

Figure 12.4

the free surface of the body. The stress components on the AACC and CCBB faces are respectively σ_x, τ_{xy} and σ_y, τ_{yx}, and AA = CC = BB = 1. The direction, of the normal to AABB is μ (Figure 12.4b).

The external forces on the boundary surface are described in terms of surface forces per unit area, with components T_x and T_y in the x and y directions, which we shall refer to as surface tractions.

For equilibrium of forces on the prism in the x direction

$$-\sigma_x(\text{AC}\times 1)-\tau_{yx}(\text{CB}\times 1)+T_x(\text{AB}\times 1) = 0$$

Substituting AC = AB $\cos\theta$ and CB = AB $\cos\phi$ we obtain

$$T_x = \sigma_x\cos\theta + \tau_{yx}\cos\phi \qquad (12.9a)$$

For equilibrium of forces in the y direction we find

$$T_y = \tau_{xy}\cos\theta + \sigma_y\cos\phi \qquad (12.9b)$$

Since θ and ϕ are also the angles which μ makes with the x and y directions, the quantities $\cos\theta$ and $\cos\phi$ appearing in equations (12.9) are the direction cosines of the normal to the boundary surface.

Equations (12.9) must be satisfied on those parts of the boundary where the surface tractions T_x and T_y are prescribed. If instead displacements were prescribed the displacement functions u and v would be required to attain the prescribed values when evaluated at the boundary.

12.2 Simple Stress Fields

Here we inquire into the existence, for a thin rectangular plate (Figure 12.5a), of certain simple states of stress which we have assumed in earlier chapters.

(a) We begin with the uniform state of stress of Article 5.2

$$\begin{pmatrix} \sigma_x & 0 & 0 \\ 0 & 0 & 0 \\ 0 & 0 & 0 \end{pmatrix} \qquad (A12.1)$$

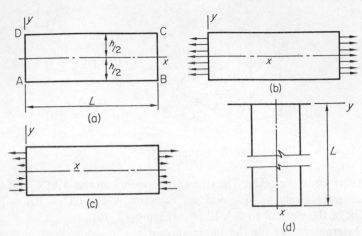

Figure 12.5

The stress equilibrium equations (12.4) are satisfied if $p_x = p_y = 0$. This is effectively true for comparatively small bodies if self-weight is the only body force. From the stress boundary conditions (12.9) we find that on BC (where $\theta = 0°$ and $\phi = 90°$) the surface tractions are

$$T_x = \sigma_x \quad \text{and} \quad T_y = 0$$

i.e. uniformly distributed normal tensile forces as in Figure 12.5b. In a similar manner we find that DC should be free of tractions. Thus the stress distribution assumed in Article 5.2 is true only if the end forces are as in Figure 12.5b. However, by St. Venant's Principle (Article 4.6) it is also true away from the ends, for any statically equivalent end force distribution.

(b) Now let us consider a state of stress of the form of (A12.1) in which

$$\sigma_x = b_0 + b_1 y \tag{12.10}$$

and b_0 and b_1 are constants. Again the equilibrium equations (12.4) are satisfied when body forces are negligible. When b_0 is zero the stress distribution corresponds to that for pure bending of a beam (equation 6.14, Section 6.3.3). The stress boundary conditions found from (12.9) for AD and BC (Figure 12.5a) are, noting that $\theta = 180°$ for AD,

$$T_x = \mp b_1 y; \quad T_y = 0$$

Similarly AB and CD are found to be free of tractions, and the boundary forces are as in Figure 12.5c. These tractions are statically equivalent to end moments as in Figure 6.11b and so away from the ends of a bar the assumed bending stress distribution is correct.

If b_0 is not zero it should be clear by reference to (a) that (12.10) would correspond to combined direct loading and bending as discussed in Article 9.1.

(c) Lastly we consider a state of stress of the form of (A12.1) when

$$\sigma_x = b_0 + b_1 x$$

The equilibrium equations (12.4) require

$$b_1 + p_x = 0; \qquad p_y = 0$$

so that

$$\sigma_x = b_0 - p_x x \tag{12.11}$$

This stress distribution is only possible if the body force acts in the x direction alone, as for example, if the plate is suspended vertically and loaded by its own weight, as in Figure 12.5d. In this problem we shall use the boundary conditions (12.9) in the normal way, i.e. to obtain information about the internal stress distribution. Thus at $x = L$ where there is no surface traction

$$T_x = 0 = \sigma_x$$

Then putting $\sigma_x = 0$ at $x = L$ in (12.11) we have $b_0 = p_x L$ and so

$$\sigma_x = p_x(L - x)$$

Thus the stress in the plate is greatest at $x = 0$ and decreases linearly to zero at $x = L$. A surface traction of magnitude $p_x L$ is of course required at the upper end. From this result we can infer the stresses in a suspended rod such as an oil well drill rod or a cable in a mine shaft. As a numerical example consider a steel rod (density 7950 kg/m³); the body force intensity is $p_x = \rho g$ and so

$$(\sigma_x)_{max} = \rho g L = 78,000 \, L \, \text{N/m}^2$$

The yield stress of 210 MN/m² for mild steel is attained when the length is 2·69 km.

Problems for Solution 12.1–12.3

12.3 Airy Stress Function

12.3.1 *Reduction of Field Equations*

Eight field equations of plane elasticity (equations 12.4, 12.6 and 3.9 or 3.12) have been derived relating the eight variables σ_x, σ_y, τ_{xy}, ϵ_x, ϵ_y, γ_{xy}, u, v. However, the compatibility condition (12.8) can be used to replace the three strain/displacement equations leaving six equations—two equilibrium, three stress/strain, one compatibility—for the six remaining unknowns σ_x, σ_y, τ_{xy}, ϵ_x, ϵ_y, γ_{xy}.

The strain/stress equations (3.9) or (3.12) can be substituted in the compatibility condition (12.8) to express it in terms of stress. Using (3.9) for plane stress we obtain

$$\frac{\partial^2}{\partial x^2}(\sigma_y - \nu\sigma_x) + \frac{\partial^2}{\partial y^2}(\sigma_x - \nu\sigma_y) = 2(1+\nu)\frac{\partial^2 \tau_{xy}}{\partial x \partial y} \qquad (12.12)$$

which leaves three equations—two equilibrium, one stress/compatibility—relating three unknowns σ_x, σ_y, τ_{xy}. An alternative form of equation (12.12) can be obtained by introducing certain functions of the equilibrium equations. Thus taking $\partial/\partial x$ of equation (12.4b) and $\partial/\partial y$ of (12.4a) and adding, we obtain, when $p_x = p_y = 0$

$$\frac{2\partial^2 \tau_{xy}}{\partial x \partial y} = -\frac{\partial^2 \sigma_x}{\partial x^2} - \frac{\partial^2 \sigma_y}{\partial y^2}$$

Substituting this in (12.12) and rearranging we have,

$$\left(\frac{\partial^2}{\partial x^2} + \frac{\partial^2}{\partial y^2}\right)(\sigma_x + \sigma_y) = 0 \qquad (12.13)$$

Any stress field which satisfies (12.4) and either (12.12) or (12.13) is a possible solution subject to the boundary conditions. In the case of plane strain, substitution of equations (3.12) in (12.8) leads ultimately in a similar manner to equation (12.13) provided the body forces are zero. It should also be noted that, since the reduced governing equations (12.4) and (12.13) do not contain material properties, the stress field is the same in objects of the same shape loaded in the same manner regardless of the material.

The system of equations may be further reduced by introducing a new variable $\Phi(x, y)$. If the stresses are derived therefrom by

$$\sigma_x = \frac{\partial^2 \Phi}{\partial y^2}; \qquad \sigma_y = \frac{\partial^2 \Phi}{\partial x^2}; \qquad \tau_{xy} = -\frac{\partial^2 \Phi}{\partial x \partial y} \qquad (12.14a, b, c)$$

then it is easily verified by substitution that the stress equilibrium equations (12.4) are always satisfied when body forces are zero. However, if Φ is to represent a problem in plane elasticity it must also satisfy the stress/compatibility equation (12.13). Substituting (12.14) in (12.13) we find

$$\frac{\partial^4 \Phi}{\partial x^4} + 2\frac{\partial^4 \Phi}{\partial x^2 \partial y^2} + \frac{\partial^4 \Phi}{\partial y^4} = 0 \qquad (12.15)$$

The solution of a problem in plane elasticity (body forces zero) is therefore reduced to that of finding a function Φ which satisfies (12.15) and the stress/boundary conditions (12.9). This function is called Airy's Stress Function.

12.3.2 *Applications*

If Φ is taken to be a polynomial in x and y some interesting solutions are obtained for a rectangular strip as in Figure 12.5a. By inspection we see that any polynomial of degree three or less automatically satisfies equation (12.15) and a first degree expression will correspond to zero stresses.

(a) *Second Degree*

$$\Phi = \tfrac{1}{2}A_2x^2 + B_2xy + \tfrac{1}{2}C_2y^2$$

Applying (12.14) we have

$$\sigma_x = C_2; \qquad \sigma_y = A_2; \qquad \tau_{xy} = -B_2$$

corresponding to a uniform stress field for which appropriate surface tractions are obtained from (12.9). In particular if $A_2 = B_2 = 0$ the stress function corresponds to a bar in direct loading as in Figure 12.5b. If $A_2 = C_2 = 0$ the stress function corresponds to a state of pure shear, as might be found in thin-panel structures.

(b) *Third Degree*

$$\Phi = \frac{A_3x^3}{3!} + \tfrac{1}{2}B_3x^2y + \tfrac{1}{2}C_3xy^2 + \frac{D_3y^3}{3!}$$

Here

$$\sigma_x = C_3x + D_3y; \qquad \sigma_y = A_3x + B_3y; \qquad \tau_{xy} = -B_3x - C_3y$$

If $A_3 = B_3 = C_3 = 0$ the system corresponds to pure bending (Figure 12.5c).

(c) *Superposition* Since equation (12.15) is a linear partial differential equation, solutions may be superimposed. As an example consider the function

$$\Phi = \tfrac{1}{2}C_2y^2 + \frac{D_3y^3}{3!}$$

which is a sum of two particular cases considered in (a) and (b). Hence

$$\sigma_x = C_2 + D_3y; \qquad \sigma_y = 0; \qquad \tau_{xy} = 0$$

which corresponds to combined direct loading and bending.

Problems for Solution 12.4, 12.5.

12.4 Thermal Effects

When thermal expansion occurs (see also Section 5.4.1) the functions u and v now represent both the displacements due to loads and those due to temperature changes. However, since the arguments of Section 12.1.3 are independent of the cause of the displacements, we can apply equations (12.6) to obtain the total strains

$$e_{xx} = \frac{\partial u}{\partial x}; \qquad e_{yy} = \frac{\partial v}{\partial y}; \qquad e_{xy} = \frac{\partial u}{\partial y} + \frac{\partial v}{\partial x} \qquad \text{(12.16a, b, c)}$$

which are the sum of the strains due to loading and thermal effects. Thus the normal strain in the x direction is

$$e_{xx} = \epsilon_x + \alpha T \qquad \qquad \text{(12.17a)}$$

where ϵ_x is the corresponding load strain, T is the temperature change, and α is the coefficient of linear expansion of the material (see equation 5.33). Similarly

$$e_{yy} = \epsilon_y + \alpha T \qquad \qquad \text{(12.17b)}$$

Since the change of temperature causes only volume expansion the shear strain is due only to the loads, i.e.

$$e_{xy} = \gamma_{xy} \qquad \qquad \text{(12.17c)}$$

The total strains can now be written in matrix form as

$$\begin{bmatrix} e_{xx} \\ e_{yy} \\ e_{xy} \end{bmatrix} = \begin{bmatrix} \epsilon_x \\ \epsilon_y \\ \gamma_{xy} \end{bmatrix} + \alpha T \begin{bmatrix} 1 \\ 1 \\ 0 \end{bmatrix} \qquad \text{(12.18)}$$

Since the load strains are related to the stresses by the strain/stress relations (3.13) and (3.15) we have for the case of plane stress

$$\begin{bmatrix} e_{xx} \\ e_{yy} \\ e_{xy} \end{bmatrix} = \frac{1}{E} \begin{bmatrix} 1 & -\nu & 0 \\ -\nu & 1 & 0 \\ 0 & 0 & 2(1+\nu) \end{bmatrix} \begin{bmatrix} \sigma_x \\ \sigma_y \\ \tau_{xy} \end{bmatrix} + \alpha T \begin{bmatrix} 1 \\ 1 \\ 0 \end{bmatrix} \qquad \text{(12.19)}$$

These equations may be solved for the stresses to give

$$\begin{bmatrix} \sigma_x \\ \sigma_y \\ \tau_{xy} \end{bmatrix} = \frac{E}{1-\nu^2} \begin{bmatrix} 1 & \nu & 0 \\ \nu & 1 & 0 \\ 0 & 0 & \frac{1-\nu}{2} \end{bmatrix} \begin{bmatrix} e_{xx} \\ e_{yy} \\ e_{xy} \end{bmatrix} - \frac{E\alpha T}{1-\nu} \begin{bmatrix} 1 \\ 1 \\ 0 \end{bmatrix} \qquad \text{(12.20)}$$

In terms of stress, the arguments of Article 12.1.2 and 12.1.4 are unchanged so that the stress equilibrium equations are (12.4) and the stress boundary conditions are (12.9).

These results will be utilized in Chapter 13 when the Finite Element Method is applied to thermal stress problems.

12.5 Virtual Work and Energy Methods

12.5.1 *Principle of Virtual Work (or Virtual Displacements)*

This basic principle of mechanics is introduced for a system of particles in Appendix I and this description can be summarized as follows. A system of particles is in equilibrium if the virtual work performed during any arbitrary virtual displacement is zero. In this context a virtual displacement is a fictitious displacement during which the forces on the particles remain unchanged; in general an *actual* displacement would result in changes of the forces. The virtual displacements can be any displacements which are continuous (so that there are no cracks or overlaps) and which do not violate the prescribed conditions of displacement at the supports (for example no translation at point A of Figure 4.5a). The virtual work is the sum of the products of each force and the virtual displacement along its line of action. The virtual displacement process may be considered as a sort of mathematical experiment to test for equilibrium.

To interpret the principle for an elastic body we begin by referring to the discussion in Section 4.1.1 of a bar in simple tension. It was found that, provided the bar is in equilibrium throughout the loading process, the work done by the external forces is stored as strain energy (equation 4.5). Then during an infinitesimal displacement δu (Figure 4.1b) the increments of work and strain energy are equal

i.e. $$\delta W_e = \delta U \tag{12.21}$$

It is clear from Figure 4.1b (compare equation 4.2) that the increment of work is given to a first approximation by $X\delta u$ which is the product of the force X (corresponding to displacement u) and the increment of displacement

i.e. $$\delta W_e = X\delta u \tag{12.22}$$

Noting that $X = A\sigma_x$ and that, since $\epsilon_x = u/L$, we can write $\delta u = L\delta\epsilon_x$ we obtain

$$X\delta u = \sigma_x\delta\epsilon_x AL \tag{12.23}$$

where $\delta\epsilon_x$ is the increment of strain due to the increment of displacement δu. Comparison with equations (12.21) and (12.22) shows that the quantity on the right-hand side of (12.23) is the increment of strain energy δU.

Now if the forces X were given a virtual displacement u^*, of magnitude equal to δu, then the virtual work of the external forces would be the product $Xu^* = X\delta u$. Thus the virtual work of the external forces is the same as the increment of actual work due to an infinitesimal actual displacement of the forces. Similarly, since we have taken u^* equal to δu, we can say that $\delta u/L = u^*/L$.

i.e. $$\delta\epsilon_x = \epsilon^* \tag{12.24}$$

where ϵ_x^* is called the virtual strain.

However, if the bar is in equilibrium subject to forces X we can apply (12.23). Then putting $\delta u = u^*$ and $\delta\epsilon_x = \epsilon_x^*$ we have

$$Xu^* = \sigma_x \epsilon_x^* AL \tag{12.25}$$

in which the left-hand side is the virtual work of the external forces denoted W_e^*. The quantity on the right is similar to the increment of strain energy and will be termed the virtual strain energy, denoted by U^*. Then (12.25) can be written

$$W_e^* = U^* \tag{12.26}$$

Thus if the bar is in equilibrium the virtual work performed by the external forces during an admissible virtual displacement is equal to the virtual strain energy. A similar conclusion can be reached for a body subject to an arbitrary set of loads (e.g. Figure 4.5a) and we can now state the Principle of Virtual Work in a form suitable for use in elasticity theory, as follows:

'An elastic body is in equilibrium, if, for any set of virtual displacements which do not violate the constraints, the virtual work of the external forces is equal to the virtual strain energy.'

For a general loading the virtual strain energy becomes

$$U^* = \int_V U_0^* \, dV \tag{12.27}$$

where

$$U_0^* = \sigma_x \epsilon_x^* + \sigma_y \epsilon_y^* + \sigma_z \epsilon_z^* + \tau_{xy} \gamma_{xy}^* + \tau_{yz} \gamma_{yz}^* + \tau_{xz} \gamma_{xz}^* \tag{12.28}$$

is the virtual strain energy density.

A formal proof for an arbitrary loading may be carried out starting from stress equilibrium equations such as (12.4). However, the mathematical complexity is outside the scope of this text and may be found in, for example, references 2 and 3.

It should be noted that the above discussion of the Principal of Virtual Work has been given primarily in preparation for its application in Chapter 13. However, some other applications can now be noted.

(a) *Minimum of the Total Potential* Introducing $\bar{V} = -W_e$ as the change of potential energy of the external loads (measured from zero relative displacement as datum), we can write (12.26) as

$$(U + \bar{V})^* = 0 \tag{12.29}$$

The quantity $(U + \bar{V})$ is known as the total potential and we see that during any admissible virtual displacement the virtual change of this quantity is zero. This result can be interpreted as requiring that when the body is in equilibrium the total potential has a stationary value (which can be shown to be a minimum). We can then state the Theorem of a Minimum of the Total Potential as:

'Of all the admissible displacement distributions, that which makes the total potential a minimum is the one which actually occurs at equilibrium.'

This theorem is of limited use for exact analysis but is of wide application when applied to an approximating procedure such as the Rayleigh–Ritz method to be described in Section 12.5.2.

As defined above, the potential of the external loads is the product of their final value and their final displacement, e.g. the area BHCD in Figure 4.4b. The strain energy actually stored in this example is the area BCH. Then the area BCD, which in Article 4.3 was defined as the complementary energy, could be interpreted as that part of the potential of the loads which cannot be converted to strain energy under equilibrium conditions. This is obviously also true for a non-linear elastic system (Figure 4.4a).

(b) *Castigliano's Theorem*: *Part 1*   Consider a body in equilibrium subject to arbitrary loads $F_1, F_2, \ldots F_i, \ldots F_n$ as in Figure 4.5a and provided with just sufficient supports to prevent rigid body displacements. Let the load F_i undergo a virtual displacement u_i^* which is a translation or rotation according as F_i is a force or a moment. Then since the body is in equilibrium we have by the Principle of Virtual Work

$$F_i u_i^* = U^* \quad \text{or} \quad F_i = \frac{U^*}{u_i^*}$$

However, as discussed earlier, a virtual displacement u_i^* is equivalent to an infinitesimal actual displacement δu_i. Therefore in the limit as u_i^* tends to zero we have

$$F_i = \frac{\partial U}{\partial u_i} \tag{12.30}$$

in which the partial derivative recognizes the dependence of U on the displacements of all the loads. This result is known as Castigliano's Theorem: Part 1 (see also the discussion in Article 4.4).

(c) *Complementary Virtual Work*   All of the above discussion was based on the equality of the real work done on a body and its strain energy (equation 12.21). It is reasonable to infer that analogous conclusions could be drawn from the corresponding statement (equations 4.21 and 4.23) for increments of complementary work and energy. A full discussion is beyond the scope of this text (see for example reference 3) but a few points can be noted for general interest.

Instead of virtual displacements, the forces are varied without changing the displacements. Equality of the complementary virtual work done by these virtual loads and the complementary virtual strain energy is interpreted as a condition that the displacements are geometrically compatible. For instance if a virtual load F_i^* is applied to the body of Figure 4.5a in the direction of displacement u_i the complementary virtual work is $u_i F_i^*$ and we have

$$u_i F_i^* = \bar{U}^*$$

In the limit as F_i^* tends to zero this becomes

$$u_i = \frac{\partial \bar{U}}{\partial F_i} \tag{12.31}$$

which is Castigliano's Theorem: Part 2 (see Article 4.4).

12.5.2 *Rayleigh–Ritz Procedure*

The displacements are expressed by suitable approximating functions, e.g. a power series with r undetermined coefficients. Using the assumed functions the total potential energy is evaluated in terms of these coefficients. The partial derivatives of this approximate total potential are taken with respect to each of the r coefficients and set equal to zero. In this way a set of r linear equations is obtained and the solution of these equations gives that set of r coefficients which minimizes the approximate total potential energy.

As an example, if $v(x)$ is the only displacement we might let

$$v = a_0 + a_1 x + a_2 x^2 + \ldots \quad \ldots + a_n x^n$$

Then we evaluate

$$\bar{P} = U + \bar{V} = \bar{P}(a_0, a_1, a_2, \ldots \quad \ldots, a_n)$$

The minimum of the total potential is then obtained by solving the equations

$$\frac{\partial \bar{P}}{\partial a_0} = 0; \qquad \frac{\partial \bar{P}}{\partial a_1} = 0; \qquad \ldots \quad \ldots \frac{\partial \bar{P}}{\partial a_n} = 0$$

i.e. the quantities $a_0, a_1, \ldots a_n$ are being selected to give the least value of $\bar{P}$ possible within the accuracy of the approximating function.

Example 12.1   Find the end deflexion of a cantilever AB of length L carrying a concentrated moment m at the free end B.

Solution   Let us try

$$v = a_0 + a_1 x + a_2 x^2 \tag{i}$$

Then

$$\frac{dv}{dx} = a_1 + 2a_2 x \tag{ii}$$

and

$$\frac{d^2 v}{dx^2} = 2a_2$$

From Section 7.2.1

$$U = \int_0^L \left(\frac{\mathrm{d}^2 v}{\mathrm{d}x^2}\right)^2 \frac{EI}{2}\,\mathrm{d}x = \int_0^L (4a_2^2)\frac{EI}{2}\,\mathrm{d}x$$
$$= 2a_2^2 EIL$$

$$\bar{V} = -m\theta_B = -m\left(\frac{\mathrm{d}v}{\mathrm{d}x}\right)_{x=L} = -m(a_1 + 2a_2 L)$$

Hence

$$\bar{P} = U + \bar{V} = 2a_2^2 EIL - ma_1 - 2ma_2 L$$

and

$$\frac{\partial \bar{P}}{\partial a_2} = 4EILa_2 - 2mL = 0$$

Hence

$$a_2 = \frac{m}{2EI} \tag{iii}$$

Also the prescribed displacements require

$$v = \frac{\mathrm{d}v}{\mathrm{d}x} = 0 \quad \text{at} \quad x = 0$$

so that from (i) and (ii)

$$a_0 = a_1 = 0 \tag{iv}$$

Substuting (iii) and (iv) in (i) and putting $x = L$ we have $v_B = mL^2/2EI$ which is the exact result listed in Figure 7.5.

In the above example the simple nature of the problem permitted an exact solution. However, the approximating functions chosen could have been improved by applying at the outset the prescribed displacements to give $a_0 = a_1 = 0$. If the assumed displacement function is not the same as the actual function the results will be more or less approximate. If the true function is different in form in different parts of the beam as in Example 7.3, a single function can never be exact. This suggests that an improved result might be obtained if the approximating functions were allowed to be different in a finite number of sub-divisions (elements) of the system. This is, in a sense, what is done in the Finite Element Method to be described in the next chapter.

Problems for Solution 12.6, 12.7

12.5.3 *The Reciprocal Theorem*

Let us consider a linear elastic body which may be subjected to concentrated loads (forces and moments) at any number of points. These forces or moments can each be resolved into three components relative to general coordinate axes so that at any point i the loads are defined by six quantities X_i, Y_i, Z_i, $m_{x,i}$, $m_{y,i}$, $m_{z,i}$. Then if there are n load points on the body a set of loads can be conveniently represented by a column matrix of size $(6n \times 1)$

$$\mathbf{F_I} = \{X_1 \quad Y_1 \quad Z_1 \quad m_{x,1} \ldots m_{y,n} \quad m_{z,n}\}$$

Corresponding to each load component at a point are displacement components u_i, v_i, w_i, ψ_i, ϕ_i, θ_i. The set of displacements due to load system I can also be represented as a $(6n \times 1)$ column matrix

$$\boldsymbol{\delta}_\mathrm{I} = \{u_{1\mathrm{I}} \quad v_{1\mathrm{I}} \quad w_{1\mathrm{I}} \quad \psi_{1\mathrm{I}} \ldots \phi_{n\mathrm{I}} \quad \theta_{n\mathrm{I}}\}$$

where the first subscript denotes location and the Roman numeral the load system causing the displacement.

It is supposed that a set of loads is applied in such a way that all components of load increase slowly and in the same proportions until the final values are attained. Then each component of displacement varies with force in the manner of Figure 4.1b and the work done is half their product. Using the matrix procedures of Section 4.1.3 we can say that the work done is $\frac{1}{2}\mathbf{F}_\mathrm{I}^\mathrm{T}\boldsymbol{\delta}_\mathrm{I}$.

Now let us suppose that a second set of loads $\mathbf{F}_\mathrm{II}$ are applied at the same points in addition to the first set and cause a set of additional displacements $\boldsymbol{\delta}_\mathrm{II}$ to be generated. As considered above these loads do work $\frac{1}{2}\mathbf{F}_\mathrm{II}^\mathrm{T}\boldsymbol{\delta}_\mathrm{II}$ in producing these displacements. However, the first set of loads also experience the additional displacements $\boldsymbol{\delta}_\mathrm{II}$. During these displacements the loads $\mathbf{F}_\mathrm{I}$ have constant magnitude so that the work done by them is $\mathbf{F}_\mathrm{I}^\mathrm{T}\boldsymbol{\delta}_\mathrm{II}$. Then the total work done during the application of the two sets of loads is

$$W_{\mathrm{I},\mathrm{II}} = \tfrac{1}{2}\mathbf{F}_\mathrm{I}^\mathrm{T}\boldsymbol{\delta}_\mathrm{I} + \tfrac{1}{2}\mathbf{F}_\mathrm{II}^\mathrm{T}\boldsymbol{\delta}_\mathrm{II} + \mathbf{F}_\mathrm{I}^\mathrm{T}\boldsymbol{\delta}_\mathrm{II}$$

where the order of subscripts of W indicates the order of application of the two sets of loads.

If the order of loading is reversed the work done during application of the $\mathbf{F}_\mathrm{II}$ set is $\frac{1}{2}\mathbf{F}_\mathrm{II}^\mathrm{T}\boldsymbol{\delta}_\mathrm{II}$. Then when the set $\mathbf{F}_\mathrm{I}$ is also applied the additional work is $(\frac{1}{2}\mathbf{F}_\mathrm{I}^\mathrm{T}\boldsymbol{\delta}_\mathrm{I} + \mathbf{F}_\mathrm{II}^\mathrm{T}\boldsymbol{\delta}_\mathrm{I})$ since the $\mathbf{F}_\mathrm{II}$ loads already present act through the displacements $\boldsymbol{\delta}_\mathrm{I}$ produced by the $\mathbf{F}_\mathrm{I}$ set. Then

$$W_{\mathrm{II},\mathrm{I}} = \tfrac{1}{2}\mathbf{F}_\mathrm{I}^\mathrm{T}\boldsymbol{\delta}_\mathrm{I} + \tfrac{1}{2}\mathbf{F}_\mathrm{II}^\mathrm{T}\boldsymbol{\delta}_\mathrm{II} + \mathbf{F}_\mathrm{II}^\mathrm{T}\boldsymbol{\delta}_\mathrm{I}$$

However, since the work done is equal to the strain energy stored, and the strain energy of an elastic material does not depend on the order of application of the loads, it follows that $W_{\mathrm{I},\mathrm{II}} = W_{\mathrm{II},\mathrm{I}}$ and so

$$\mathbf{F}_\mathrm{I}^\mathrm{T}\boldsymbol{\delta}_\mathrm{II} = \mathbf{F}_\mathrm{II}^\mathrm{T}\boldsymbol{\delta}_\mathrm{I} \qquad (12.32)$$

This expression of the Reciprocal Theorem was first given by Betti and can be stated as:

'The work done by the loads of a first load system acting over the displacements due to a second load system, is equal to the work done by the second system of loads acting over the displacements due to the first system.'

Maxwell was in fact the first to describe a Reciprocal Theorem but in the particular form obtained when the two load systems consist of the same force applied at different locations. The original papers are somewhat obscure but an account has been given by Niles.[4]

As a simple example two load locations have been chosen on a beam (Figures 12.6a and b) with load systems

$$F_I = \{Y_1 \quad 0 \quad 0 \quad 0\} \quad \text{and} \quad F_{II} = \{0 \quad 0 \quad 0 \quad m_2\}$$

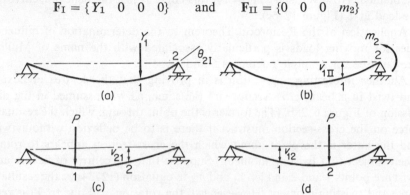

Figure 12.6

The displacements are

$$\delta_I = \{v_{1I} \quad \theta_{1I} \quad v_{2I} \quad \theta_{2I}\} \quad \text{and} \quad \delta_{II} = \{v_{1II} \quad \theta_{1II} \quad v_{2II} \quad \theta_{2II}\}$$

Then applying equation (12.32) we obtain

$$Y_1 v_{1II} = m_2 \theta_{2I} \tag{12.33}$$

Results of this type are often used experimentally when for example the angle θ_{2I} could be deduced from the more easily measured deflexion v_{1II}. This method can also be used to deduce deflexion or slope at an inaccessible location. Since there is only one component in each of the above load systems the terminology can be simplified to

$$Y_1 v_{12} = m_2 \theta_{21} \tag{12.34}$$

Another important result is obtained when each load system contains the same load component but at different locations. Then referring to Figures 12.6c and d

$$F_I = \{P \quad 0\}; \quad F_{II} = \{0 \quad P\}; \quad \delta_I = \{v_{11} \quad v_{21}\}; \quad \delta_{II} = \{v_{12} \quad v_{22}\}$$

and from (12.32)

$$v_{12} = v_{21} \tag{12.35}$$

which can be stated as: 'The deflexion at location 1 due to a load at location 2 is the same as the deflexion at 2 due to the same load acting at 1.' If in Figure

12.6c the load location is constant while 2 is allowed to move along the beam then v_{21} traces out its deflected shape. Now let the deflexion at location 1 be measured while the load location 2 (Figure 12.6d) is allowed to move along the beam. If the succession of values of v_{12} so obtained is plotted against the load position 2 a curve known as the influence line for deflexion at position 1 is obtained. However, since $v_{12} = v_{21}$ this curve is just the deflexion curve for the load at 1 (Figure 12.6c).

Application of the Reciprocal Theorem to the determination of influence lines for moving loads is particularly associated with the name of Muller-Breslau (see for example references 5 and 6).

Another interesting application is in proving that the centres of flexure and twist in a beam cross-section are the same, as was assumed in the discussion of Figure 6.28b. (The former is the point through which the resultant force on the cross-section must act if there is to be deflexion without twist and the latter is the point about which the cross-section appears to rotate when subject to a twisting moment.) Suppose that the centres of flexure and twist are points 1 and 2 and let Y_1 and m_2 in equation (12.34) be the resultant force and twisting moment. However θ_{21}, the rotation at 2 due to Y_1 is zero by virtue of the definition of the centre of flexure. Consequently from equation (12.34) v_{12} the deflexion at point 1 due to a twisting moment, is also zero; this can only be true if point 1, i.e. the centre of flexure, is also the centre of twist.

Lastly, referring to equation (12.34) we note that the deflexion v_{12} being caused by m_2 (the load at location 2) could be expressed as (m_2/s_{21}) where s_{21} is a 'stiffness' coefficient defined as the load at 2 to cause unit deflexion at location 1. Similarly we can write (Y_1/s_{12}) for θ_{21} and so substituting in (12.34) we obtain the result

$$s_{12} = s_{21} \tag{12.36}$$

which explains the observed symmetry of stiffness matrices.

12.6 References

(a) *Cited*

1. Chou, P. C. and Pagano, N. J., *Elasticity*, Van Nostrand, Princeton, N.J., 1967.
2. Argyris, J. H., *Energy Theorems and Structural Analysis*, Butterworth, London, 1960.
3. Przemieniecki, J. S., *Theory of Matrix Structural Analysis*, McGraw-Hill, N.Y., 1968.
4. Niles, A. S., 'Clerk Maxwell and the theory of indeterminate structures', *Engineering*, **170**, 194 (1950).
5. Preece, B. W. and Davies, J. D., *Models for Structural Concrete*, C.R. Books, London, 1964.
6. Charlton, T. M., *Model Analysis of Plane Structures*, Pergamon, Oxford, 1966.

(b) *General*

Timoshenko, S. P. and Goodier, J. N., loc. cit., Article 1.15.
Boley, B. A. and Weiner, J. H., *Theory of Thermal Stresses*, Wiley, N. Y., 1960.
Johns, D. J., loc. cit., Article 5.7.

12.7 Problems for Solution

12.1. Interpret the stress boundary conditions for a rectangular plate if σ_x and σ_y are constants and $\tau_{xy} = 0$.

12.2. Interpret the condition of a rectangular plate in which $\sigma_x = b_1xy$, $\sigma_y = 0$, $\tau_{xy} = b_2(1 - (4y^2/h^2))$ where h is as in Figure 12.5a.

12.3. A plate tapers uniformly in width from $2h$ to h over length L. The only boundary forces are normal and uniformly distributed over the parallel ends. If body forces are negligible find whether or not the state of stress can be of the form of (A12.1).

12.4. The stress function for a rectangular plate in plane stress has the form $\Phi = \frac{1}{2}Ax^2 + \frac{1}{2}By^2$. Interpret the stress distribution and state the maximum shear stress when (a) $A = B$, (b) $B = -A$.

12.5. The stress function for a plate as in Figure 12.5a and in plane stress, has the form $\Phi = Bxy^3 - Cxy$. Find the stress distribution in terms of C if AB and DC are free surfaces. State the maximum shear stress when (a) $y = 0$, (b) $y = \pm h/2$. To what bending problem does this stress function correspond?

12.6. A simply supported beam, length L, is subject to concentrated moments m at $x = 0$, $-m$ at $x = L$. Find the central deflexion and slope.

12.7. A cantilever, length L, carries a concentrated force P acting down at the free end. Find the end deflexion.

The Finite Element Method for Plane Elasticity

13.1 Introduction

In Chapters 5, 7, 8 and 9 methods were developed for the analysis of collinear structures such as continuous beams by considering them to consist of an assembly of **n** bars joining a finite number n of points called nodes (Figure 5.3). Loads were considered to act only at nodes and the displacement of the system was characterized by the displacements of the nodes. From a knowledge of the distribution of stress and strain in the bar linear stiffness equations (such as 7.14 and 7.29) were derived expressing end loads on one bar in terms of its end displacements. Then, by considering compatibility of displacement and equilibrium at the nodes, stiffness equations for the assembly can be formed from those for the individual bars, and have the general form (compare equations 7.42),

$$F = K\delta \tag{13.1}$$

Here

$$F = \{F_1 \quad F_2 \ldots F_i \ldots F_n\} \tag{13.2}$$

and

$$\delta = \{\delta_1 \quad \delta_2 \ldots \delta_i \ldots \delta_n\} \tag{13.3}$$

are respectively, vectors of nodal loads and displacements, and K is a square matrix of stiffness coefficients.

Representation of the collinear structure in terms of its behaviour at a finite number of points constitutes a very considerable simplification. This arises from the fact that the complete structure has of course a continuous stress/strain distribution, which from Chapter 12 we can expect to be governed by a set of partial differential equations. However, the need for a complete solution of these for the overall boundary conditions has been avoided by the assembly process. The differential equations are solved only within each bar in order to relate its end loads and displacements by algebraic equations (compare Section 7.4.3), which are then used to form the assembly stiffness equations. Since these are easy to form and solve using a digital computer very large systems can be analysed relatively quickly. We might call this technique a process of 'discrete representation' since the behaviour of the structure is expressed in terms of loads and displacements at a number of discrete points (the nodes) by first considering the behaviour of a number of discrete regions, or finite elements, of the structure.

It is possible to apply the concept of discrete representation to problems of plane elasticity. To this end we choose a finite number of nodes numbered 1, 2, ... *n*, and distribute them over the plane area (Figure 13.1). The

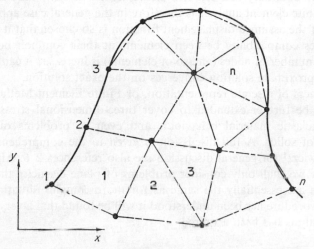

Figure 13.1

choice of locations is less obvious than for a structure composed of bars (Figure 5.3), and will depend on the judgement of the analyst; boundaries must be adequately covered, and nodes may be concentrated in regions where stresses are expected to be relatively large. Unlike assemblies of bars there are no natural sub-divisions and we therefore imagine the plane divided into polygonal finite elements (numbered **1, 2, ... n**) by straight lines joining each node to at least two others (Figure 13.1). Curved boundaries are then usually represented with sufficient accuracy by a series of straight lines as shown (while it is possible to have elements with curved sides this is outside the scope of an introductory treatment).

Following the previous discussion for an assembly of bars we now need to develop stiffness equations relating nodal loads and displacements for each finite element. Unfortunately exact expressions comparable to those of equations (7.14 and 7.29) cannot be obtained for a polygonal element. This difficulty is overcome by postulating the form of the relation between displacement and position within an element. The equations of Chapter 12 can then be used to deduce the element stiffness equations from which the assembly stiffness equations (13.1) are obtained by considering compatibility of displacement and equilibrium at the nodes.

Comparing the plane body procedure to that for an assembly of bars we note two sources of approximation. In the first place the arbitrary choice of displacement function restricts the ability of the discrete system to represent the stress/strain behaviour of the plane body. Secondly, compatibility and equilibrium between elements are imposed only at a finite number of points (the nodes) whereas in the actual body these conditions must be met over the whole linear boundary between two elements, i.e. at an infinite number of points. A finite element analysis is therefore in the general case approximate. However, if the assumed displacement function is so chosen that it automatically ensures compatibility between elements at their common boundaries, then as the number of nodes (and so of elements) is increased it can be shown that the approximate solution converges on the exact solution.

This process of discrete representation, or Finite Element Method as it is called, can be further extended to cover three-dimensional stress analysis, cases of inelastic material behaviour, and even to problems outside the mechanics of solids. A full discussion is given in the comprehensive work by Zienkiewicz;[1] for general discussion see also references 2–6. Although in this chapter we shall only consider problems of plane elasticity the steps of the analysis are essentially the same as for more complex situations. Once this basic procedure has been understood it will be found that a very powerful tool for analysis has been acquired.

13.2 Finite Element Representation of Plane Elasticity

As discussed in Article 12.1 a body in a state of plane elasticity in the xy plane can be considered by reference to a thin slice of thickness t with middle surface in the xy plane and subject to forces parallel to the xy plane and uniformly distributed over t (Figure 12.1). If σ_z is zero everywhere the body is in plane stress, whereas if ϵ_z is everywhere zero, we have a condition of plane strain. In either case the elastic stress distribution can be found if the displacements u and v of the body parallel to the x and y directions are known at every point in the xy plane.

Reference to equation (3.11c) shows that in plane strain σ_z is not zero in general. There must therefore be normal forces on the end surfaces of the slice normal to the z direction. However, because there is zero displacement in the z direction ($\epsilon_z = 0$) these forces need not be taken into account. Further justification of this statement will be indicated at a later state in the development.

In the discrete representation method the behaviour of the body will be described in terms of the loads and displacements at n nodes (Figure 13.1).

At any node i external forces X_i, Y_i act parallel to the x, y axes; displacements u_i, v_i of the node occur parallel to the x, y axes. The nodal load and displacement vectors are therefore

$$\mathbf{F}_i = \{X_i \quad Y_i\} \qquad \text{and} \qquad \boldsymbol{\delta}_i = \{u_i \quad v_i\} \qquad (13.4\text{a, b})$$

Referring to equations (13.2) and (13.3) we see that for the whole body there are $2n$ components of nodal load, and of nodal displacement. Since the object of our investigation is to express any load component as a function of all the displacement components, typical relations would be

$$X_i = s_{2i-1,1}u_1 + s_{2i-1,2}v_1 + \ldots + s_{2i-1,2i-1}u_i + s_{2i-1,2i}v_i + \ldots$$
$$Y_i = s_{2i,1}u_1 \mid s_{2i,2}v_1 + \ldots + s_{2i,2i-1}u_i + s_{2i,2i}v_i + \ldots \qquad (13.5)$$

where $s_{2i-1,1}$ etc. are stiffness coefficients. In matrix form these equations become

$$
\begin{bmatrix}
X_1 \\
Y_1 \\
\cdot \\
\cdot \\
\cdot \\
X_n \\
Y_n
\end{bmatrix}
=
\begin{bmatrix}
s_{11} & s_{12} & & \cdot\cdot & \\
s_{21} & s_{22} & & \cdot\cdot & \\
\cdot & \cdot & \cdot & & \\
\cdot & \cdot & \cdot & & \\
\cdot & \cdot & \cdot & & \\
\cdot\cdot & & & s_{2n-1,2n} & \\
\cdot\cdot & & & s_{2n,2n} &
\end{bmatrix}
\begin{bmatrix}
u_1 \\
v_1 \\
\cdot \\
\cdot \\
\cdot \\
u_n \\
v_n
\end{bmatrix}
\qquad (13.6)
$$

The column matrices are the assembly load and displacement vectors (see equations 13 2–13.4). The square matrix of stiffness coefficients is called the assembly stiffness matrix and denoted $\mathbf{K}$. Equations (13.6) could then be written in the shorthand form of equation (13.1).

As discussed in Article 13.1 the assembly stiffness equations are found by considering the behaviour of finite elements of the body. Here we shall only consider triangular elements, as these are relatively simple to analyse and yet extremely versatile in representing plane bodies of irregular shape. However, the general sequence of operations is the same for any other shape of element. Figure 13.2 shows a triangular finite element numbered **j** and having its vertices at nodes numbered i, j, m (conventionally read in anticlockwise order round the element). The components of force exerted on the element at the ith node are shown as $X_{i;j}$, $Y_{i;j}$. The notation is the same as for bars (see Section 5.3.1) the first subscript being the nodal number, while that following the semicolon is the number of the element.

In Article 13.3 we shall develop explicitly the simplest form of assembly stiffness equations for a plane body using triangular elements. We shall

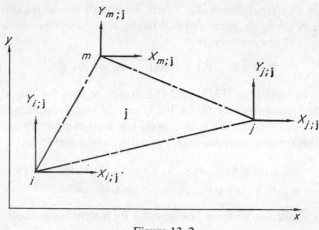

Figure 13.2

then in Article 13.4 redevelop these results in a more general fashion which indicates how more complex systems are dealt with.

13.3 Assembly Stiffness Equations for Plane Elasticity (Explicit Approach)

13.3.1 *Linear Displacement Functions*

In order to develop stiffness equations for a triangular finite element we begin by assuming that the variation of displacement within the element is expressed by the linear functions

$$u = \alpha_1 + \alpha_2 x + \alpha_3 y$$
$$v = \alpha_4 + \alpha_5 x + \alpha_6 y$$
(13.7)

in which the six coefficients α_1 etc. may be called the amplitudes, and are constant within the element.

If we were to plot displacement u or v at any point in an element on a base representing the element we would obtain a 'displacement surface' which in the case of the linear displacement assumption (equations 13.7) is a plane (Figure 13.3a). Since the actual surface will in general be curved the assumed displacement variation is in general only approximate, as was pointed out in Article 13.1. However, if the element is divided into two by addition of a node p as in Figure 13.3b then the original displacement surface is replaced by two planes which ought in general to be a better approximation to the true displacement surface. Thus although, as noted in Article 13.1, the method is approximate the approximation will improve as the number of elements is increased, subject to the practical limitation of the accuracy of the associated calculations.

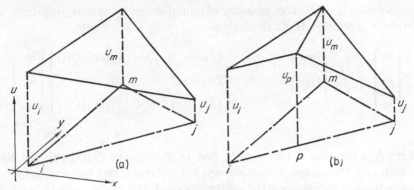

Figure 13.3

We may summarize by saying that using the linear functions (13.7) the displacement surfaces u and v of the body are represented by a series of plane facets, and from Figure 13.3b it is clear that there is continuity of displacement at interelement boundaries. Within an element the slopes of the u surface are

$$\frac{\partial u}{\partial x} = \alpha_2 \quad \text{and} \quad \frac{\partial u}{\partial y} = \alpha_3$$

in the x and y directions. Similarly for the v surface

$$\frac{\partial v}{\partial x} = \alpha_5 \quad \text{and} \quad \frac{\partial v}{\partial y} = \alpha_6$$

Hence, referring to the strain/displacement equations (12.6) we see that

$$\epsilon_x = \alpha_2; \qquad \epsilon_y = \alpha_6; \qquad \gamma_{xy} = \alpha_2 + \alpha_3 \tag{13.8}$$

Thus the strain components are uniform within the element for this type of displacement function. If these strains are to be related to the nodal displacements we must first express the amplitudes in these terms. To do so we substitute the nodal displacements u_i, u_j, u_m (which occur at points with coordinates x_i, y_i etc.) in the first of equations (13.7) obtaining

$$u_i = \alpha_1 + \alpha_2 x_i + \alpha_3 y_i$$
$$u_j = \alpha_1 + \alpha_2 x_j + \alpha_3 y_j$$
$$u_m = \alpha_1 + \alpha_2 x_m + \alpha_3 y_m$$

These can also be written in matrix form as

$$\begin{bmatrix} u_i \\ u_j \\ u_m \end{bmatrix} = \begin{bmatrix} 1 & x_i & y_i \\ 1 & x_j & y_j \\ 1 & x_m & y_m \end{bmatrix} \begin{bmatrix} \alpha_1 \\ \alpha_2 \\ \alpha_3 \end{bmatrix} \tag{13.9}$$

and solved either by conventional elimination or by matrix inversion, as shown in Appendix II, to obtain

$$
\begin{bmatrix} \alpha_1 \\ \alpha_2 \\ \alpha_3 \end{bmatrix} = \frac{1}{2\Delta} \begin{bmatrix} (x_j y_m - x_m y_j) & (x_m y_i - x_i y_m) & (x_i y_j - x_j y_i) \\ (y_j - y_m) & (y_m - y_i) & (y_i - y_j) \\ (x_m - x_j) & (x_i - x_m) & (x_j - x_i) \end{bmatrix} \begin{bmatrix} u_i \\ u_j \\ u_m \end{bmatrix}
$$

$$(13.10)$$

where Δ is the area of the triangle *ijm*. In the square matrix we see that in each row the component in the second column could be obtained from that in the first by changing the subscript(s) one step in the cyclic order *ijmij*, e.g. in the second row the second-column component $(y_m - y_i)$ has subscripts m, i whereas the first-column component has subscripts j, m. Also in each column one subscript is absent. We now adopt for convenience the following shorthand notation

$$
\begin{bmatrix} \alpha_1 \\ \alpha_2 \\ \alpha_3 \end{bmatrix} = \frac{1}{2\Delta} \begin{bmatrix} a_A & a_B & a_C \\ b_A & b_B & b_C \\ c_A & c_B & c_C \end{bmatrix} \begin{bmatrix} u_i \\ u_j \\ u_m \end{bmatrix}
$$

$$(13.10a)$$

Substituting from these in the first of equations (13.7) we have

$$
u = \frac{1}{2\Delta} \left((a_A u_i + a_B u_j + a_C u_m) + (b_A u_i + b_B u_j + b_C u_m)\, x + (c_A u_i + c_B u_j + c_C u_m)\, y \right)
$$

$$
= \frac{1}{2\Delta} \left((a_A + b_A x + c_A y)\, u_i + (a_B + b_B x + c_B y)\, u_j + (a_C + b_C x + c_C y)\, u_m \right)
$$

$$(13.11)$$

Referring to equations (13.8) and (13.10a) we can write

$$
\epsilon_x = \alpha_2 = \frac{1}{2\Delta} (b_A u_i + b_B u_j + b_C u_m)
$$

Similarly by substituting the values v_i, v_j, v_m in the second of equations (13.7) we have

$$
\begin{bmatrix} v_i \\ v_j \\ v_m \end{bmatrix} = \begin{bmatrix} 1 & x_i & y_i \\ 1 & x_j & y_j \\ 1 & x_m & y_m \end{bmatrix} \begin{bmatrix} \alpha_4 \\ \alpha_5 \\ \alpha_6 \end{bmatrix}
$$

Since the square matrix is the same as in (13.9) it is clear that the coefficients α_4, α_5, α_6 are related to the nodal displacements v_i, v_j, v_m by the same square matrix as in equations (13.10). Then referring also to equations (13.8) we have

$$\epsilon_y = \alpha_6 = \frac{1}{2\Delta} (c_A v_i + c_B v_j + c_C v_m)$$

$$\gamma_{xy} = \alpha_3 + \alpha_5 = \frac{1}{2\Delta} (c_A u_i + c_B u_j + c_C u_m + b_A v_i + b_B v_j + b_C v_m)$$

In matrix form the element strains are related to the nodal displacements as

$$
\begin{bmatrix} \epsilon_x \\ \epsilon_y \\ \gamma_{xy} \end{bmatrix} = \frac{1}{2\Delta}
\begin{bmatrix} b_A & 0 & b_B & 0 & b_C & 0 \\ 0 & c_A & 0 & c_B & 0 & c_C \\ c_A & b_A & c_B & b_B & c_C & b_C \end{bmatrix}
\begin{bmatrix} u_i \\ v_i \\ u_j \\ v_j \\ u_m \\ v_m \end{bmatrix}
\tag{13.12}
$$

Symbolically

$$\epsilon = B\delta_j \tag{13.12a}$$

where $\epsilon = \{\epsilon_x \ \epsilon_y \ \gamma_{xy}\}$ is a vector of element strain components, and $\delta_j = \{u_i \ v_i \ u_j \ v_j \ u_m \ v_m\}$ is a vector of nodal displacements for the triangular element

The stresses in the element can now be obtained by using the appropriate stress/strain equations (3.14) or (3.16) according to whether the body is in plane stress or plane strain. Symbolically

$$\sigma = D\epsilon \tag{13.13}$$

where $\sigma = \{\sigma_x \ \sigma_y \ \tau_{xy}\}$ is the vector of stress components for the element and D is the stress/strain operator. The particular form (3.14) for plane stress is requoted below for easy reference

$$
\begin{bmatrix} \sigma_x \\ \sigma_y \\ \tau_{xy} \end{bmatrix} = \frac{E}{1-\nu^2}
\begin{bmatrix} 1 & \nu & 0 \\ \nu & 1 & 0 \\ 0 & 0 & \frac{1-\nu}{2} \end{bmatrix}
\begin{bmatrix} \epsilon_x \\ \epsilon_y \\ \gamma_{xy} \end{bmatrix}
\tag{13.13a}
$$

Substituting for ϵ from (13.12a) in (13.13) we have

$$\sigma = DB\delta \tag{13.14}$$

Taking **D** from (13.13a) and **B** from (13.12) we have for plane stress

$$
\begin{bmatrix} \sigma_x \\ \sigma_y \\ \tau_{xy} \end{bmatrix} = \frac{E}{2\Delta(1-\nu^2)} \begin{bmatrix} b_A & \nu c_A & b_B & \nu c_B & b_C & \nu c_C \\ \nu b_A & c_A & \nu b_B & c_B & \nu b_C & c_C \\ \Omega c_A & \Omega b_A & \Omega c_B & \Omega b_B & \Omega c_C & \Omega b_C \end{bmatrix} \begin{bmatrix} u_i \\ v_i \\ u_j \\ v_j \\ u_m \\ v_m \end{bmatrix}
$$

(13.14a)

where $\Omega = (1-\nu)/2$.

Thus when linear displacement functions are assumed the states of strain and stress are uniform within an element.

13.3.2 *Element Stiffness Equations*

The forces exerted on the nodes of an element are shown in Figure 13.2 and were described in Article 13.2. They may be written as a vector of forces for the **j**th element as

$$\mathbf{F}_j = \{X_{i;j} \quad Y_{i;j} \quad X_{j;j} \quad Y_{j;j} \quad X_{m;j} \quad Y_{m;j}\} \tag{13.15a}$$

Similarly a vector of nodal displacements for that element is

$$\boldsymbol{\delta}_j = \{u_i \quad v_i \quad u_j \quad v_j \quad u_m \quad v_m\} \tag{13.15b}$$

The stiffness equations for the element express each force component in terms of all the nodal displacements of the element. Typically

$$X_{i;j} = a_{11}u_i + a_{12}v_i + \ldots \quad \ldots + a_{16}v_m$$

$$\cdot \qquad \cdot \qquad \cdot \qquad \qquad \cdot$$
$$\cdot \qquad \cdot \qquad \cdot \qquad \qquad \cdot \qquad\qquad\qquad (13.16)$$
$$\cdot \qquad \cdot \qquad \cdot \qquad \qquad \cdot$$

$$Y_{m;j} = a_{61}u_i + \ldots \qquad \ldots + a_{66}v_m$$

If these are written in matrix form we obtain the shorthand expression

$$\mathbf{F}_j = \mathbf{k}_j \boldsymbol{\delta}_j \tag{13.16a}$$

where $\mathbf{F}_j$, $\boldsymbol{\delta}_j$ are the (6×1) column matrices defined above and $\mathbf{k}_j$ is a square matrix (size 6×6) of stiffness coefficients which we call the element stiffness matrix.

In order to find these relations from the assumed states of stress and strain we investigate the equilibrium of the element using the Principle of Virtual Work (see Section 12.5.1). A set of virtual displacements are imposed and if the virtual work of the external forces is equal to the virtual strain energy the system is in equilibrium. Virtual displacements are ones which do not change the magnitudes of the forces (both internal and external) and do not violate any constraints on the system.

We begin by imposing virtual displacements

$$\boldsymbol{\delta}_j^* = \{u_i^* \quad v_i^* \quad u_j^* \quad v_j^* \quad u_m^* \quad v_m^*\}$$

at the nodes i, j, m of the element. (Note the use of $*$ to indicate a virtual quantity.) The virtual work of the force $X_{i;j}$ is $u_i^* X_{i;j}$ since the magnitude of the force does not change during the virtual displacement. Similar expressions can be written for the virtual work of the other components of $\mathbf{F}_j$ and the total virtual work of the external forces $\mathbf{F}_j$ in moving through the virtual displacements $\boldsymbol{\delta}_j^*$ is (compare equation 4.16)

$$W_{\text{ext}}^* = [u_i^* \quad v_i^* \quad u_j^* \quad v_j^* \quad u_m^* \quad v_m^*] \begin{bmatrix} X_{i;j} \\ Y_{i;j} \\ X_{j;j} \\ Y_{j;j} \\ X_{m;j} \\ Y_{m;j} \end{bmatrix} = (\boldsymbol{\delta}_j^*)^{\text{T}} \mathbf{F}_j \qquad (13.17)$$

where the superscript ^{T} indicates the transpose of a matrix.

The virtual displacements of the nodes will cause virtual strains $\boldsymbol{\epsilon}^*$ in the element, in accordance with equations (13.12a)

i.e.
$$\boldsymbol{\epsilon}^* = \mathbf{B}\boldsymbol{\delta}_j^* \qquad (13.18)$$

where

$$\boldsymbol{\epsilon}^* = \{\epsilon_x^* \quad \epsilon_y^* \quad \gamma_{xy}^*\}$$

The virtual strain energy density is obtained from equation (12.28) as

$$U_0^* = \epsilon_x^* \sigma_x + \epsilon_y^* \sigma_y + \gamma_{xy}^* \tau_{xy} = [\epsilon_x^* \quad \epsilon_y^* \quad \gamma_{xy}^*] \begin{bmatrix} \sigma_x \\ \sigma_y \\ \tau_{xy} \end{bmatrix} = (\boldsymbol{\epsilon}^*)^{\text{T}} \boldsymbol{\sigma} \qquad (13.19)$$

Then if V is the volume of the element the virtual strain energy is

$$U^* = \int_V U_0^* \, \mathrm{d}V = \int_V ((\boldsymbol{\epsilon}^*)^{\text{T}} \boldsymbol{\sigma}) \, \mathrm{d}V$$

By the Principle of Virtual Work for equilibrium of the element we require (see 12.26)

$$W_{\text{ext}}^* = U^*$$

i.e.
$$(\boldsymbol{\delta}_j^*)^{\text{T}} \mathbf{F}_j = \int_V ((\boldsymbol{\epsilon}^*)^{\text{T}} \boldsymbol{\sigma}) \, \mathrm{d}V$$

The virtual strains $\boldsymbol{\epsilon}^*$ are given by (13.18) and so (see Appendix II)

$$(\boldsymbol{\epsilon}^*)^{\mathrm{T}} = (\boldsymbol{\delta}_j^*)^{\mathrm{T}}\mathbf{B}^{\mathrm{T}}$$

Using this and substituting also for $\boldsymbol{\sigma}$ from (13.14) we obtain

$$(\boldsymbol{\delta}_j^*)^{\mathrm{T}}\mathbf{F}_j = \int_V (\boldsymbol{\delta}_j^*)^{\mathrm{T}}\mathbf{B}^{\mathrm{T}}\mathbf{D}\mathbf{B}\boldsymbol{\delta}_j \, \mathrm{d}V$$

Both $\boldsymbol{\delta}_j^*$ and $\boldsymbol{\delta}_j$ are independent of the integration and so we have

$$\mathbf{F}_j = \left(\int_V \mathbf{B}^{\mathrm{T}}\mathbf{D}\mathbf{B} \, \mathrm{d}V\right) \boldsymbol{\delta}_j \qquad (13.20)$$

Comparing this with equation (13.16a) we see that the element stiffness matrix is

$$\mathbf{k}_j = \int_V \mathbf{B}^{\mathrm{T}}\mathbf{D}\mathbf{B} \, \mathrm{d}V \qquad (13.21)$$

In Section 13.3.1 it was seen that for the particular system under discussion the components of the $\mathbf{B}$ and $\mathbf{D}$ matrices are constants. Therefore (13.21) can be written as

$$\mathbf{k}_j = \mathbf{B}^{\mathrm{T}}\mathbf{D}\mathbf{B} \int_V \mathrm{d}V = (\mathbf{B}^{\mathrm{T}}\mathbf{D}\mathbf{B}) \, t\Delta \qquad (13.22)$$

where t is the thickness, and Δ the area, of the element.

It should be noted that equations (13.21) and (13.22) are valid for plane stress or plane strain. In the latter case it is a condition of the system that there are no displacements in the z direction; consequently no virtual displacements can be imposed in this direction and there are no contributions to virtual work or virtual strain energy from external or internal forces acting in this direction. Equations (13.17) and (13.19) are therefore valid for both plane stress and plane strain.

The element stiffness equations were derived for brevity in terms of symbolic operations. Examples will now be given to illustrate the actual numerical evaluation of these quantities.

Example 13.1 Calculate the stiffness matrix for the element numbered **2** in Figure 13.4a. Assume plane stress:

Solution (a) The information required comprises the coordinates of the nodes i, j, m (given in the figure), the thickness t and the properties of the material of the element. Here $t = 0\cdot01$ m; $E = 70 \, \mathrm{GN/m^2}$; $\nu = 0\cdot35$.

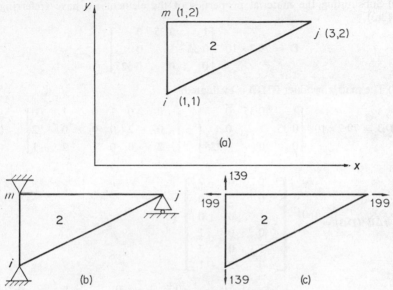

Figure 13.4 Units of force kN

(b) Taking nodal coordinates from Figure 13.4a and referring to (13.9) we form

$$\begin{bmatrix} 1 & x_i & y_i \\ 1 & x_j & y_j \\ 1 & x_m & y_m \end{bmatrix} = \begin{bmatrix} 1 & 1 & 1 \\ 1 & 3 & 2 \\ 1 & 1 & 2 \end{bmatrix} \quad \text{(i)}$$

(c) The inverse of this matrix is calculated in practice by a computer. Here we may use the explicit form in equations (13.10) to obtain

$$\frac{1}{2\Delta} \begin{bmatrix} 4 & -1 & -1 \\ 0 & 1 & -1 \\ -2 & 0 & 2 \end{bmatrix} \quad \text{(ii)}$$

where

$$\Delta = 1 \quad \text{(iii)}$$

Comparing equation (ii) with equation (13.10a) we see that

$$\begin{aligned} a_A &= 4; & a_B &= -1; & a_C &= -1 \\ b_A &= 0; & b_B &= 1; & b_C &= -1 \\ c_A &= -2; & c_B &= 0; & c_C &= 2 \end{aligned} \quad \text{(iv)}$$

(d) Matrix **B** (defined in (13.12)) is formed by substituting from (iv) above

$$\mathbf{B} = \frac{1}{2\Delta} \begin{bmatrix} b_A & 0 & b_B & 0 & b_C & 0 \\ 0 & c_A & 0 & c_B & 0 & c_C \\ c_A & b_A & c_B & b_B & c_C & b_C \end{bmatrix} = \tfrac{1}{2} \begin{bmatrix} 0 & 0 & 1 & 0 & -1 & 0 \\ 0 & -2 & 0 & 0 & 0 & 2 \\ -2 & 0 & 0 & 1 & 2 & -1 \end{bmatrix} \quad \text{(v)}$$

Matrix $\mathbf{B}^T$ is formed at the same time.

(e) Substituting the material properties of the element we have (referring to (13.13a))

$$D = 79.7 \times 10^9 \begin{bmatrix} 1 & 0.35 & 0 \\ 0.35 & 1 & 0 \\ 0 & 0 & 0.325 \end{bmatrix} \tag{vi}$$

(f) The matrix product $\mathbf{B^T DB}$ is evaluated.

$$\mathbf{DB} = 79.7 \times 10^9 \begin{bmatrix} 1 & 0.35 & 0 \\ 0.35 & 1 & 0 \\ 0 & 0 & 0.325 \end{bmatrix} \tfrac{1}{2} \begin{bmatrix} 0 & 0 & 1 & 0 & -1 & 0 \\ 0 & -2 & 0 & 0 & 0 & 2 \\ -2 & 0 & 0 & 1 & 2 & -1 \end{bmatrix} \tag{vii}$$

Then

$$\mathbf{k_j} = t\Delta \mathbf{B^T(DB)} = \frac{0.01}{2} \begin{bmatrix} 0 & 0 & -2 \\ 0 & -2 & 0 \\ 1 & 0 & 0 \\ 0 & 0 & 1 \\ -1 & 0 & 2 \\ 0 & 2 & -1 \end{bmatrix} \times$$

$$39.9 \times 10^9 \begin{bmatrix} 0 & -0.7 & 1 & 0 & -1 & 0.7 \\ 0 & -2 & 0.35 & 0 & -0.35 & 2 \\ -0.65 & 0 & 0 & 0.325 & 0.65 & -0.325 \end{bmatrix} \tag{viii}$$

i.e. $\mathbf{k_2} = 199 \times 10^6 \begin{bmatrix} 1.3 & 0 & 0 & -0.65 & -1.3 & 0.65 \\ 0 & 4 & -0.7 & 0 & 0.7 & -4 \\ 0 & -0.7 & 1 & 0 & -1 & 0.7 \\ -0.65 & 0 & 0 & 0.325 & 0.65 & -0.325 \\ -1.3 & 0.7 & -1 & 0.65 & 2.3 & -1.35 \\ 0.65 & -4 & 0.7 & -0.325 & -1.35 & 4.325 \end{bmatrix} \tag{ix}$

When all the nodal displacements $\boldsymbol{\delta_j}$ of the element j are known the element stiffness equations (13.16a) can be used directly to determine the loads $\mathbf{F_j}$ on the vertices of the element. The state of stress in the element may also be obtained using equations (13.14a).

Example 13.2 The element **2** of Example 13.1 is supported as shown in Figure 13.4b and subjected to a displacement $u_j = 10^{-3}$ m. Determine the loads at the vertices and the state of stress.

Solution From (13.16a) $\mathbf{F_2} = \mathbf{k_2}\boldsymbol{\delta_2}$ where $\mathbf{k_2}$ has been calculated in Example 13.1. From Figure 13.4b we see that the supports prevent displacements in the x and y directions at i and m, and in the y direction at j. Hence

$$\boldsymbol{\delta_2} = \{0 \quad 0 \quad 10^{-3} \quad 0 \quad 0 \quad 0\}\,\text{m}$$

and

$$
\begin{bmatrix} X_{i;2} \\ Y_{i;2} \\ X_{j;2} \\ Y_{j;2} \\ X_{m;2} \\ Y_{m;2} \end{bmatrix} = 199 \times 10^6 \begin{bmatrix} 1\cdot3 & 0 & 0 & -0\cdot65 & -1\cdot3 & 0\cdot65 \\ 0 & 4 & -0\cdot7 & 0 & 0\cdot7 & -4 \\ 0 & -0\cdot7 & 1 & 0 & -1 & 0\cdot7 \\ -0\cdot65 & 0 & 0 & 0\cdot325 & 0\cdot65 & -0\cdot325 \\ -1\cdot3 & 0\cdot7 & -1 & 0\cdot65 & 2\cdot3 & -1\cdot35 \\ 0\cdot65 & -4 & 0\cdot7 & -0\cdot325 & -1\cdot35 & 4\cdot325 \end{bmatrix}
$$

$$
\times \begin{bmatrix} 0 \\ 0 \\ 10^{-3} \\ 0 \\ 0 \\ 0 \end{bmatrix} = 10^3 \begin{bmatrix} 0 \\ -139 \\ 199 \\ 0 \\ -199 \\ 139 \end{bmatrix} \text{ N \quad (i)}
$$

Figure 13.4c illustrates these loads. From (13.14) $(\sigma)_2 = \mathbf{DB\delta}_2$ where $\mathbf{DB}$ was determined in Example 13.1 (see equations (vii) and (viii) of that example). Hence

$$
\begin{bmatrix} \sigma_x \\ \sigma_y \\ \tau_{xy} \end{bmatrix} = 39\cdot9 \times 10^9 \begin{bmatrix} 0 & -0\cdot7 & 1 & 0 & -1 & 0\cdot7 \\ 0 & -2 & 0\cdot35 & 0 & -0\cdot35 & 2 \\ -0\cdot65 & 0 & 0 & 0\cdot325 & 0\cdot65 & -0\cdot325 \end{bmatrix}
$$

$$
\times \begin{bmatrix} 0 \\ 0 \\ 10^{-3} \\ 0 \\ 0 \\ 0 \end{bmatrix} = 10^6 \begin{bmatrix} 39\cdot9 \\ 14 \\ 0 \end{bmatrix} \text{ N/m}^2 \quad \text{(ii)}
$$

i.e. the state of stress throughout the element is

$$
\begin{pmatrix} 39\cdot9 & 0 & 0 \\ 0 & 14 & 0 \\ 0 & 0 & 0 \end{pmatrix} \text{ MN/m}^2
$$

(iii)

As illustrated in Example 13.1 the element stiffness matrix is in practice evaluated numerically. However, it is of course possible to obtain an explicit expression by performing the matrix operations of equation (13.22) using the explicit expressions for $\mathbf{B}$ (equation 13.12) and $\mathbf{D}$ (equation 3.14 or 3.16). The procedure is routine but tedious even for a triangular element, but can be made slightly easier by partitioning the matrices $\mathbf{B}^T$, $\mathbf{D}$, $\mathbf{B}$ to take advantage of the zeros in $\mathbf{D}$ (see equations 3.14 and 3.16)

$$
\mathbf{k}_j = \begin{bmatrix} \mathbf{B}_I^T & \vdots & \mathbf{B}_{II}^T \\ 6\times2 & \vdots & 6\times1 \\ & \vdots & \end{bmatrix} \begin{bmatrix} \mathbf{D}_I & \vdots & \mathbf{0} \\ 2\times2 & \vdots & 2\times1 \\ \hline 0 & \vdots & \mathbf{D}_{II} \\ 1\times2 & \vdots & 1\times1 \end{bmatrix} \begin{bmatrix} \mathbf{B}_I \\ 2\times6 \\ \hline \mathbf{B}_{II} \\ 1\times6 \end{bmatrix} t\Delta
$$

$$
= (\mathbf{B}_I^T\mathbf{D}_I\mathbf{B}_I + \mathbf{B}_{II}^T\mathbf{D}_{II}\mathbf{B}_{II})\, t\Delta = \mathbf{k}_I + \mathbf{k}_{II} \qquad (13.23)
$$

These matrices $\mathbf{k}_I$ and $\mathbf{k}_{II}$ consist of the contributions of normal stress and shear stress respectively to the element stiffness matrix, as can be seen by considering the compositions of $\mathbf{D}_I$ and $\mathbf{D}_{II}$. Evaluating (13.23) for the case of plane stress we have, taking $\mathbf{B}$ and $\mathbf{D}$ from (13.12) and (13.13a).

$$
\mathbf{k}_I = \frac{Et}{4\Delta(1-\nu^2)}
\begin{bmatrix}
b_A b_A & \nu b_A c_A & b_A b_B & \nu b_A c_B & b_A b_C & \nu b_A c_C \\
\nu b_A c_A & c_A c_A & \nu c_A b_B & c_A c_B & \nu c_A b_C & c_A c_C \\
b_A b_B & \nu c_A b_B & b_B b_B & \nu b_B c_B & b_B b_C & \nu b_B c_C \\
\nu b_A c_B & c_A c_B & \nu b_B c_B & c_B c_B & \nu c_B b_C & c_B c_C \\
b_A b_C & \nu c_A b_C & b_B b_C & \nu c_B b_C & b_C b_C & \nu b_C c_C \\
\nu b_A c_C & c_A c_C & \nu b_B c_C & c_B c_C & \nu b_C c_C & c_C c_C
\end{bmatrix}
$$

$$
\mathbf{k}_{II} = \frac{Et}{8\Delta(1+\nu)}
\begin{bmatrix}
c_A c_A & b_A c_A & c_A c_B & c_A b_B & c_A c_C & c_A b_C \\
b_A c_A & b_A b_A & b_A c_B & b_A b_B & b_A c_C & b_A b_C \\
c_A c_B & b_A c_B & c_B c_B & b_B c_B & c_B c_C & c_B b_C \\
c_A b_B & b_A b_B & b_B c_B & b_B b_B & b_B c_C & b_B b_C \\
c_A c_C & b_A c_C & c_B c_C & b_B c_C & c_C c_C & b_C c_C \\
c_A b_C & b_A b_C & c_B b_C & b_B b_C & b_C c_C & b_C b_C
\end{bmatrix}
$$

$$(13.24)$$

For plane strain $\mathbf{k}_{II}$ is as above. To obtain $\mathbf{k}_I$ an additional scalar multiplier $(1-\nu)/(1-2\nu)$ is applied to the expression given above, and, in addition, the terms in the array not multiplied by ν are multiplied by $(1-\nu)$.

It is clear from equations (13.24) that the element stiffness matrix is symmetric, which is in accordance with previous findings for bar-type finite elements.

Lastly, we note that the element load vector (equation 13.15a) may be written

$$
\mathbf{F_j} = \{\mathbf{F}_{i;j} \quad \mathbf{F}_{j;j} \quad \mathbf{F}_{m;j}\}
$$

where for example $\mathbf{F}_{i;j} = \{X_{i;j} \quad Y_{i;j}\}$ is a sub-matrix or vector of loads on the jth element at the ith node. Similarly we can write the element displacement vector as

$$
\boldsymbol{\delta_j} = \{\boldsymbol{\delta}_i \quad \boldsymbol{\delta}_j \quad \boldsymbol{\delta}_m\}
$$

where for example $\boldsymbol{\delta}_i = \{u_i \quad v_i\}$. Then the element stiffness equations (13.16) may be written in the form

$$
\begin{bmatrix}
\mathbf{F}_{i;j} \\
\mathbf{F}_{j;j} \\
\mathbf{F}_{m;j}
\end{bmatrix}
=
\begin{bmatrix}
\mathbf{k}_{ii;j} & \mathbf{k}_{ij;} & \mathbf{k}_{im;j} \\
\mathbf{k}_{ji;j} & \mathbf{k}_{jj;j} & \mathbf{k}_{jm;j} \\
\mathbf{k}_{mi;j} & \mathbf{k}_{mj;j} & \mathbf{k}_{mm;j}
\end{bmatrix}
\begin{bmatrix}
\boldsymbol{\delta}_i \\
\boldsymbol{\delta}_j \\
\boldsymbol{\delta}_m
\end{bmatrix}
\qquad (13.25)
$$

where the components of the stiffness matrix are (2×2) sub-matrices. The sub-matrix $\mathbf{k}_{mi;j}$ for example, relates forces on element $\mathbf{j}$ at node m to displacement at node i. In the particular case of the element of Example 13.1 this sub-matrix was, referring to (ix) of that example,

$$\mathbf{k}_{mi;2} = 199 \times 10^6 \begin{bmatrix} -1\cdot3 & 0\cdot7 \\ 0\cdot65 & -4 \end{bmatrix}$$

Equations (13.25) should be compared with the general form (7.37) for a bar, which is an element with only two nodes. Finally we note from (13.25) that for example

$$\mathbf{F}_{i;j} = \mathbf{k}_{ii;j}\boldsymbol{\delta}_i + \mathbf{k}_{ij;j}\boldsymbol{\delta}_j + \mathbf{k}_{im;j}\boldsymbol{\delta}_m \tag{13.26}$$

Problems for Solution   13.1–13.3

13.3.3 *Assembly Stiffness Equations*

As in the case of bar-type finite elements (see Sections 5.3.2 and 7.4.5) the assembly stiffness equations are obtained from element stiffness equations by application of the conditions of compatibility of deformation, and of equilibrium, at the nodes. The first condition requires that all the elements remain united at the nodes when the system is under load. This requirement has been automatically satisfied by writing the element stiffness equations (13.16) and (13.25) in terms of nodal displacements. The second condition requires that the external forces exerted on a node are the resultants of the forces exerted on the elements meeting at that node. For example, considering the two-element assembly in Figure 13.5a, the equilibrium conditions at node 4, where elements 1 and 2 meet, are in matrix form (compare equations 7.43)

$$\begin{bmatrix} X_4 \\ Y_4 \end{bmatrix} = \begin{bmatrix} X_{4,1} \\ Y_{4;1} \end{bmatrix} + \begin{bmatrix} X_{4;2} \\ Y_{4;2} \end{bmatrix} \tag{13.27}$$

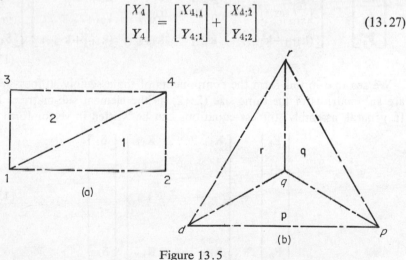

Figure 13.5

and in vector form

$$F_4 = F_{4;1} + F_{4;2} \qquad (13.27a)$$

Similarly we can write down the vector equilibrium conditions for node q in Figure 13.5b as

$$F_q = F_{q;p} + F_{q;q} + F_{q;r} \qquad (13.28)$$

An equation of the type of (13.27a) or (13.28) can be written for each node of the assembly. Each element meeting at the node contributes one term to the right-hand side of the equilibrium condition, and each of these terms can be expressed in terms of nodal displacements by an element stiffness equation of the type of (13.26). In this way the required relations (13.1) are obtained between the assembly nodal loads and displacements.

As an example of this we consider now the two-element assembly in Figure 13.5a. At node 1 the equilibrium condition is

$$F_1 = F_{1;1} + F_{1;2}$$

By setting $i = 1, j = 2, m = 4, \mathbf{j} = 1$ in equations (13.25) we can write

$$F_{1;1} = k_{11;1}\delta_1 + k_{12;1}\delta_2 + k_{14;1}\delta_4$$

Proceeding in a similar manner for element 2 to obtain an expression for $F_{1;2}$ we can rewrite the nodal equilibrium condition as the stiffness equation

$$F_1 = (k_{11;1} + k_{11;2})\,\delta_1 + k_{12;1}\delta_2 + k_{13;2}\delta_3 + (k_{14;1} + k_{14;2})\,\delta_4$$

After carrying out a similar procedure for the other nodes we can write the assembly stiffness equations as

$$\begin{bmatrix} F_1 \\ F_2 \\ F_3 \\ F_4 \end{bmatrix} = \begin{bmatrix} (k_{11;1} + k_{11;2}) & k_{12;1} & k_{13;2} & (k_{14;1} + k_{14;2}) \\ k_{21;1} & k_{22;1} & 0 & k_{24;1} \\ k_{31;2} & 0 & k_{33;2} & k_{34;2} \\ (k_{41;1} + k_{41;2}) & k_{42;1} & k_{43;2} & (k_{44;1} + k_{44;2}) \end{bmatrix} \begin{bmatrix} \delta_1 \\ \delta_2 \\ \delta_3 \\ \delta_4 \end{bmatrix}$$

$$(13.29)$$

We see that in this form the components of the assembly stiffness matrix are sub-matrices of the same size (2×2) as the element sub-matrices $k_{pq;j}$. In general, assembly stiffness equations can be written in vector form as

$$\begin{bmatrix} F_1 \\ \cdot \\ \cdot \\ \cdot \\ \cdot \\ F_n \end{bmatrix} = \begin{bmatrix} K_{11} & \cdots & K_{1n} \\ \cdot & \cdot & \cdot \\ \cdot & \cdot & \cdot \\ \cdot & \cdot & \cdot \\ \cdot & \cdot & \cdot \\ K_{n1} & \cdots & K_{nn} \end{bmatrix} \begin{bmatrix} \delta_1 \\ \cdot \\ \cdot \\ \cdot \\ \cdot \\ \delta_n \end{bmatrix} \qquad (13.30)$$

In the case of the two-element assembly n is 4 and we see that, for example,

$$K_{41} = k_{41;1} + k_{41;2}$$

This sub-matrix relates the nodal load vector F_4 to the nodal displacement vector δ_1, and is seen to be the sum of the element sub-matrices relating loads at node 4 to displacements at node 1. This is the basis of the general method for the formation of the assembly stiffness matrix.

Using K_{pq} to represent the sub-matrix in row p column q of K, then K_{pq} is the sum of all element sub-matrices having the same nodal subscripts,

i.e.
$$K_{pq} = \sum_{j=1}^{n} k_{pq;j} \tag{13.31}$$

In this formal expression (cf. Section 7.4.8) all elements $1, 2, \ldots n$ are considered. However, in practice only a limited number will have a contribution, the remainder making zero contribution. For instance in the two-element example considered above, element **2** makes no contribution to K_{12} and neither element contributes to K_{32}. The latter result is due to the fact that nodes 2 and 3 are not nodes of any one element (Figure 13.5a). In contrast nodes 4 and 1 are nodes of both elements and so there are two contributions to K_{41} and K_{14}.

Thus we can form the assembly stiffness equations for a system with n nodes in vector form by direct generation as follows. The assembly load and displacement vectors F and δ are $(n \times 1)$ column matrices of nodal loads F_i and displacements δ_i as in equations (13.2) and (13.3). The assembly stiffness matrix K is formed as an $(n \times n)$ null matrix. The sub-matrices of each element stiffness matrix are inserted in the sub-matrix of K having the same nodal subscripts—typically $k_{pq;j}$ is added to K_{pq}.

When we come to consider the direct generation of the assembly stiffness equations in their actual numerical form it is necessary to allow for the fact that each nodal force and displacement vector has two components. Therefore both F and δ in scalar form are column matrices of size $(2n \times 1)$.

The component K_{pq} of the assembly stiffness matrix is formed of element sub-matrices $k_{pq;j}$ which are of size (2×2). Therefore K is a $(2n \times 2n)$ matrix of scalar quantities, and the (2×2) components of $k_{pq;j}$ are placed in rows $2p-1$, $2p$ and columns $2q-1$, $2q$ of the scalar form of K. The following Table 13.1 summarizes these rules for the (6×6) components a_{gh} of the scalar stiffness matrix for an element with nodes i, j, m.

Example 13.3 Form the plane stress stiffness equations for the assembly in Figure 13.5a where the coordinates of nodes 1 to 4 are respectively $(1, 1)$, $(3, 1)$, $(1, 2)$, $(3, 2)$ (units metres). Take the nodal numbers of elements 1 and 2 to be 1, 2, 4 and 1, 4, 3 respectively. Each element is 0·01 m thick and the material has $E = 70$ GN/m², $\nu = 0·35$.

Table 13.1   Comparative row and column addresses
in element and assembly stiffness matrices

	Row		Column
In k_j (i.e. g)	In K	In k_j (i.e. h)	In K
1	$2i-1$	1	$2i-1$
2	$2i$	2	$2i$
3	$2j-1$	3	$2j-1$
4	$2j$	4	$2j$
5	$2m-1$	5	$2m-1$
6	$2m$	6	$2m$

Solution   The assembly load and displacement vectors are column matrices of size $2n \times 1$, i.e. 8×1, and have components as follows

$$\mathbf{F} = \{X_1 \quad Y_1 \quad X_2 \quad Y_2 \quad X_3 \quad Y_3 \quad X_4 \quad Y_4\} \tag{i}$$

$$\boldsymbol{\delta} = \{u_1 \quad v_1 \quad u_2 \quad v_2 \quad u_3 \quad v_3 \quad u_4 \quad v_4\} \tag{ii}$$

The stiffness matrices of the two elements are calculated either from the general equation (13.22) as illustrated in Example 13.1 or by direct substitution in the explicit expressions (13.23) and 13.24).

It will be noted that element **2** corresponds exactly to the element considered in Example 13.1 (Figure 13.4a). The stiffness matrix is therefore equation (ix) of that example, and will not be rewritten here. For element **1** the square matrices of equations (13.9) and (13.10a), are respectively

$$\begin{bmatrix} 1 & 1 & 1 \\ 1 & 3 & 1 \\ 1 & 3 & 2 \end{bmatrix} \quad \text{and} \quad \tfrac{1}{2} \begin{bmatrix} 3 & 1 & -2 \\ -1 & 1 & 0 \\ 0 & -2 & 2 \end{bmatrix} \tag{iii}$$

Then evaluating (13.22) or substituting in (13.23) and (13.24) we find

$$\mathbf{k_1} = \Omega \begin{bmatrix} 1 & 0 & -1 & 0.7 & 0 & -0.7 \\ 0 & 0.325 & 0.65 & -0.325 & -0.65 & 0 \\ -1 & 0.65 & 2.3 & -1.35 & -1.3 & 0.7 \\ 0.7 & -0.325 & -1.35 & 4.325 & 0.65 & -4 \\ 0 & -0.65 & -1.3 & 0.65 & 1.3 & 0 \\ -0.7 & 0 & 0.7 & -4 & 0 & 4 \end{bmatrix} \tag{iv}$$

where

$$\Omega = 199 \times 10^6 \, \text{N/m}.$$

The assembly stiffness matrix is formed as a null matrix of size $2n \times 2n$ (i.e. 8×8) with general component s_{pq}. The components of k_1 are now placed in locations in **K** determined from the rules of Table 13.1, with $i = 1, j = 2, m = 4$, e.g.

$a_{16} = -0.7 \, \Omega$ is placed in row $(2 \times 1)-1$; column (2×4); i.e. adds to s_{18}

$a_{35} = -1.3 \, \Omega$ is placed in row $(2 \times 2)-1$; column $(2 \times 4)-1$; i.e. adds to s_{37}

The components of k_2 are then inserted in **K** in a similar way. Here $i = 1, j = 4$, $m = 3$, and for example, referring to (ix) of Example 13.1

$a_{23} = -0.7 \, \Omega$ is placed in row (2×1); column $(2 \times 4)-1$; i.e. adds to s_{27}

$a_{46} = -0.325 \, \Omega$ is placed in row (2×4); column (2×3); i.e. adds to s_{86}

The resulting assembly stiffness matrix is incorporated in (v) below.

$$
\begin{bmatrix} X_1 \\ Y_1 \\ X_2 \\ Y_2 \\ X_3 \\ Y_3 \\ X_4 \\ Y_4 \end{bmatrix} = \Omega
$$

$$
\times
\begin{bmatrix}
2{\cdot}3 & 0 & -1 & 0{\cdot}7 & -1{\cdot}3 & 0{\cdot}65 & 0 & -1{\cdot}35 \\
0 & 4{\cdot}325 & 0{\cdot}65 & -0{\cdot}325 & 0{\cdot}7 & -4 & -1{\cdot}35 & 0 \\
-1 & 0{\cdot}65 & 2{\cdot}3 & -1{\cdot}35 & 0 & 0 & -1{\cdot}3 & 0{\cdot}7 \\
0{\cdot}7 & -0{\cdot}325 & -1{\cdot}35 & 4{\cdot}325 & 0 & 0 & 0{\cdot}65 & -4 \\
-1{\cdot}3 & 0{\cdot}7 & 0 & 0 & 2{\cdot}3 & -1{\cdot}35 & -1 & 0{\cdot}65 \\
0{\cdot}65 & -4 & 0 & 0 & -1{\cdot}35 & 4{\cdot}325 & 0{\cdot}7 & -0{\cdot}325 \\
0 & -1{\cdot}35 & -1{\cdot}3 & 0{\cdot}65 & -1 & 0{\cdot}7 & 2{\cdot}3 & 0 \\
-1{\cdot}35 & 0 & 0{\cdot}7 & -4 & 0{\cdot}65 & -0{\cdot}325 & 0 & 4{\cdot}325
\end{bmatrix}
\begin{bmatrix} u_1 \\ v_1 \\ u_2 \\ v_2 \\ u_3 \\ v_3 \\ u_4 \\ v_4 \end{bmatrix}
$$

$$(v)$$

These are the scalar stiffness equations (13.6) for the assembly of Figure 13.5a. Note the zero components corresponding to the null sub-matrices in equations (13.29). The vectors **F** and **δ** are taken from equations (i) and (ii).

13.3.4 *Applications*

If all the displacements **δ** are known for an assembly, then the loads **F** can be calculated directly.

Example 13.4 If for the assembly of Example 13.3 all nodal displacements are zero except $u_2 = u_4 = -10^{-3}$ m (see Figure 13.6a) calculate the nodal loads.

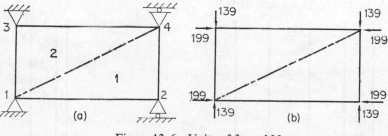

Figure 13.6   Units of force kN

Solution   Here

$$\delta = \{0 \quad 0 \quad -10^{-3} \quad 0 \quad 0 \quad 0 \quad -10^{-3} \quad 0\} \, \text{m}. \tag{i}$$

Substituting for δ in equations (v) of Example 13.3 we obtain

$$F = 10^3 \{199 \quad 139 \quad -199 \quad 139 \quad 199 \quad -139 \quad -199 \quad -139\} \, \text{N} \tag{ii}$$

The nodal loads are shown in Figure 13.6b. The system can be interpreted as equivalent to a slab of material subject to uniform compression in the x direction and completely restrained in the lateral (y) direction.

As was pointed out in Sections 5.3.3 and 7.4.6 if only nodal loads are known rigid body displacements of an assembly may occur. There is therefore no unique set of nodal displacements corresponding to the loads. Mathematically this takes the form of singularity of the matrix **K** so that the equations (13.1) cannot be solved for the assembly displacements δ.

The plane body has three degrees of freedom of rigid body displacement, viz. translation parallel to the x and y directions and rotation in the xy plane. These are eliminated by prescribing one nodal value of each of the displacements in the x and y directions and one other nodal displacement. The effect of introducing the prescribed displacements is now examined.

Let us introduce a single prescribed displacement, say $v_1 = a$ in equations (13.5). Then typically

$$Y_1 = s_{21}u_1 + s_{22}a + s_{23}u_2 + \ldots + s_{2,2i-1}u_i + s_{2,2i}v_i + \ldots$$

$$\cdot \qquad \cdot \qquad \cdot$$
$$\cdot \qquad \cdot \qquad \cdot$$
$$\cdot \qquad \cdot \qquad \cdot$$

$$X_i = s_{2i-1,1}u_1 + s_{2i-1,2}a + \ldots + s_{2i-1,2i-1}u_i + \ldots$$

$$\cdot \qquad \cdot \qquad \cdot$$
$$\cdot \qquad \cdot \qquad \cdot$$
$$\cdot \qquad \cdot \qquad \cdot$$

We note that since v_1 is prescribed Y_1 is a reactive load, which can be determined from the other displacements $u_1, u_2, v_2, \ldots$. The equation for Y_1 can therefore be eliminated leaving $2n - 1$ equations. Secondly we note that the term involving a in each equation is now a constant with dimension of force which can be transferred to the left-hand sides of the equations. Then in matrix form we can write

$$
\begin{bmatrix} X_1 \\ X_2 \\ Y_2 \\ \cdot \\ \cdot \\ \cdot \\ Y_n \end{bmatrix}
- a
\begin{bmatrix} s_{12} \\ s_{32} \\ s_{42} \\ \cdot \\ \cdot \\ \cdot \\ s_{2n,2} \end{bmatrix}
=
\begin{bmatrix} s_{11} & s_{13} & \cdot\cdot & & \cdot\cdot \\ s_{31} & s_{33} & \cdot\cdot & & \cdot\cdot \\ s_{41} & \cdot\cdot & \cdot\cdot & & \cdot\cdot \\ \cdot & & & & \\ \cdot & & & & \\ \cdot & & & & \\ \cdot\cdot & \cdot\cdot & & \cdot\cdot & s_{2n,2n} \end{bmatrix}
\begin{bmatrix} u_1 \\ u_2 \\ v_2 \\ \cdot \\ \cdot \\ \cdot \\ v_n \end{bmatrix}
$$

Thus we see that a prescribed displacement is introduced by deleting the appropriate row, say p, from each of $\mathbf{F}$, $\mathbf{K}$ and $\boldsymbol{\delta}$, and subtracting from the left-hand side the products of the prescribed displacement with the components of column p of $\mathbf{K}$. The matrix equations are thus reduced in size. Introduction of for example three nodal displacements reduces $\mathbf{F}$ and $\boldsymbol{\delta}$ to column matrices $\mathbf{F}'_R$ and $\boldsymbol{\delta}_R$, of size $(2n-3) \times 1$ and $\mathbf{K}$ to a $(2n-3) \times (2n-3)$ square matrix $\mathbf{K}_{RR}$, i.e. equations (13.1) are reduced to

$$\mathbf{F}'_R = \mathbf{K}_{RR}\boldsymbol{\delta}_R \qquad (13.32)$$

where $\mathbf{F}'_R$ incorporates the prescribed displacement products. Solving for the unknown displacements we have

$$\boldsymbol{\delta}_R = \mathbf{K}_{RR}^{-1}\mathbf{F}'_R \qquad (13.33)$$

If, as is often the case, the prescribed displacements are all zero, reduction proceeds by direct elimination of rows and the corresponding columns of $\mathbf{K}$. Once the unknown displacements have been determined, the reactive loads can be found using the complete stiffness equations.

Example 13.5 The assembly of Example 13.3 is supported as in Figure 13.7a, and subject to nodal forces $X_2 = X_4 = -100$ kN; $Y_3 = Y_4 = 0$. Determine the rest of the nodal loads and displacements, and the stress in element **2**.

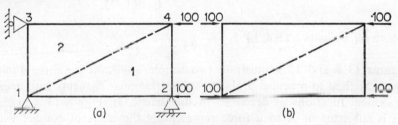

Figure 13.7   Units of force kN

Solution From Figure 13.7 we see that $u_1 = v_1 = v_2 = u_3 = 0$ and so

$$\boldsymbol{\delta} = \{0 \quad 0 \quad u_2 \quad 0 \quad 0 \quad v_3 \quad u_4 \quad v_4\} \qquad (i)$$

Introducing known loads in the assembly load vector

$$\mathbf{F} = \{X_1 \quad Y_1 \quad -10^5 \quad Y_2 \quad X_3 \quad 0 \quad -10^5 \quad 0\} \qquad (ii)$$

Using (i) and (ii) for $\mathbf{F}$ and $\boldsymbol{\delta}$ in the assembly stiffness equations ((v) of Example 13.3) we note that to reduce the equations to the form of (13.32) rows 1, 2, 4 and 5 must be eliminated. Since the prescribed displacements are all zeros we directly eliminate columns 1, 2, 4 and 5 of $\mathbf{K}$. Then the reduced equations are

$$\begin{bmatrix} -10^5 \\ 0 \\ -10^5 \\ 0 \end{bmatrix} = 199 \times 10^6 \begin{bmatrix} 2\cdot3 & 0 & -1\cdot3 & 0\cdot7 \\ 0 & 4\cdot325 & 0\cdot7 & -0\cdot325 \\ -1\cdot3 & 0\cdot7 & 2\cdot3 & 0 \\ 0\cdot7 & -0\cdot325 & 0 & 4\cdot325 \end{bmatrix} \begin{bmatrix} u_2 \\ v_3 \\ u_4 \\ v_4 \end{bmatrix} \qquad (iii)$$

Solving we find $u_2 = u_4 = -0.575 \times 10^{-3}$ m; $v_3 = v_4 = 0.1 \times 10^{-3}$ m, so that

$$\delta = 10^{-3}\{0 \quad 0 \quad -0.575 \quad 0 \quad 0 \quad 0.1 \quad -0.575 \quad 0.1\}\,\text{m} \qquad \text{(iv)}$$

Using this in equations (v) of Example 13.3 we find

$$\mathbf{F} = \{10^5 \quad 0 \quad -10^5 \quad 0 \quad 10^5 \quad 0 \quad -10^5 \quad 0\}\,\text{N} \qquad \text{(v)}$$

These loads (shown in Figure 13.7b) compress the assembly in the x direction. Having determined the nodal displacements of the assembly the stress components in the elements are determined by application of equations (13.14). Thus for element **2**

$$\delta_2 = \{\delta_1 \quad \delta_4 \quad \delta_3\} = 10^{-3}\{0 \quad 0 \quad -0.575 \quad 0.1 \quad 0 \quad 0.1\}$$

and **(DB)** has been evaluated in Example 13.1 (equations (vii) and (viii)). Hence

$$
\begin{bmatrix} \sigma_x \\ \sigma_y \\ \tau_{xy} \end{bmatrix} = 39.9 \times 10^9
\begin{bmatrix} 0 & -0.7 & 1 & 0 & -1 & 0.7 \\ 0 & -2 & 0.35 & 0 & -0.35 & 2 \\ -0.65 & 0 & 0 & 0.325 & 0.65 & -0.325 \end{bmatrix}
$$

$$
\times 10^{-3}
\begin{bmatrix} 0 \\ 0 \\ -0.575 \\ 0.1 \\ 0 \\ 0.1 \end{bmatrix} = 10^6
\begin{bmatrix} 20 \\ 0 \\ 0 \end{bmatrix}\,\text{N/m}^2
$$

Problems for Solution   13.4, 13.5

Figures 13.8 and 13.9 illustrate two simple applications of the Finite Element Method to practical problems using triangular elements with linear displacement functions as described in this article. In Figure 13.8 a gear tooth is subjected at A to a force representing the effect of contact with another tooth. Displacements of the tooth are assumed to be zero along the boundary BCDE. At the centroid of each element lincs have been drawn parallel to the principal axes of stress and proportional to the magnitudes of the principal stresses. The flow of internal force (i.e. stress) across the element is clearly seen. Apart from the high contact stresses around A there are regions of tensile and compressive stresses around B and E due to bending action.

In Figure 13.9a the top of the slab has been subjected to a set of loads statically equivalent to zero load. The purpose is to illustrate the application of St. Venant's Principle (see Article 4.6) and the computed stress distribution is shown to a larger scale in Figure 13.9b. It is clear that within a distance from the top equal to the width of the slab the stresses have become negligible. If more information were required about the stress distribution in the upper part of the slab more elements could be used giving a finer sub-division.

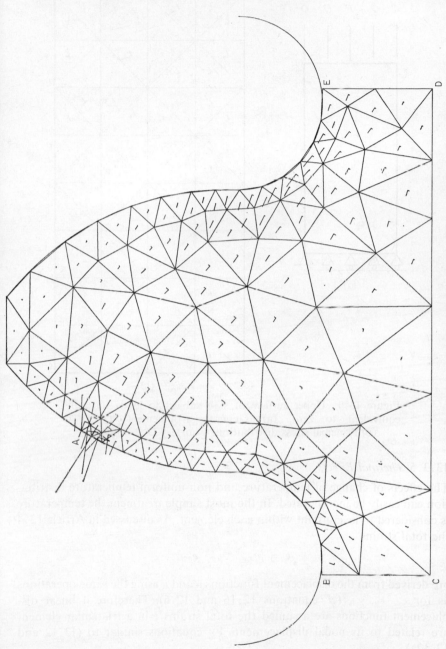

Figure 13.8 Principal stress distribution in Novikov all addendum gear tooth. Courtesy P. N. Bissell[7]

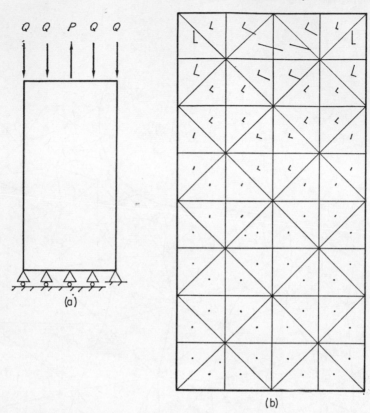

(a)

(b)

Figure 13.9   Upper surface of slab subject to loads statically equivalent to zero. (a) General arrangement $Q = P/4$. (b) Principal stress distribution. Courtesy D. J. Maggs[8]

13.3.5 *Thermal Effects*

The effects of change of temperature and non-uniform temperature distribution can easily be incorporated. In the most simple treatment the temperature is considered to be constant within each element. As discussed in Article 12.4 the total strains

$$\mathbf{e} = \{e_{xx} \quad e_{yy} \quad e_{xy}\}$$

are derived from the displacement functions u and v using the same operations as for ϵ_x, ϵ_y, γ_{xy} (cf. equations 12.16 and 12.6). Therefore if linear displacement functions are assumed the total strains $\mathbf{e}$ in a triangular element are related to its nodal displacements by equations similar to (13.12 and 13.12a).

i.e. $\mathbf{e} = \mathbf{B}\boldsymbol{\delta}_j$ (13.34)

Stresses can be related to total strains in the manner

$$\sigma = De + D_T \tag{13.35}$$

For the particular case of plane stress the explicit form of these is given by equations (12.20) from which we see D is as in equation (13.13a) and

$$D_T = \frac{E\alpha T}{1-\nu}\{-1 \quad -1 \quad 0\} \tag{13.36}$$

As in Section 13.3.2 element stiffness equations are derived by introducing a set of virtual nodal displacements δ_j^* for the element. The virtual work of the external forces is $(\delta_j^*)^T F_j$ as in equation (13.17). Following equation (13.19) the virtual strain energy can be stated as

$$U^* = \int_V (e^*)^T \sigma \, dV$$

where from (13.34) $e^* = B\delta_j^*$ are the virtual total strains. Substituting for σ from (13.35) and noting that $(\delta_j^*)^T$ is independent of the integration, we have

$$U^* = (\delta_j^*)^T \int_V B^T(DB\delta_j + D_T) \, dV$$

Equating the virtual work of the external forces to the virtual strain energy, and simplifying, we obtain the element equilibrium condition in the form

$$F_j = \left(\int_V B^T DB \, dV \right) \delta_j + \int_V B^T D_T \, dV \tag{13.37}$$

$$= k_j \delta_j + Q_j \tag{13.37a}$$

where k_j is as previously defined, and a 'thermal force' matrix is defined by

$$Q_j = \int_V B^T D_T \, dV \tag{13.38}$$

By comparing equations (13.16), (13.25) and (13.37a) it is seen that Q_j has six scalar components or three vector components; if i, j, m are the element nodes we have

$$Q_j = \{c_1 \quad c_2 \quad c_3 \quad c_4 \quad c_5 \quad c_6\}$$
$$= \{Q_{i;j} \quad Q_{j;j} \quad Q_{m;j}\} \tag{13.39}$$

Following equation (13.26) a typical row of (13.37a) is

$$F_{i;j} = k_{ii;j}\delta_i + k_{ij;j}\delta_j + k_{im;j}\delta_m + Q_{i;j} \tag{13.40}$$

To obtain an explicit form of Q_j for plane stress we substitute in (13.38) from (13.12) and (13.36) as follows

$$Q_j = \frac{1}{2\Delta} \int_V dV \begin{bmatrix} b_A & 0 & c_A \\ 0 & c_A & b_A \\ b_B & 0 & c_B \\ 0 & c_B & b_B \\ b_C & 0 & c_C \\ 0 & c_C & b_C \end{bmatrix} \frac{E\alpha T}{1-v} \begin{bmatrix} -1 \\ -1 \\ 0 \end{bmatrix} = -\frac{Et\alpha T}{2(1-v)} \begin{bmatrix} b_A \\ c_A \\ b_B \\ c_B \\ b_C \\ c_C \end{bmatrix}$$

(13.41)

Comparing this result with (13.39) we note that for example

$$Q_{i;j} = -\frac{Et\alpha T}{2(1-v)} \{b_A \quad c_A\}$$

Assembly stiffness equations are now determined as in Section 13.3.3 by considering nodal equilibrium. Then at the ith node of an assembly we have

$$F_i = F_{i;p} + F_{i;q} + \dots$$

in which there is a term on the right-hand side from each element meeting at this node. Each of these terms can be expanded as in (13.40) giving terms involving nodal displacements, and in addition terms of the type $Q_{i;q} + Q_i; + \dots$. Denoting the sum of these as a nodal thermal force vector

i.e. $$Q_i = \sum_{j=1}^{n} Q_{i;j}$$ (13.42)

we obtain one row of the assembly stiffness equations as

$$F_i = K_{i1}\delta_1 + \dots + K_{in}\delta_n + Q_i$$

Thus thermal effects are allowed for by the addition of a column vector

$$Q = \{Q_1 \quad Q_2 \quad \dots \quad \dots \quad Q_n\}$$ (13.43)

of nodal thermal forces, i.e. (13.1) becomes

$$F = K\delta + Q$$ (13.44)

When the nodal displacements are all zero the nodal forces are equal to the thermal forces. If the nodal displacements are all known the nodal forces are obtained from (13.44). Otherwise sufficient displacements must be prescribed to eliminate rigid body degrees of freedom. Then following the discussion in Section 13.3.4 we obtain reduced equations which can be solved for the unknown displacements as

$$\delta_R = K_{RR}^{-1}(F_R' - Q_R) \tag{13.45}$$

To generate the scalar form (size $2n \times 1$) of Q we must allow for the fact that each term of the type $Q_{i;j}$ has two scalar components. The typical term c_g of Q_j (equation 13.39) is therefore placed in a row of Q given by the row addresses of Table 13.1.

Example 13.6 The assembly of Example 13.3 (Figure 13.5a) experiences a rise of temperature of 20°C. (a) Write down the thermal force matrix. (b) Find the nodal loads to maintain complete restraint and the stresses in the elements. $\alpha = 23 \times 10^{-6}/°C$ for both elements.

Solution (a) For both elements

$$\frac{E\alpha T}{2(1-\nu)} = \frac{70 \times 10^9 \times 0.01 \times 23 \times 10^{-6} \times 20}{2(1-0.35)} = 248 \times 10^3 \text{ N/m}$$

For element 1 the quantities b_A, c_A, etc., are obtained from (iii) of Example 13.3. Then using (13.41)

$$Q_1 = -248 \times 10^3 \{-1 \quad 0 \quad 1 \quad -2 \quad 0 \quad 2\} \tag{i}$$

Similarly, referring to equations (iv) of Example 13.1

$$Q_2 = -248 \times 10^3 \{0 \quad -2 \quad 1 \quad 0 \quad -1 \quad 2\} \tag{ii}$$

Node 1 is the i node of both elements so that using Table 13.1

$$Q_1 = -248 \times 10^3 \{(-1+0) \quad (0-2)\} = -248 \times 10^3 \{-1 \quad -2\}$$

Node 2 is the j node of 1 and

$$Q_2 = -248 \times 10^3 \{1 \quad -2\}$$

Finally

$$Q = -248 \times 10^3 \{-1 \quad -2 \quad 1 \quad -2 \quad -1 \quad 2 \quad 1 \quad 2\} \text{ N} \tag{iii}$$

(b) Here $\delta = 0$ in (13.44) and so $F = Q$ where Q is given by equation (iii) above. Since $\delta = 0$ then $\delta_1 = \delta_2 = 0$ and using (13.34) to (13.36) the stresses in both elements are

$$-49.6 \{1 \quad 1 \quad 0\} \text{ MN/m}^2$$

Problem for Solution 13.6

13.4 Plane Elasticity (Formal Approach)

Here we shall develop in a more formal manner the particular results obtained in Article 13.3. We began by selecting the number and locations of nodes.

Triangular elements were adopted and a suitable displacement function chosen. In Section 13.3.1 the linear functions (13.7) which were used had six coefficients and these can be written as a column matrix

$$\alpha = \{\alpha_1 \quad \alpha_2 \quad \alpha_3 \quad \alpha_4 \quad \alpha_5 \quad \alpha_6\} \qquad (13.46)$$

The first major step is to relate these coefficients (or amplitudes as they were called) to the nodal displacements for an element by substituting the nodal displacements in the displacement functions. Thus if node q has coordinates x_q, y_q then, substituting in equations (13.7) we have

$$u_q = \alpha_1 + \alpha_2 x_q + \alpha_3 y_q; \quad v_q = \alpha_4 + \alpha_5 x_q + \alpha_6 y_q$$

For an element with nodes i, j, m as in Figure 13.2 we have

$$
\begin{bmatrix} u_i \\ v_i \\ u_j \\ v_j \\ u_m \\ v_m \end{bmatrix}
=
\begin{bmatrix}
1 & x_i & y_i & 0 & 0 & 0 \\
0 & 0 & 0 & 1 & x_i & y_i \\
1 & x_j & y_j & 0 & 0 & 0 \\
0 & 0 & 0 & 1 & x_j & y_j \\
1 & x_m & y_m & 0 & 0 & 0 \\
0 & 0 & 0 & 1 & x_m & y_m
\end{bmatrix}
\begin{bmatrix} \alpha_1 \\ \alpha_2 \\ \alpha_3 \\ \alpha_4 \\ \alpha_5 \\ \alpha_6 \end{bmatrix}
\qquad (13.47)
$$

The column vectors are from (13.15b) and (13.46) δ_j and α. Then, denoting the square matrix as H, we can write the above as

$$\delta_j = H\alpha \qquad (13.47a)$$

Inverting we obtain an expression for α as

$$\alpha = H^{-1}\delta_j \qquad (13.48)$$

This expression summarizes information such as that in equations (13.10) and (13.10a).

The next step is to express the element strains in terms of element nodal displacements. We begin by applying the strain/displacement equations (12.6) to the assumed displacement functions, obtaining equations (13.8). These express the strains in terms of the amplitudes of the displacement functions and may be written in matrix form as

$$
\begin{bmatrix} \epsilon_x \\ \epsilon_y \\ \gamma_{xy} \end{bmatrix}
=
\begin{bmatrix}
0 & 1 & 0 & 0 & 0 & 0 \\
0 & 0 & 0 & 0 & 0 & 1 \\
0 & 0 & 1 & 0 & 1 & 0
\end{bmatrix}
\begin{bmatrix} \alpha_1 \\ \alpha_2 \\ \alpha_3 \\ \alpha_4 \\ \alpha_5 \\ \alpha_6 \end{bmatrix}
\qquad (13.49)
$$

Denoting the (3×6) matrix by $\mathbf{b}$ we have in shorthand form

$$\boldsymbol{\epsilon} = \mathbf{b}\boldsymbol{\alpha} \tag{13.49a}$$

and clearly a similar relation holds for total strains e when thermal effects are present. Substituting for $\boldsymbol{\alpha}$ from (13.48) we have

$$\boldsymbol{\epsilon} = \mathbf{b}\mathbf{H}^{-1}\boldsymbol{\delta}_j \tag{13.50}$$

Comparing this with equation (13.12a) we note that

$$\mathbf{B} = \mathbf{b}\mathbf{H}^{-1} \tag{13.51}$$

The distinction between these two forms will be found to be important when more general displacement functions are considered in Article 13.5.

Having found the strains in terms of nodal displacements the stresses in the element are obtained as in Section 13.3.1 by applying the stress/strain equations (13.13). Thus

$$\boldsymbol{\sigma} = \mathbf{D}\boldsymbol{\epsilon} = \mathbf{D}\mathbf{b}\mathbf{H}^{-1}\boldsymbol{\delta}_j \tag{13.52}$$

This brings us to the end of Section 13.3.1. In computational terms we need only have supplied nodal coordinates, and the nodal numbers and E and ν for each element. The computer would be programmed to generate from this information the matrix $\mathbf{H}$ for any element; matrix $\mathbf{b}$ is the same for all elements, while $\mathbf{D}$ is generated from the plane stress or plane strain standard form (equations 3.14 and 3.16) using the values of E and ν for the element.

The forces exerted on the nodes of the jth element are shown in Figure 13.2 and were defined as a vector $\mathbf{F}_j$ by equation (13.15a). The element stiffness equations relating these forces to the nodal displacements $\boldsymbol{\delta}_j$ were derived in Section 13.3.2 by considering equilibrium of the element using the Principle of Virtual Work. Introducing a set of virtual nodal displacements $\boldsymbol{\delta}_j^*$ the virtual work of the external forces is as in (13.17) and the virtual strain energy follows from (13.19). Equating these for equilibrium we have, as before

$$(\boldsymbol{\delta}_j^*)^{\mathrm{T}}\mathbf{F}_j = \int_V (\boldsymbol{\epsilon}^*)^{\mathrm{T}}\boldsymbol{\sigma} \, \mathrm{d}V$$

Using equation (13.50) and the rule for transpose of a matrix product we have

$$(\boldsymbol{\epsilon}^*)^{\mathrm{T}} = (\mathbf{b}\mathbf{H}^{-1}\boldsymbol{\delta}_j^*)^{\mathrm{T}} = (\boldsymbol{\delta}_j^*)^{\mathrm{T}}(\mathbf{H}^{-1})^{\mathrm{T}}\mathbf{b}^{\mathrm{T}}$$

Then substituting also for $\boldsymbol{\sigma}$ from (13.52) and cancelling the virtual displacements, we find

$$\mathbf{F}_j = \int_V (\mathbf{H}^{-1})^{\mathrm{T}}\mathbf{b}^{\mathrm{T}}\mathbf{D}\mathbf{b}\mathbf{H}^{-1}\boldsymbol{\delta}_j \, \mathrm{d}V$$

Now the nodal displacements are independent of the integration, as are the matrices H^{-1} which were formed from nodal coordinates. Therefore we can rewrite the equilibrium condition as

$$F_j = (H^{-1})^T \left(\int_V b^T Db \, dV \right) H^{-1} \delta_j \qquad (13.53)$$

$$= k_j \delta_j$$

where

$$k_j = (H^{-1})^T \left(\int_V b^T Db \, dV \right) H^{-1} \qquad (13.54)$$

is the element stiffness matrix. For the particular system discussed in Article 13.3 matrices b and D (see equations 13.49, 3.14 and 3.16) are composed of constants for any one element and so (13.54) can be further simplified to

$$k_j = (H^{-1})^T b^T Db H^{-1} t \Delta \qquad (13.55)$$

where t is the thickness and Δ the area of the element.

The above results correspond to the set of equations (13.20–13.22). As far as computation is concerned all the matrices in (13.55) have already been formed and so the element stiffness matrix is obtained by evaluating the matrix products utilizing the appropriate computer routines which are available as standard facilities.

The assembly stiffness equations are now deduced by application of the twin conditions of compatibility of displacement, and equilibrium, at the nodes. The first condition has been satisfied by using nodal displacements to define element behaviour. The second condition requires that at any node q the external load F_q must be the resultant of the loads exerted on the elements meeting at the node, and is expressed for the qth node as

$$F_q = \sum_{j=1}^{n} F_{q;j} \qquad (13.56)$$

Here $F_{q;j}$ is the vector of loads exerted at node q on the jth element. The summation has been taken formally over all n elements; those elements not having q as a node make zero contribution.

As shown by (13.26) vectors of the form $F_{q;j}$ can be expressed in terms of nodal displacements and this form can be generalized as

$$F_{q;j} = k_{q1;j} \delta_1 + k_{q2;j} \delta_2 + \ldots + k_{qp;j} \delta_p + \ldots + k_{qn;j} \delta_n \qquad (13.57)$$

Again the summation has been taken formally over all nodes, but, for many of the nodes, $k_{qp;j}$ will be a null matrix. Then substituting in (13.56) and using (13.31) we obtain

$$F_q = K_{q1} \delta_1 + \ldots + K_{qp} \delta_p + \ldots + K_{qn} \delta_n \qquad (13.58)$$

There are n of these equations comprising the required assembly stiffness equations (13.1). The components $\mathbf{K}_{qp}$ of the assembly stiffness matrix are the summation of all element sub-matrices with the same nodal subscripts (equation 13.31). In computational terms each sub-matrix is (2×2) for plane elasticity and the components are transferred to addresses in $\mathbf{K}$ in accordance with the rules of Table 13.1.

When thermal effects are present total strains $\mathbf{e}$ are used and stresses are related to the nodal displacements by

$$\sigma = \mathbf{De} + \mathbf{D_T} = \mathbf{DbH^{-1}\delta_j} + \mathbf{D_T} \qquad (13.59)$$

Applying the Principle of Virtual Work as before the modified element stiffness equations are

$$\mathbf{F_j} = \mathbf{k_j\delta_j} + \mathbf{Q_j} \qquad (13.60)$$

where

$$\mathbf{Q_j} = (\mathbf{H^{-1}})^\mathbf{T} \int_V \mathbf{b^T D_T} \, dV \qquad (13.61)$$

is the matrix of thermal loads for the element. The component of $\mathbf{Q_j}$ at any node of the element is typically $\mathbf{Q}_{q;j}$ (equation 13.39). Adding this term to the right-hand side of (13.57) and proceeding as before we find that a term

$$\mathbf{Q}_q = \sum_{j=1}^{n} \mathbf{Q}_{q;j} \qquad (13.62)$$

must be added to the right-hand side of (13.58), giving the general assembly equations

$$\mathbf{F} = \mathbf{K\delta} + \mathbf{Q} \qquad (13.63)$$

If all displacements and temperatures are known the nodal loads can at once be calculated. More usually the required unknowns are certain of the displacements, comprising a vector $\mathbf{\delta}_R$. Partitioning $\mathbf{\delta}$ we have

$$\mathbf{\delta} = \{\mathbf{\delta}_R \quad \mathbf{\delta}_P\} \qquad (13.64)$$

where $\mathbf{\delta}_P$ are known displacements. As discussed in Section 13.3.4 there must always be sufficient of these to prevent rigid body displacement. If this condition is not met it will be found that the matrix equations (13.63) cannot be solved for the unknown displacements.

Partitioning equations (13.63) to correspond with (13.64) we have

$$\begin{bmatrix} \mathbf{F_R} \\ \mathbf{F_P} \end{bmatrix} = \begin{bmatrix} \mathbf{K_{RR}} & \mathbf{K_{RP}} \\ \mathbf{K_{PR}} & \mathbf{K_{PP}} \end{bmatrix} \begin{bmatrix} \mathbf{\delta_R} \\ \mathbf{\delta_P} \end{bmatrix} + \begin{bmatrix} \mathbf{Q_R} \\ \mathbf{Q_P} \end{bmatrix} \qquad (13.65)$$

where $\mathbf{F_R}$ is a vector of known forces. Then expanding the first row and rearranging we have

$$\mathbf{F'_R} = (\mathbf{F_R} - \mathbf{K_{RP}\delta_P} - \mathbf{Q_R}) = \mathbf{K_{RR}\delta_R} \qquad (13.66)$$

which reduces to the particular expression (13.32) when thermal effects are absent. These reduced stiffness equations may be inverted giving the unknown displacements

$$\boldsymbol{\delta}_{R} = \mathbf{K}_{RR}^{-1}\mathbf{F}_{R}' \tag{13.67}$$

The computational procedure for these equations has already been described in Section 13.3.4. Once $\boldsymbol{\delta}_R$ has been found the nodal displacement vector $\boldsymbol{\delta}_i$ (equation 13.15b) for any element can be written, and the stresses found using (13.59).

13.5 Further Applications

We begin by examining the possibility of using alternative displacement functions for a triangular element in plane elasticity. As shown in Section 13.3.1 the linear displacement function makes strain and stress uniform within each element. Exact results are then obtained only in such simple situations as direct loading (see Example 13.5). When the actual state of stress varies with position satisfactory results can still be obtained by using a large number of elements so that the actual displacement surface is represented with sufficient accuracy (cf. Figures 13.3a and b).

As an alternative to using many elements with corresponding increase of data and other computational difficulties, an assumed displacement function could be chosen which would allow a more accurate representation of the state of strain. For example let

$$u = \alpha_1 + \alpha_2 x + \alpha_3 y + \alpha_4 x^2 + \alpha_5 xy + \alpha_6 y^2$$
$$v = \alpha_7 + \alpha_8 x + \alpha_9 y + \alpha_{10} x^2 + \alpha_{11} xy + \alpha_{12} y^2 \tag{13.68}$$

Applying the strain/displacement equations (12.6) we have

$$\epsilon_x = \alpha_2 + 2\alpha_4 x + \alpha_5 y; \qquad \epsilon_y = \alpha_9 + \alpha_{11} x + 2\alpha_{12} y$$
$$\gamma_{xy} = \alpha_3 + \alpha_5 x + 2\alpha_6 y + \alpha_8 + 2\alpha_{10} x + \alpha_{11} y \tag{13.69}$$

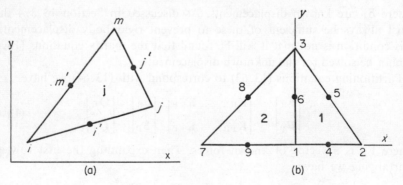

(a)   (b)

Figure 13.10

Clearly a system such as this can represent the strain distribution more accurately for a given number of elements. However, a price has to be paid in the greater complexity of the calculations. As there are twelve amplitudes the formation of equations (13.47a) for this case will require the substitution in equations (13.68) of the displacements at six nodes. The usual practice is to take the additional nodes at the midpoints of the sides of the triangle (Figure 13.10a). Then equations (13.47a) have in this case the form

$$
\begin{bmatrix} u_i \\ v_i \\ 0 \\ \cdot \\ \cdot \\ v_m \end{bmatrix} = \begin{bmatrix} 1 & x_i & y_i & x_i^2 & x_i y_i & y_i^2 & 0 & 0 & 0 & 0 & 0 & 0 \\ 0 & 0 & 0 & 0 & 0 & 0 & 1 & x_i & \cdots \\ & & \cdot & & & \cdot & & & \cdot \\ & & \cdot & & & \cdot & & & \cdot \\ & & \cdot & & & \cdot & & & \cdot \end{bmatrix} \begin{bmatrix} \alpha_1 \\ \alpha_2 \\ \cdot \\ \cdot \\ \alpha_{12} \end{bmatrix}
$$

Since the square matrix **H** is composed of numerical quantities derived from nodal coordinates it can be inverted to give equations of the form of (13.48) expressing amplitudes in terms of nodal displacements.

The strains are related to the amplitudes by equations (13.69) which when written in the form of (13.49a) are

$$
\begin{bmatrix} \epsilon_x \\ \epsilon_y \\ \cdot \\ \cdot \\ \gamma_{xy} \end{bmatrix} = \begin{bmatrix} 0 & 1 & 0 & 2x & y & 0 & 0 & 0 & 0 & 0 & 0 & 0 \\ 0 & 0 & 0 & 0 & 0 & 0 & 0 & 0 & 1 & 0 & x & 2y \\ & & \cdot & & & \cdot & & & \cdot \\ & & \cdot & & & \cdot & & & \cdot \\ 0 & 0 & 1 & 0 & x & 2y & 0 & 1 & 0 & 2x & y & 0 \end{bmatrix} \begin{bmatrix} \alpha_1 \\ \alpha_2 \\ \cdot \\ \cdot \\ \alpha_{12} \end{bmatrix}
$$

The stresses are obtained from these strains using equations (13.13) in which **D** is as in (3.14) or (3.16). Since matrix **b** does not consist of constants the general expression (13.54) for the element stiffness matrix has to be used and the integral of the matrix product $\mathbf{b^T D b}$ evaluated over the volume of the element. This may be done by deriving explicit forms analogous to (13.23) and (13.24), or by numerical integration. The details of the latter are beyond the scope of this text and the reader is referred to the comprehensive work of Zienkiewicz.[1]

The assembly stiffness matrix is derived from the element stiffness matrices by application of equation (13.31). Since the typical sub-matrix $\mathbf{k}_{pq;j}$ is again of size (2×2) its components are placed in rows $(2p-1)$, $2p$ and columns $(2q-1)$, $2q$ of **K**. All other steps of the analysis are as previously described.

It is interesting to note that the displacement surfaces derived from a second-order polynomial are continuous in value across element boundaries, and so meet the condition for convergence. For example in the particular case shown in Figure 13.10b the displacement functions for element 1 reduce, along line 1, 6, 3 to

$$
u = \alpha_1 + \alpha_3 y + \alpha_6 y^2; \qquad v = \alpha_7 + \alpha_9 y + \alpha_{12} y^2 \tag{13.70}
$$

As we have seen the values of the amplitudes α_1 etc. are so chosen that at nodes 1, 6 and 3 the displacements are u_1, u_6, u_3, v_1, v_6, v_3. Now for element 2 the displacement functions along line 1, 6, 3 have the *form* of (13.70) and the amplitudes must also be such that the displacements at 1, 6 and 3 are u_1, etc. Consequently the displacements along the line 1, 6, 3 are the same for both elements. A similar result can be established for any boundary between elements, and the displacement function is said to be conforming.

By use of higher-order polynomials the displacement surfaces can be represented still more accurately, but the increased complexity of the computations may not be justified on economic grounds. Another possibility is the use of other element shapes such as rectangle or quadrilateral. The procedure is the same as described starting with the choice of a conforming displacement function, followed by generation of **H** and **b** matrices, and application of equations (13.54) and (13.31). However, there is usually little to be gained as a triangle is eminently suited to the representation of complex plane shapes.

Lastly we may quickly illustrate how the method can be applied to three-dimensional problems in elasticity for which it is necessary to consider displacements u, v and w in the x, y and z directions. The basic element is a tetrahedron (Figure 13.11) for which we choose linear assumed displacement functions

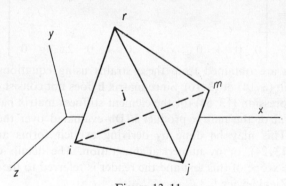

Figure 13.11

$$u = \alpha_1 + \alpha_2 x + \alpha_3 y + \alpha_4 z$$
$$v = \alpha_5 + \alpha_6 x + \alpha_7 y + \alpha_8 z \tag{13.71}$$
$$w = \alpha_9 + \alpha_{10} x + \alpha_{11} y + \alpha_{12} z$$

The strain displacement equations can be shown to be a generalization of those derived in Section 12.1.3, see for example reference 9 and we can write

$$\epsilon_x = \alpha_2; \qquad \epsilon_y = \alpha_7; \qquad \epsilon_z = \alpha_{12}$$
$$\gamma_{xy} = \alpha_3 + \alpha_6; \qquad \gamma_{yz} = \alpha_8 + \alpha_{11}; \qquad \gamma_{xz} = \alpha_4 + \alpha_{10} \tag{13.72}$$

Introducing the nodal coordinates x_i, y_i, z_i, etc. in equations (13.71) we obtain equations (13.47a) in the form

$$\begin{bmatrix} u_i \\ v_i \\ \cdot \\ \cdot \\ \cdot \\ w_r \end{bmatrix} = \begin{bmatrix} 1 & x_i & y_i & z_i & 0 & 0 & 0 & 0 & 0 & 0 & 0 & 0 \\ 0 & 0 & 0 & 0 & 1 & x_i & y_i & z_i & 0 & 0 & 0 & 0 \\ & & & & & \cdot & & & & & & \\ & & & & & \cdot & & & & & & \\ & & & & & \cdot & & & & & & \\ 0 & 0 & 0 & 0 & 0 & 0 & 0 & 0 & 1 & x_r & y_r & z_r \end{bmatrix} \begin{bmatrix} \alpha_1 \\ \alpha_2 \\ \cdot \\ \cdot \\ \cdot \\ \alpha_{12} \end{bmatrix}$$

The square matrix is inverted to give $\mathbf{H}^{-1}$. By writing equations (13.72) in matrix form corresponding to (13.49a) we find that $\mathbf{b}$ consists of constants and the element stiffness matrix is therefore obtained using equation (13.55) with $t\Delta$ replaced by the volume of the tetrahedron. The assembly stiffness matrix is obtained formally by applying (13.31). However, since there are three displacements at each node, e.g. u_p, v_p, w_p, the components of the typical element sub-matrix $\mathbf{k}_{pq;}$ are placed in rows $3p-2$, $3p-1$, $3p$ and columns $3q-2$, $3q-1$, $3q$ of the scalar assembly stiffness matrix.

For a survey of the vast further possibilities of the method the reader is referred to the Bibliography.

13.6 References

1. Zienkiewicz, O. C., *The Finite Element Method in Engineering Science*, McGraw-Hill, 1971.
2. Zienkiewicz, O. C. and Cheung, Y. K., *The Finite Element Method in Structural and Continuum Mechanics*, McGraw-Hill, 1967.
3. Holand, I. and Bell, K. (Eds.), *Finite Element Methods in Stress Analysis*, Tapir (Trondheim), 1968.
4. *Proc. Conf. on Matrix Methods in Structural Analysis*, Wright-Patterson Air Force Base, 1965. AFFDL-TR-66-80, 1966.
5. *Proc. Conf. on Matrix Methods in Structural Analysis*, Wright-Patterson Air Force Base, AFFDL-TR-68-150, 1968.
6. Przemieniecki, J. S., loc. cit., Article 12.6.
7. Bissell, P. N., Private communication, Univ. of Dundee, 1972.
8. Maggs, D. J., Private communication, Univ. of Dundee, 1972.
9. Timoshenko, S. P. and Goodier, J. N., loc. cit., Article 1.15.

13.7 Problems for Solution

13.1. Calculate the forces required at j of element of Example 13.2 when displacement is prevented at i and m and j is displaced 1 mm upwards only. Find the state of stress.

13.2. Calculate the plane stress stiffness matrix for an element with nodes at points $(1, 1)$, $(3, 1)$, $(3, 2)$. $E = 70$ GN/m², $v = 0.35$, $t = 0.01$ m.

13.3. Calculate the plane stress stiffness matrix for steel elements ($E = 200$ GN/m², $v = 0.3$) with coordinates (in cm) (a) $(0, 0)$, $(1, 0)$, $(1, 2)$, (b) $(1, 2)$, $(1, 0)$, $(2, 0)$.

13.4 The assembly of Example 13.3 is supported as in Figure 13.7a but subjected to forces $X_2 = -100 \, \text{kN} = -X_4$. Find the unknown displacements and the stresses in element **2**.

13.5. Form assembly stiffness equations for the two elements in Problem 13.3 when the given nodes are numbered 1, 2, 4 and 4, 2, 3 respectively. If a force $X_4 = +4 \, \text{kN}$ is imposed and displacement is prevented at nodes 1, 2, 3 find the remaining displacements and the stresses in the elements.

13.6. If the system of Example 13.6 is supported as in Figure 13.6a, find the unknown displacements.

14

Framed Structures

14.1 Description

14.1.1 *Introduction*

Frames are structures formed by connecting together bars (Figure 14.1). If a frame is extensive only in one plane and is subject only to loads acting in that plane then it will be termed a plane frame (Figure 14.1a). Any more general arrangement will be referred to as a space frame (Figure 14.1b). The behaviour of a frame is determined by the properties of the bars, the configuration of the frame, the properties of the joints, and the arrangement of loads and supports.

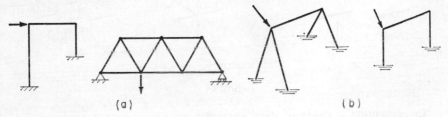

(a)   (b)

Figure 14.1

Joints in a plane frame are considered to take one of two forms, viz. (a) pinned-joints (Figure 14.2a) which offer no resistance to relative rotation of the bars and so can only transmit forces and (b) rigid joints (Figure 14.2b) which prevent *relative* rotation of the bars and can therefore transmit both

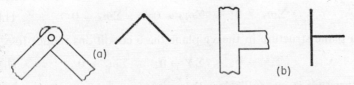

(a)   (b)

Figure 14.2 (a) Pinned, (b) rigid, joints, with schematic
representation also shown

moments and forces between the bars. This distinction has important consequences as will be shown later. The two types of joint so defined are, of course, idealizations, but provide a basis of approximation to actual conditions. Joints which permit or prevent relative rotation can also be described for space frames.

The supports for a plane frame are idealized into the three forms shown in Figure 14.3a. The roller support prevents translation normal to the surface on which it rests. The hinged-type support completely prevents translation, which is equivalent to preventing translation in any two perpendicular directions in the plane. The fixed support totally prevents both translation and rotation in the plane. These supports therefore eliminate respectively, one, two and three degrees of freedom of rigid body displacement of a plane structure. Supports imposing corresponding restrictions for a space frame eliminate one, three and six degrees of freedom (Figure 14.3b).

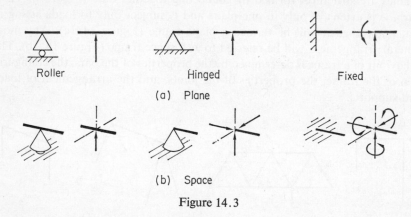

Roller   Hinged   Fixed

(a) Plane

(b) Space

Figure 14.3

In preventing a mode of displacement a support exerts a corresponding reactive load (Figure 14.3). Restraint of a component of translation requires the development of a concentrated force in the same line of action. Similarly, prevention of rotation in any plane requires the development of a concentrated moment.

A frame is in equilibrium if for all external loads (applied and reactive)

$$\Sigma X = 0; \qquad \Sigma Y = 0; \qquad \Sigma Z = 0;$$

$$\Sigma m_x = 0; \qquad \Sigma m_y = 0; \qquad \Sigma m_z = 0 \qquad (14.1a\text{--}f)$$

For a plane structure in the xy plane these conditions reduce to

$$\Sigma X = 0; \qquad \Sigma Y = 0; \qquad \Sigma m_z = 0 \qquad (14.2a\text{--}c)$$

If a structure is in equilibrium then so are all its parts. In particular the loads exerted by the bars meeting at any joint, together with any external loads on the joint must satisfy the appropriate equilibrium condition—(14.1) or (14.2).

The displacements imposed on a frame must be such that it remains continuous. In particular any displacements proposed for the bars which meet at a particular joint must be such that the bars can continue to meet at that joint. This is a condition of geometric compatibility.

14.1.2 *Plane Frames (Pin-joints)*

Plane frames with only *pin-joints* between the bars are also known as trusses or braced frames. In the analysis of these frames it is first assumed that loads are applied only at joints, any distributed loads on a bar being apportioned to its ends. Since no moments can be sustained by a pin-joint, the external load on it can be described by force components in two perpendicular directions. Similarly the forces exerted on a joint by the bars meeting there can also be described by two perpendicular components. Since the joints are to be in equilibrium, we have at the ith joint

$$(\Sigma X)_i = 0; \qquad (\Sigma Y)_i = 0$$

There are $2n$ of these equilibrium equations where n is the number of joints. However, the number of external forces which can be independently specified is $(2n-3)$ since the remaining three must be so chosen as to satisfy the three equilibrium equations (14.2). These three force components are supplied by the reactive loads at the supports which must be so disposed as to prevent rigid body displacement of the frame. The arrangements in Figures 14.4a and c are satisfactory while that in Figure 14.4d is not, although there are three reactive forces.

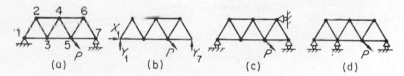

Figure 14.4

When there are just sufficient support constraints to maintain equilibrium the reactive loads can be determined from the external equilibrium conditions (14.2) alone. The frame is then said to be *statically determinate*. If there are more support constraints than this minimum the frame is said to be *statically indeterminate*, or *hyperstatic*. For instance if the roller at joint 7 of Figure 14.4a were replaced by a hinge the frame would become statically indeterminate since the number of reactive loads would be increased from three to four. The reactive loads could then be found only by considering also the deformation of the frame.

The internal forces acting on the ends of a pin-jointed bar are in general as in Figure 14.5a. Taking moments about joint i for equilibrium we see that the sum can only be zero if the resultant of X_j and Y_j acts along the centre line of the bar. Since the same argument can be applied for joint j it follows that the resultant end forces must be collinear with the bar, which is therefore in a state of direct loading (Figure 14.5b)—see Article 5.2.

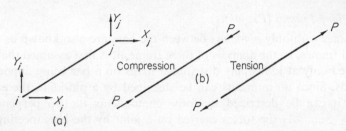

Figure 14.5

We shall now consider the development of a plane frame, starting from a statically determinate bar (line 1, 2, Figure 14.6a). To add joint 3 we clearly require two more bars (dashed lines in Figure 14.6a) as can be proved by considering equilibrium of the added joint. Since the external load F on the added joint (Figure 14.6b) can have any direction the triangle of forces

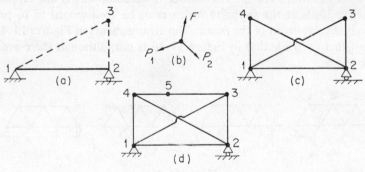

Figure 14.6

requires for equilibrium two other forces P_1 and P_2 to act on the joint, i.e. two bars are required. Addition of another joint (4 in Figure 14.6c) requires two more bars and so on. (Note that bars 1,3 and 2,4 are supposed to be independent as indicated by the loop sign.) Thus the total number of bars required to form in this way a frame with n joints is

$$n = 1+2(n-2) = 2n-3 \tag{14.3}$$

Such a frame will be statically determinate internally since the triangle of forces can be applied at each joint. For example the frame of Figure 14.6c has four joints and five bars and that of Figure 14.4a has seven joints and eleven bars. If the number of bars is less than that given by (14.3) the frame is unstable since there will be one joint at which no triangle of forces can be drawn.

It may be noted that the number subtracted from $2n$ in equation (14.3) is equal to the number of support constraints in the example previously discussed (Figure 14.6c). That this is not fortuitous can be shown by removing bar 2,4 of the frame of Figure 14.4a and observing that stability could then

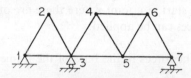

Figure 14.7

be restored by, for example, introduction of a roller support as in Figure 14.7. Thus we might conclude that a bar is equivalent to a single support constraint and rewrite (14.3) as

$$2n = n + c \tag{14.4}$$

where c is the number of support constraints. Although it has four constraints, the frame of Figure 14.7 is statically determinate, the fourth equation of external equilibrium being obtained from the condition that zero moment is transmitted by joint 3, i.e. for either of the two substructures joined by 3 the moments about 3 are independently zero.

We therefore conclude that (14.4) is a general condition for statical determinacy of a pin-jointed frame. However, care is necessary in its application as can be seen by considering the classic example in Figure 14.8a in which there are six joints, three constraints and nine bars. Although equation (14.4) is satisfied the frame is quite obviously unstable against the side force. Thus equation (14.4) is a *necessary* but not a *sufficient* condition for a frame to be

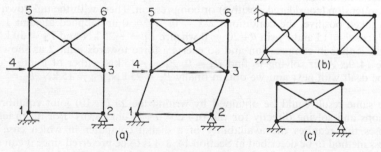

Figure 14.8

statically determinate and stable. The configuration must be obtained by addition of two bars per joint in accordance with the triangle of forces. This condition also prohibits the addition of a joint supported by two collinear bars, e.g. joint 5 in Figure 14.6d. If the number of bars exceeds that required by equation (14.4) the frame may be unstable, e.g. Figure 14.8b, or statically indeterminate, e.g. Figure 14.8c.

For statically determinate frames the basic method of analysis is to find the reactive loads from equilibrium of the external forces, and then to consider internal equilibrium at each joint in turn to obtain the forces on the bars. It is necessary to start at a joint where there are only two unknowns so that the triangle of forces can be applied.

Example 14.1   Find the forces in the bars of the frame of Figure 14.9.

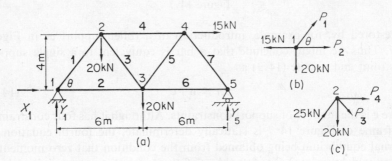

Figure 14.9

Solution   For external equilibrium we find

$$X_1 = -15 \text{ kN}; \quad Y_1 = 20 \text{ kN}; \quad Y_5 = 20 \text{ kN}$$

At joints 2, 3, 4 there are three or more unknown bar forces, but a start can be made at 1 or 5. For example joint 1 is subject to the reactive forces X_1, Y_1 and forces P_1, P_2 (Figure 14.9b) exerted by bars 1 and 2. It will be recalled that these forces are collinear with the bars and act away from the joint, or towards it, according as the bar is in tension (considered positive) or compression. Then, with the unknown bar forces taken positive, the conditions $\Sigma X = 0$, $\Sigma Y = 0$ are applied at joint 1, i.e. $P_2 + P_1 \cos \theta = 15$ and $P_1 \sin \theta + 20 = 0$ whence $P_1 = -25$ kN and $P_2 = 30$ kN.

Bar 1 is thus in compression and so exerts a force towards joint 2 as shown in Figure 14.9c. After solving to find $P_3 = 0$, $P_4 = -15$ kN, either of joints 3 or 4 may be dealt with next and we obtain finally $P_5 = 25$ kN, $P_6 = 15$ kN, $P_7 = -25$ kN.

The same result could be obtained by writing the $2n$ (= 10) joint equilibrium equations and solving directly for the (n+c = 7+3) unknowns. For all practical purposes this requires the availability of a digital computer in which case the stiffness method to be described in Section 14.3.4 is to be preferred since it can also solve statically indeterminate frames and supply the deflexions of the joints.

Problems for Solution 14.1–14.5

14.1.3 *Plane Frames* (*Rigid Joints*)

The external equilibrium and determinacy of these frames are governed by the same rules as in Section 14.1.2. However, since each joint can transmit a moment as well as the two force components it follows that there are $3n$ equations of joint equilibrium. Because of the moments exerted by the rigid joints it is possible to add a joint using only one bar. For example in Figure 14.10a joint 3 is added using bar 2,3. Any external loads exerted on this bar

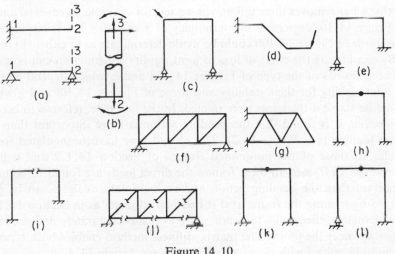

Figure 14.10

at 3 can be maintained in equilibrium by the actions transmitted by the joint 2 (Figure 14.10b). This figure also shows that the bars are in a state of combined direct and bending loading (cf. Figure 9.13). Thus the addition of one bar adds three unknowns, viz. the direct and transverse forces and the moment transmitted by the bar. (Compare Section 14.1.2 where each bar added one unknown, viz. the direct force). Since the first bar (Figure 14.10a) also contributes three unknowns there are $3n$ unknowns due to the bars. To these must be added the number of unknown support constraints c, making the total of unknowns $(3n+c)$. Following the discussion in Section 14.1.2 we recognize the following. If $3n > (3n+c)$ the frame is in an unstable condition; if $3n - (3n+c)$ the frame is statically determinate; if $3n < (3n+c)$ the frame is statically indeterminate, the quantity $((3n+c)-3n)$ being called the degree of indeterminacy (or redundancy). For example in Figure 14.10 case (c) is unstable, (d) is statically determinate, and (e), (f) and (g) are statically indeterminate. In particular the frame of Figure 14.10f has $3n = 24$ and $(3n+c) = 43$ so that the degree of indeterminacy is 19.

The above method of assessing indeterminacy becomes more difficult to apply when internal pin-joints are also present in the frame (Figure 14.10h). Since no moment is transmitted by a pin-joint the sum of moments to either side of it is zero, and so the number of available equilibrium equations becomes $(3n+c)$ where c is the number of such internal pin-joints. Comparing $(3n+c)$ with $(3\mathbf{n}+\mathbf{c})$ we see that there are 2 degrees of indeterminacy in case (h) of Figure 14.10.

It is generally held that for frames with rigid joints a more reliable procedure for assessing the degree of indeterminacy is to cut bars and remove support constraints until a statically determinate system is obtained. Since cutting a bar removes three unknowns we see, for example, in cases (i) and (j) of Figure 14.10, degrees of indeterminacy of 3 and $((6 \times 3)+1)$ respectively. The frame of Figure 14.10h could be made determinate as in either (k) or (l).

By considering the effect of loss of joint rigidity a distinction can be made between frames of the type of Figures 14.10d and e, which depend entirely on joint rigidity for their stability and those of Figures 14.10f and g which would be stable if the joints were pinned. In the first type, referred to here as *non-braced*, it is found that the bending action is more important than the direct loading (see Article 14.2). The second type have triangulated forms similar to those of the pin-jointed frames of Section 14.1.2 and will be termed *braced* frames. In these frames the direct loads are found to be more important than the bending action and so a primary analysis can be performed by treating the frame as if it had pinned-joints as in Section 14.1.2. The secondary effects due to bending are evaluated separately and are most important near the joints. The matrix stiffness method enables both types of frame to be solved with the same procedure, see Article 14.3.

Problem for Solution 14.6

14.2 Moment Distribution

14.2.1 *Introduction*

Introduced by Hardy Cross,[1] moment distribution is a method of successive approximations for determining end moments on the bars of continuous beams and rigid jointed (especially non-braced) frames. Before describing the procedure it is convenient to define the various special terms used.

The moments exerted on the ends of a beam supported by two fixed supports are known as *fixed end moments* (FEM). These can be determined for various loads by the methods described in Chapter 7, and some common examples are given in Figure 7.20. The sign convention used is that an anti-clockwise moment on a bar is positive. An additional case of some importance, illustrated in Figure 14.11a, can be obtained from the stiffness matrix of a

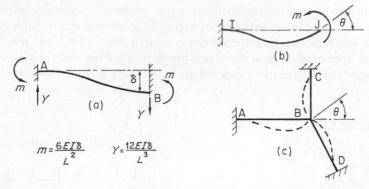

Figure 14.11

beam finite element (equation 7.28). If B were displaced upwards the moments and forces would have opposite senses. The order of subscripts in the designation FEM$_{BA}$ indicates that the moment acts on the B end of the bar AB. The rotational stiffness at end J of the bar IJ in Figure 14.11b can also be obtained from the stiffness matrix (7.28) and is

$$k_{JI} = \frac{4EI}{L} \tag{14.5}$$

Then the moment required for rotation θ at J without translation is $m = k_{JI}\theta$. If two or more bars meet at a rigid joint (Figure 14.11c) then the moment to rotate this joint by an angle θ is the sum of the moments to rotate each bar separately through θ. This is so, because, by definition, the rigid joint rotates as a whole, maintaining unchanged the relative orientation of the bars. Note that for the present we assume that the bars are fixed at their other ends. Then for the system in Figure 14.11c the moment to be exerted at B is $m = (k_{BA}+k_{BC}+k_{BD})\theta$. The proportion of this moment absorbed in bending of bar AB is called the *distribution factor* (DF) for AB and is $k_{BA}\theta/((k_{BA}+k_{BC}+k_{BD})\theta)$, i.e. $k_{BA}/\Sigma k_B$, where Σk_B is the sum of the rotational stiffnesses of the bars meeting at B. In general

$$DF_{JI} = \frac{k_{JI}}{\Sigma k_J} \tag{14.6}$$

It follows from this definition that the sum of the distribution factors at a joint is unity.

Again referring to equation (7.28) we note that when rotation without translation is imposed at J in Figure 14.11b reactive moment is developed at the fixed support. This carry-over-moment has the same sense as the moment at J but half its magnitude. We say that the bar has a *carry-over-factor* (COF) of $\frac{1}{2}$ from J to I.

The first step of the moment distribution procedure is to lock all joints not already fixed against rotation. In Figure 14.12a for example, joint B has to be locked, indicated by a special clamp symbol (Figure 14.12b). When the loads are applied all bars of the frame are then in a fixed end condition and the moments calculated using expressions from Figure 7.20 can be considered

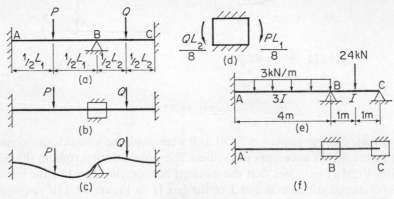

Figure 14.12

a first approximation to the moments in the actual structure. The nature of the approximation in Figure 14.12b is that the locked beam is horizontal at B whereas in the actual structure it would generally be inclined. To maintain this condition the clamp supplies the out-of-balance moment equal to the algebraic sum of the fixed end moments at the joint. From Figure 14.12d this moment is $(-PL_1 + QL_2)/8$.

To improve the approximation we release the clamp at B whereupon rotation occurs (Figure 14.11c) until the consequent bending of the bars absorbs the out-of-balance moment previously exerted by the clamp. Alternatively we could think of clamp removal as equivalent to the superposition of an equal and opposite moment, i.e. $(PL_1 - QL_2)/8$. The products of this moment with the distribution factors determine the moments absorbed at B by each bar. In turn the products of these quantities with the carry-over-factors give the changes in the moments at the other ends of the two bars. For example the end moments for AB are,

at B

$$\left(\left(\frac{-PL_1}{8}\right) + DF_{BA} \times \frac{(PL_1 - QL_2)}{8}\right)$$

and at A

$$\left(\left(\frac{PL_1}{8}\right) + COF_{BA} \times DF_{BA} \times \frac{(PL_1 - QL_2)}{8}\right)$$

If the procedure has been correctly carried out the end moments at B sum to zero and the actual form of the beam has been attained.

If more than one joint has to be locked as in Figures 14.12e and f we proceed to unlock a joint, say C, and distribute the out-of-balance moment as described above. However, before B can be balanced C must be relocked since the distribution factors for a joint are calculated on the basis that the far ends of the bars (i.e. A and C) are held against rotation. Then when joint B is balanced a carry-over-moment is developed at C so that after one balancing of each joint there is still some out-of-balance moment. This first cycle of moment distribution gives an improved approximation to the actual moments. Further cycles may be carried out until the remaining out-of-balance moments are as small as is acceptable for the purpose in hand.

14.2.2 *Organization of Calculations*

Moment distribution is best carried out using a tabular presentation as illustrated by the following example.

Example 14.2 Determine the end moments for the steel beam in Figure 14.12e. The second moments of area of AB and BC are in the ratio 3 : 1 and the support at C subsides 2 mm. Take $E = 200 \text{ GN/m}^2$, $I = 500 \text{ cm}^4$.

Solution The first step is to calculate the distribution factors and fixed end moments. At joint B using equations (14.5) and (14.6)

$$DF_{BA} = k_{BA}/(k_{BA}+k_{BC}) = \frac{4E(3I)}{4} \Big/ \left(\frac{4E(3I)}{4} + \frac{4EI}{2}\right) = \tfrac{3}{5}$$

Similarly $DF_{BC} = \tfrac{2}{5}$ and $\Sigma DF_B = 1$ as a check. Since the fixing at A prevents any rotation $DF_{AB} = (k_{AB}/\infty) = 0$. At joint C $DF_{CB} = k_{CB}/k_{CB} = 1$.

Using expressions from Figure 7.20 and Figure 14.11a we calculate fixed end moments as follows. Bar AB is subject to distributed load so that $FEM_{AB} = wL^2/12 = 4 \text{ kNm}$; at end B $FEM_{BA} = -wL^2/12 = -4 \text{ kNm}$. Bar BC is subject to a central concentrated load (fixed end moments $\pm PL/8$) and support displacement at C for which the moments are $6EI\delta/L^2$. Now $PL/8 = 6 \text{ kNm}$ and $6EI\delta/L^2 = 6 \times 200 \times 10^9 \times 500 \times 10^{-8} \times 2 \times 10^{-3}/4 \text{ Nm} = 3 \text{ kNm}$. Then at end B $FEM_{BC} = 6+3 = 9 \text{ kNm}$ and at end C $FEM_{CB} = -6+3 = -3 \text{ kNm}$.

The working table is now constructed as in Table 14.1

The carry-over-factors are required because the other joints are locked when B or C is being balanced. We see that B and C are initially out of balance by $+5$ and -3 kNm respectively. If we start by balancing joint B, this is equivalent to adding -5 kNm of which (0.6×-5) is absorbed by BA and (0.4×-5) by BC. Carry-over-moments of half these amounts are developed at the other ends of the bars, i.e. columns AB and CB. Joint B is now balanced (total moment zero) and is reclamped before balancing C by adding $+4 \text{ kNm}$. Relocking of C completes the first cycle of distribution with only B unbalanced (due to carry-over from C).

The process is repeated as often as considered necessary. After three cycles the remaining out-of-balance moment of 0.02 kNm would be small enough for most purposes. The end moments on the bars are obtained by summing each column and, as would be expected, the moment at joint C is zero. The two moments at B should

Table 14.1

Joint	A	B		C	
Bar	AB	BA	BC	CB	
DF	0	0·6	0·4	1	
COF	—	$\frac{1}{2}$	$\frac{1}{2}$	$\frac{1}{2}$	
FEM	+4	−4	+9	−3	
	−1·5	−3	−2	−1	
			+2	+4	First cycle
	−0·6	−1·2	−0·8	−0·4	
			+0·2	+0·4	Second cycle
	−0·06	−0·12	−0·08	−0·04	
			+0·02	+0·04	Third cycle
Summation	+1·84	−8·32	+8·34	0	Units kNm

sum to zero, the difference being the remaining unbalanced moment. It is worth noting the substantial fixed end moments developed due to the small displacement of joint C which could easily arise due to settlement or misalignment.

Problems for Solution 14.7–14.9

14.2.3. *Shear Force and Bending Moment Diagrams*

Each bar is considered as a free body subject to end forces (unknown) and end moments (taken from the moment distribution table), together with the actual external loads. For the beam of Example 14.2 these free body diagrams are shown in Figures 14.13a and b. At each end two components of force and a moment are required, the moments being obtained from Table 14.1. Note that at end C of BC there is zero moment.

To determine shear force and bending moment we only need to find the transverse forces Y_{AB} etc. which can be done by considering static equilibrium. Hence for example, taking moments about B for AB we find

$$Y_{AB} = ((3 \times 4) \times 2 + 1 \cdot 84 - 8 \cdot 33)/4 = 4 \cdot 38 \text{ kN}$$

Similarly

$$Y_{BA} = 7 \cdot 62 \text{ kN}; \quad Y_{BC} = 16 \cdot 17 \text{ kN}; \quad Y_{CB} = 7 \cdot 83 \text{ kN}$$

The shear force and bending moment diagrams can now be constructed

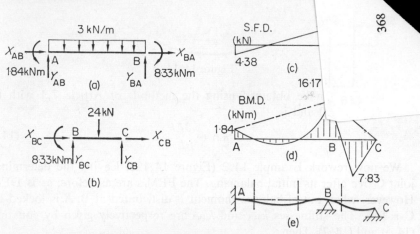

Figure 14.13 (c) and (d) Positive up, negative down

in the usual way (see Article 6.2) and are shown in Figures 14.13c and d; the deflexion curve is sketched in Figure 14.13e.

A useful short cut of the basic procedure for the bending moment diagram is obtained by recognizing that the loads on bar BC in Figure 14.13b are equivalent to those on a simply supported beam subject to an end moment and a central concentrated force. The bending moment diagram is then obtained by superposition of separate bending moment diagrams for these loadings. In Figure 14.13d that for the moment of 8·33 kNm is the chain-dashed line while that due to the concentrated force is triangular (see Figure 6.5, Example 6.1) with maximum value $-PL/4$, i.e. -12 kNm. Combining these we obtain the net bending moment as the shaded portion shown. For a uniformly distributed load the bending moment varies parabolically with central value $-wL^2/8$ (Example 6.3, Figure 6.7a).

Problems for Solution 14.10–14.13

14.2.4 *Determinate Joints*

The moments at outer joints which are free to rotate can be determined from statics. For example in Figure 14.12e there is zero moment at C, and in Figure 14.25c the moments at A and C are zero and 4·5 kNm respectively. During the first cycle of a moment distribution (Table 14.1) such joints attain their correct moments but subsequently become unbalanced due to carry-overs from other joints. The rate of convergence is enhanced if these determinate joints are left free to rotate after their first balancing. This is because when a moment is applied to end J of a bar (Figure 14.14) and vertical displacement of J is prevented then no moment is developed at the other end I provided it is free to rotate, i.e. the carry-over-factor is zero. The rotational

Figure 14.14

stiffness at J can be obtained using the methods of Article 7.1 with the condition $v_J = 0$ and is

$$k_{JI} = \frac{3EI}{L} \qquad (14.7)$$

We now rework Example 14.2 (Figure 14.12e) leaving the determinate joint C free after its initial balancing. The FEM's are as before, as is DF_{CB}. However, when out-of-balance moment is distributed at B, A is locked but C is free. The stiffnesses k_{BA} and k_{BC} are respectively given by equations (14.5) and (14.7). Then

$$DF_{BC} = \frac{3EI}{2} \bigg/ \left(\frac{4E(3I)}{4} + \frac{3EI}{2} \right) = \tfrac{1}{3}; \qquad DF_{BA} = \tfrac{2}{3}$$

There is COF of $\tfrac{1}{2}$ from C to B since B is locked when C is balanced but zero COF from B to C since C is free when B is balanced—see Table 14.2.

Table 14.2

Joint	A	B		C
Bar	AB	BA	BC	CB
DF	0	$\tfrac{2}{3}$	$\tfrac{1}{3}$	1
COF	—	$\tfrac{1}{2}$	0	$\tfrac{1}{2}$
FEM	+4	−4	+9	−3
			+1·5	+3
	−2·17	−4·33	−2·17	
Summation	+1·83	−8·33	+8·33	0 kNm

Thus the same result is obtained as in Table 14.1 but after only one cycle of distribution. Note that the sequence of operations must be decided in advance.

Problems for Solution 14.14–14.16

14.2.5 Frames

Some examples are now given of moment distribution for frames. It is assumed that the translations of the joints are known.

Example 14.3 Obtain the bending moment diagram for the steel frame in Figure 14.15.

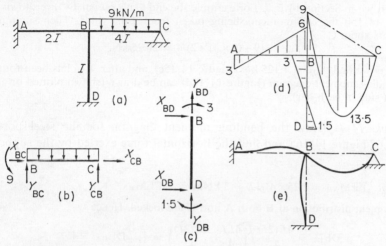

Figure 14.15   AB = BC = BD = 4 m, units of moment kNm,
(e) schematic deflexion curve

Solution   In finding end moments on the bars by moment distribution the solution is obtained more quickly if the determinate joint C is balanced first and then left unclamped.

FEM$_{BC}$ = $9 \times 4^2/12$ = 12 kNm = $-$FEM$_{CB}$; the other bars have zero FEM. At C we have DF$_{CB}$ = 1 and COF = $\frac{1}{2}$ from C to the locked joint B. At B we have A and D locked but C is free to rotate. Hence using equations (14.5) to (14.7) we find

$$\text{DF}_{BC} = \frac{3E(4I)}{4} \bigg/ \left(\frac{4E(2I)}{4} + \frac{3E(4I)}{4} + \frac{4EI}{4} \right) - \tfrac{1}{2}; \quad \text{DF}_{BA} = \tfrac{1}{3}; \quad \text{DF}_{BD} = \tfrac{1}{6}$$

There is zero COF from B to C, but COF = $\frac{1}{2}$ from B to the locked joints A and D.

Table 14.3

Joint	A		B		C	D
Bar	AB	BA	BC	BD	CB	DB
DF	0	$\frac{1}{3}$	$\frac{1}{2}$	$\frac{1}{6}$	1	0
COF	—	$\frac{1}{2}$	0	$\frac{1}{2}$	$\frac{1}{2}$	—
FEM	0	0	+12	0	−12	0
			+6		+12	
	−3	−6	−9	−3		−1·5
Summation	−3	−6	+9	−3	0	−1·5 kNm

In Table 14.3 joint C is balanced first and then left unclamped. As there is then no carry-over to C from B, distribution is complete after one cycle with both B and C balanced. The tranverse forces at the ends of each bar are determined from statics as described in Section 14.2.3. For example the end loads on bar BC are shown in Figure 14.15b the end moments being taken from Table 14.3. Then summing moments about C

$$Y_{BC} = (9+(9\times4)\times2)/4 = 20\cdot25 \text{ kN}.$$

Similarly we find $X_{DB} = 1\cdot125$ kN (Figure 14.15c), and after Y_{AB} has been found the bending moment diagram (Figure 14.15d) can be drawn (shown plotted on the tension side of each bar).

Example 14.4 Draw the bending moment diagram for the steel portal frame of Figure 14.16a and find the horizontal force exerted by the support at C.

Solution $FEM_{AB} = 2\cdot25 \times 4^2/12 = 3$ kNm $= -FEM_{BA}$.

For moment distribution at B both A and C are locked. Hence

$$DF_{BC} = \frac{4E(2I)}{4} \bigg/ \left(\frac{4EI}{4} + \frac{4E(2I)}{4}\right) = \tfrac{2}{3}; \quad DF_{BA} = \tfrac{1}{3}$$

and COF $= \tfrac{1}{2}$ for BA and BC. Similarly $DF_{CB} = 0\cdot6$ etc. and Table 14.4 is obtained.

Table 14.4

Joint	A	B		C		D
Bar	AB	BA	BC	CB	CD	DC
DF	0	$\tfrac{1}{3}$	$\tfrac{2}{3}$	0·6	0·4	0
COF	—	$\tfrac{1}{2}$	$\tfrac{1}{2}$	$\tfrac{1}{2}$	$\tfrac{1}{2}$	—
FEM	+3	−3	0	0	0	0
	+0·5	+1	+2	+1		
Example 14.4			−0·3	−0·6	−0·4	−0·2
	+0·05	+0·1	+0·2	+0·1		
			−0·03	−0·06	−0·04	−0·02
Summation	+3·55	−1·90	+1·87	+0·44	−0·44	−0·22 kNm
		(−1·89)	(+1·89)			
FEM	+4·5	+4·5	0	0	+8	+8
Example 14.5	−0·75	−1·5	−3	−1·5		

After three complete cycles the following summation is obtained

	+4·11	+3·72	−3·72	−5·11	+5·11	+6·56 kNm

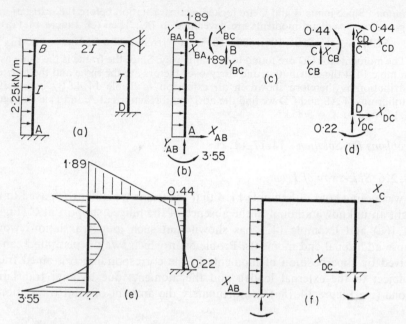

Figure 14.16 AB = BC = 4 m, CD = 3 m, units of moment kNm

The values in BA and BC in brackets are averages. From Figures 14.16b, c and d we find $X_{AB} = -4\cdot915$ kN, $Y_{BC} = 0\cdot583$ kN, $X_{DC} = 0\cdot22$ kN, which allows the bending moment diagram to be drawn (Figure 14.16e). Referring to Figure 14.16f the horizontal force X_C exerted by support C is found from the sum of the external forces in this direction on the complete frame viz.

$$X_C + X_{AB} + X_{DC} + (2\cdot25 \times 4) = 0$$

Hence $X_C = -4\cdot305$ kN, i.e. a force acting towards the left. Without this force joints B and C would translate to the right.

Example 14.5 Find the force required to translate joints B and C (Figure 14.17a) horizontally through $u = 10$ mm. Take $E = 200$ GN/m² and $I = 600$ cm⁴.

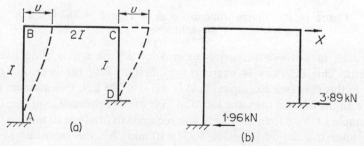

Figure 14.17 AB = BC = 4 m, CD = 3 m

13

Solution Since joints B and C are locked against rotation before the translation is imposed, no fixed end moments are developed in BC. Then (cf. Figure 14.11a)

$$\text{FEM}_{AB} = 6 \times 200 \times 10^9 \times 600 \times 10^{-8} \times 10^{-2}/16 \text{ Nm} = 4\cdot5 \text{ kNm} = \text{FEM}_{BA}$$

The moments for CD are found in the same way. Since the frame is the same as in Example 14.4 the distribution and carry-over-factors are the same and the moment distribution is therefore shown in an extension of Table 14.4. By considering equilibrium of AB and CD we find the horizontal reactions at A and D as in Figure 14.17b and so $X = 5\cdot85$ kN.

Problems for Solution   14.17–14.21

14.2.6 *Side-sway of Frames*

It was pointed out in Example 14.4 that the frame would have swayed to the right an unknown amount in the absence of the hinged support at C (Figure 14.16a) and Example 14.5 has shown that such joint translations would cause additional end moments. Problems in which sway is permitted can be solved by superposition of moments in the corresponding restrained frame (subject to the external loads), and the moments due to joint translation alone (see Figure 14.18a). Unfortunately the amount of joint translation is

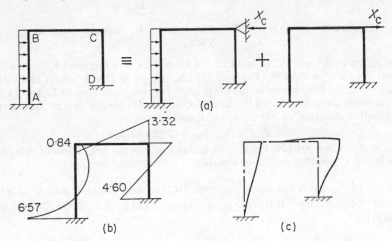

Figure 14.18   Frame dimensions as in Figure 14.16. Units of moment kNm

not known in advance to permit calculation of the appropriate fixed end moments. This difficulty is overcome by first finding the restraining force which in this case (see Example 14.4) is $X_C = 4\cdot305$ kN. Comparison is then made with the force at the same location to cause an arbitrary joint translation. In Example 14.5 we found $5\cdot85$ kN was required to produce 10 mm and so the actual joint translation is $(4\cdot305/5\cdot85) \times 10$ mm. All the moments in Table 14.4 (part 2) should therefore be multiplied by $(4\cdot305/5\cdot85)$ to find those due to the actual side-sway. Adding these moments to those in Table 14.4 (part 1) we obtain Table 14.5.

Table 14.5

Example 14.5 ($\times 0.735$)	$+3.02$	$+2.73$	-2.73	-3.76	$+3.76$	$+4.82$
Example 14.4	$+3.55$	-1.89	$+1.89$	$+0.44$	-0.44	-0.22
kNm	$+6.57$	$+0.84$	-0.84	-3.32	$+3.32$	$+4.60$

The bending moment and deflexion diagrams are shown in Figures 14.18b and c.

Problems for Solution 14.22, 14.23

14.3 Matrix Stiffness Method for Frames

14.3.1 *Introduction*: *Local and Global Coordinates*

In extending to frames the stiffness method of analysis used in earlier chapters we can consider the joints as nodes and the bars as discrete (i.e. finite) elements. The system is then to be described in terms of the loads exerted, and the displacements occurring, at the nodes. The stiffness equations relating these quantities will be developed from the elastic properties of the bars and their geometrical arrangement.

In Chapters 5, 7, 8 and 9 stiffness equations were derived for bars relative to coordinate axes based on the longitudinal axis of the bar (Figure 5.1). These axes will henceforth be called *local* axes, while the coordinate axes used for an assembly of bars will be termed *global* axes (datum is also used). Since for a collinear assembly (Figure 5.3) the components of load and displacement relative to global and local axes are parallel, the assembly stiffness equations can be formed, as previously described (e.g. Section 5.3.2) directly from element equations in local coordinates. However, for a frame (e.g. Figure 14.1) it is clear that the global axes must be inclined to the local axes of some of the bars. Consequently components of nodal load and displacement appearing in element stiffness equations may have different directions from those in the assembly stiffness equations.

There are two ways of dealing with this difficulty. Element stiffness equations could be derived relative to global axes (effectively the procedure adopted for triangular elements of a plane body in Article 13.3). Alternatively we must devise means of transforming element stiffness equations from local to global axes. Whichever procedure is adopted the same assembly stiffness equations are obtained for a particular choice of global axes. It is a matter of convenience which procedure is adopted, the second being usually more convenient for bars.

In what follows the notation previously defined is retained for global axes, quantities relative to local coordinates being distinguished by overlaid lines, e.g. $\bar{X}$, $\bar{x}$, $\bar{u}$.

14.3.2 *Coordinate Transformation Matrices*

We start by discussing the simple case of a plane pin-jointed frame in which (see Section 14.1.2) the bars are subject to direct loading only. In local coordinates the end loads are forces $\bar{X}_{i;j}$ and $\bar{X}_{j;j}$, and the end displacements are translations $\bar{u}_i$ and $\bar{u}_j$ (see Figure 14.19a). In global coordinates we

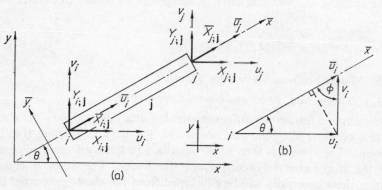

Figure 14.19

require two components to specify each end force, viz. $X_{i;j}$, $Y_{i;j}$, $X_{j;j}$, $Y_{j;j}$, and there are two components of each nodal displacement, viz. u_i, v_i, u_j, v_j. Expressing end forces in global coordinates in terms of end forces in local coordinates we find at node i

$$\bar{X}_{i;j} = X_{i;j} \cos\theta + Y_{i;j} \cos\phi = X_{i;j} \cos\theta + Y_{i;j} \sin\theta$$

where θ and ϕ are the angles which the x and y axes make with the $\bar{x}$ axis. Similarly

$$\bar{X}_{j;j} = X_{j;j} \cos\theta + Y_{j;j} \sin\theta$$

Combining these results the end forces in the two systems are related by

$$\begin{bmatrix} \bar{X}_{i;j} \\ \bar{X}_{j;j} \end{bmatrix} = \begin{bmatrix} \cos\theta & \sin\theta & 0 & 0 \\ 0 & 0 & \cos\theta & \sin\theta \end{bmatrix} \begin{bmatrix} X_{i;j} \\ Y_{i;j} \\ X_{j;j} \\ Y_{j;j} \end{bmatrix} \quad (14.8)$$

i.e.

$$\bar{\mathbf{F}}_j = \boldsymbol{\lambda} \mathbf{F}_j \quad (14.9)$$

where

$$\bar{F}_j = \{\bar{X}_{i;j} \quad \bar{X}_{j;i}\} \quad \text{and} \quad F_j = \{X_{i;j} \quad Y_{i;j} \quad X_{j;i} \quad Y_{j;i}\}$$

are end load vectors in local and global coordinates. The matrix

$$\lambda = \begin{bmatrix} \cos\theta & \sin\theta & 0 & 0 \\ 0 & 0 & \cos\theta & \sin\theta \end{bmatrix} \tag{14.10}$$

is called a transformation matrix.

Referring to Figure 14.19b we see that the local nodal displacement $\bar{u}_i$ has components u_i, v_i relative to the global axes, and

$$\bar{u}_i = u_i \cos\theta + v_i \cos\phi = u_i \cos\theta + v_i \sin\theta$$

A similar result can be obtained for node j and we can then write

$$\bar{\delta}_j = \lambda\delta_j \tag{14.11}$$

analogous to (14.9). Here the nodal displacement vectors in the two systems are

$$\bar{\delta}_j = \{\bar{u}_i \quad \bar{u}_j\} \quad \text{and} \quad \delta_j = \{u_i \quad v_i \quad u_j \quad v_j\}$$

and λ is the transformation matrix defined by equation (14.10).

Now let us consider a more general loading (Figure 14.20) with direct and transverse end forces and end moments in the plane of the forces. This

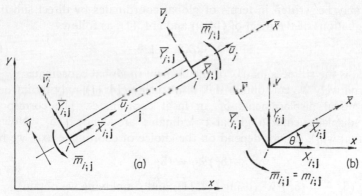

Figure 14.20 (a) Loads and displacements in local coordinates.
(b) Loads at i in local and global coordinates

type of loading corresponds to that on a bar of a plane frame with rigid joints (Section 14.1.3). Introducing $l_{\bar{x}x}$, $l_{\bar{x}y}$ for the cosines of the angles between the $\bar{x}$ axis and the x and y axes, we obtain (referring to Figure 14.20)

$$\bar{X}_{i;j} = X_{i;j}l_{\bar{x}x} + Y_{i;j}l_{\bar{x}y} = X_{i;j} \cos\theta + Y_{i;j} \sin\theta$$

Similarly

$$\bar{Y}_{i;j} = X_{i;j}l_{\bar{y}x} + Y_{i;j}l_{\bar{y}y} = -X_{i;j} \sin\theta + Y_{i;j} \cos\theta$$

where for example $l_{\bar{y}x} = \cos{(90° + \theta)}$ is the cosine of the angle between the $\bar{y}$ and x axes. It is apparent from Figure 14.20b that $\bar{m}_{i;j} = m_{i;j}$. Similar results can be obtained for node j, and writing $c = \cos{\theta}$, $s = \sin{\theta}$ we have

$$
\begin{bmatrix} \bar{X}_{i;j} \\ \bar{Y}_{i;j} \\ \bar{m}_{i;j} \\ \bar{X}_{j;j} \\ \bar{Y}_{j;j} \\ \bar{m}_{j;j} \end{bmatrix} = \begin{bmatrix} c & s & 0 & 0 & 0 & 0 \\ -s & c & 0 & 0 & 0 & 0 \\ 0 & 0 & 1 & 0 & 0 & 0 \\ 0 & 0 & 0 & c & s & 0 \\ 0 & 0 & 0 & -s & c & 0 \\ 0 & 0 & 0 & 0 & 0 & 1 \end{bmatrix} \begin{bmatrix} X_{i;j} \\ Y_{i;j} \\ m_{i;j} \\ X_{j;j} \\ Y_{j;j} \\ m_{j;j} \end{bmatrix} \qquad (14.12)
$$

i.e.

$$\bar{F}_j = \lambda F_j$$

A similar result can be derived for nodal displacements.

14.3.3 *Stiffness Equations*

In Article 9.4 it was shown that for combined direct loading and bending of a bar the stiffness equations (9.11a) in local coordinates have the form equivalent to

$$\bar{F}_j = \bar{k}_j \bar{\delta}_j \qquad (14.13)$$

which may be written in terms of global coordinates by direct substitution from equations of the form of (14.9) and (14.11), as follows

$$F_j = \lambda^{-1} \bar{k}_j \lambda \delta_j = k_j \delta_j \qquad (14.14)$$

where k_j is the stiffness matrix of the element in global coordinates.

Alternatively we can obtain this result using (14.11) only by introducing virtual nodal displacements $\bar{\delta}_j^*$ in local coordinates. The corresponding virtual displacements in global coordinates are denoted δ_j^*. Now since the virtual work cannot depend on the choice of coordinate axes we have

$$(\delta_j^*)^T F_j = (\bar{\delta}_j^*)^T \bar{F}_j$$

Putting $(\bar{\delta}_j^*)^T = (\delta_j^*)^T \lambda^T$ (using (14.11)) and simplifying we obtain

$$F_j = \lambda^T \bar{F}_j$$

Substituting from (14.11) and (14.13) we have

$$F_j = \lambda^T \bar{k}_j \lambda \delta_j \qquad (14.15)$$

Comparing equations (14.14) and (14.15) we find

$$k_j = \lambda^T \bar{k}_j \lambda = \lambda^{-1} \bar{k}_j \lambda \qquad (14.16)$$

Thus the transformation matrix λ has its transpose equal to its inverse,

.e. it is an example of an *orthogonal* matrix. The formation of assembly stiffness equations now proceeds as described in general in Section 9.4.3.

14.3.4 *Plane Frames with Pin-joints*

As already noted the bars of this type of frame are in a state of direct loading, for which the element stiffness matrix in local coordinates $\bar{k}_j$ was derived in Section 5.3.1 and the coordinate transformation matrix λ in Section 14.3.2 (equation 14.10). To transform to global coordinates $\bar{k}_j$ and λ are substituted directly in equation (14.16). Alternatively the matrix product of (14.16) can be evaluated formally to give the following explicit expression

$$\mathbf{k}_j = k' \begin{bmatrix} \mathbf{k}_0 & -\mathbf{k}_0 \\ -\mathbf{k}_0 & \mathbf{k}_0 \end{bmatrix} \qquad (14.17)$$

where

$$\mathbf{k}_0 = \begin{bmatrix} \cos^2 \theta & \cos \theta \sin \theta \\ \cos \theta \sin \theta & \sin^2 \theta \end{bmatrix} \quad \text{and} \quad k' = \frac{AE}{L}$$

and θ is the inclination of $\bar{x}$ to the x axis.

Example 14.6 Find the displacement of node 2 and the stresses in the bars of the pin-jointed frame in Figure 14.21a. The bars are each 2·5 m long and have cross-sectional areas of 800 and 1600 mm² respectively. $E = 210$ GN/m².

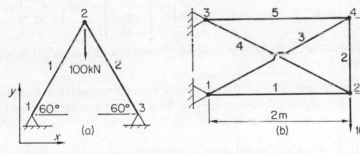

Figure 14.21

Solution Putting $\theta = 60°$ for bar **1** and $\theta = -60°$ for bar **2** in equation (14.17) we find

$$\mathbf{k}_1 = \frac{\alpha}{4} \begin{bmatrix} 1 & \sqrt{3} & -1 & -\sqrt{3} \\ \sqrt{3} & 3 & -\sqrt{3} & -3 \\ -1 & -\sqrt{3} & 1 & \sqrt{3} \\ -\sqrt{3} & -3 & \sqrt{3} & 3 \end{bmatrix} \text{(i)} \quad \mathbf{k}_2 = \frac{\beta}{4} \begin{bmatrix} 1 & -\sqrt{3} & -1 & \sqrt{3} \\ -\sqrt{3} & 3 & \sqrt{3} & -3 \\ -1 & \sqrt{3} & 1 & -\sqrt{3} \\ \sqrt{3} & -3 & -\sqrt{3} & 3 \end{bmatrix} \text{(ii)}$$

where $\alpha = 67·2 \times 10^6$ N/m and $\beta = 2\alpha$.

The assembly stiffness matrix **K** is of size (6×6). Then for example putting $i = 2, j = 3$ in rules (9.16), with $D_f = 2$, the components of the first row of $\mathbf{k}_2$ are placed in row 3 and columns 3, 4, 5, 6. Placing zeros in unoccupied locations we have finally

$$
\begin{bmatrix} X_1 \\ Y_1 \\ 0 \\ -10^5 \\ X_3 \\ Y_3 \end{bmatrix} = \frac{\alpha}{4} \begin{bmatrix} 1 & \sqrt{3} & -1 & -\sqrt{3} & 0 & 0 \\ \sqrt{3} & 3 & -\sqrt{3} & -3 & 0 & 0 \\ -1 & -\sqrt{3} & 3 & -\sqrt{3} & -2 & 2\sqrt{3} \\ -\sqrt{3} & -3 & -\sqrt{3} & 9 & 2\sqrt{3} & -6 \\ 0 & 0 & -2 & 2\sqrt{3} & 2 & -2\sqrt{3} \\ 0 & 0 & 2\sqrt{3} & -6 & -2\sqrt{3} & 6 \end{bmatrix} \begin{bmatrix} 0 \\ 0 \\ u_2 \\ v_2 \\ 0 \\ 0 \end{bmatrix}
\tag{iii}
$$

where $\alpha = 67 \cdot 2 = 10^6$ N/m. Introducing the support conditions $u_1 = v_1 = u_3 = v_3 = 0$ we obtain the reduced stiffness equations using (i)

$$
\begin{bmatrix} 0 \\ -10^5 \end{bmatrix} = \frac{\alpha}{4} \begin{bmatrix} 3 & -\sqrt{3} \\ -\sqrt{3} & 9 \end{bmatrix} \begin{bmatrix} u_2 \\ v_2 \end{bmatrix}
\tag{iv}
$$

from which $u_2 = -0 \cdot 43 \times 10^{-3}$ m; $v_2 = -0 \cdot 744 \times 10^{-3}$ m. The end forces on each bar are found using equations (14.14). For bar **1**

$$
\begin{bmatrix} X_{1;1} \\ Y_{1;1} \\ X_{2;1} \\ Y_{2;1} \end{bmatrix} = \frac{\alpha}{4} \begin{bmatrix} 1 & \sqrt{3} & -1 & -\sqrt{3} \\ \sqrt{3} & 3 & -\sqrt{3} & -3 \\ -1 & -\sqrt{3} & 1 & \sqrt{3} \\ -\sqrt{3} & -3 & \sqrt{3} & 3 \end{bmatrix} 10^{-3} \begin{bmatrix} 0 \\ 0 \\ -0 \cdot 43 \\ -0 \cdot 744 \end{bmatrix} = 10^3 \begin{bmatrix} 28 \cdot 9 \\ 50 \\ -28 \cdot 9 \\ -50 \end{bmatrix} \text{N}
$$

The end forces in local coordinates are found using the coordinate transformation equations (14.8), with $\theta = 60°$, i.e.

$$
\begin{bmatrix} \bar{X}_{1;1} \\ \bar{X}_{2;1} \end{bmatrix} = \begin{bmatrix} 0 \cdot 5 & 0 \cdot 866 & 0 & 0 \\ 0 & 0 & 0 \cdot 5 & 0 \cdot 866 \end{bmatrix} 10^3 \begin{bmatrix} 28 \cdot 9 \\ 50 \\ -28 \cdot 9 \\ -50 \end{bmatrix} = 10^3 \begin{bmatrix} 57 \cdot 7 \\ -57 \cdot 7 \end{bmatrix} \text{N}
$$

Hence from equation (5.5) $(\sigma_{\bar{x}})_1 = -57 \cdot 7 \times 10^3 / (800 \times 10^{-6})$, i.e. $-72 \cdot 1$ MN/m²; similarly $(\sigma_{\bar{x}})_2 = -36 \cdot 05$ MN/m².

The complete sequence of operations is easily arranged for automatic processing on a digital computer, so that large frames are easily solved for displacements, stresses and reactive forces.

Example 14.7 Find the nodal forces and displacements, and the stresses in the bars for the frame in Figure 14.21b. All the bars are steel with cross-sectional area 800 mm². $E = 210$ GN/m².

Solution To save space only the sub-matrices $\mathbf{k}_0$ (equation 14.17) are quoted.

$$
(\mathbf{k}_0)_1 = (\mathbf{k}_0)_5 = \begin{bmatrix} 1 & 0 \\ 0 & 0 \end{bmatrix}; \quad (\mathbf{k}_0)_2 = \begin{bmatrix} 0 & 0 \\ 0 & 1 \end{bmatrix};
$$

$$
(\mathbf{k}_0)_3 = \tfrac{1}{5} \begin{bmatrix} 4 & 2 \\ 2 & 1 \end{bmatrix}; \quad (\mathbf{k}_0)_4 = \tfrac{1}{5} \begin{bmatrix} 4 & -2 \\ -2 & 1 \end{bmatrix}
$$

After forming **K**, using rules (9.16), and introducing the prescribed displacements $u_1 = v_1 = u_3 = v_3 = 0$, we obtain the reduced stiffness equations

$$
\begin{bmatrix} 0 \\ -10^5 \\ 0 \\ 0 \end{bmatrix} = 10^6 \begin{bmatrix} 14\cdot4 & -30 & 0 & 0 \\ -30 & 183 & 0 & -168 \\ 0 & 0 & 144 & 30 \\ 0 & -168 & 30 & 183 \end{bmatrix} \begin{bmatrix} u_2 \\ v_2 \\ u_4 \\ v_4 \end{bmatrix}
$$

Hence $\delta = 10^{-3}\{0 \quad 0 \quad -1\cdot22 \quad -5\cdot85 \quad 0 \quad 0 \quad 1\cdot16 \quad -5\cdot56\}$ m.

From the complete stiffness equations we can obtain

$$F = 10^3\{51 \quad 200 \quad 0 \quad -10^5 \quad 49 \quad -200 \quad 0 \quad 0\}\,N$$

Using element stiffness and transformation equations we find (units MN/m^2)
$(\sigma_x)_1 = -128;\ (\sigma_x)_2 = 61\cdot2;\ (\sigma_x)_3 = -135;\ (\sigma_x)_4 = -143;\ (\sigma_x)_5 = 122$

Comparing Examples 14.6 and 14.7 we see that the procedure is the same whether or not the frame is statically determinate, which is one of the advantages of the matrix stiffness method.

Problem for Solution 14.24

14.3.5 *Plane Frames with Rigid Joints*

As discussed in Section 14.1.3 these frames are in a condition of combined bending and direct loading for which the element stiffness matrix in global coordinates is determined from equations (7.28) and (14.16) using the coordinate transformation matrix in equation (14.12). An explicit form of k_i can be derived as was done in Section 14.3.4 but is rather unwieldy. Since there are three degrees of freedom at each node the element stiffness matrix is of size 6×6.

Similarly the assembly stiffness matrix **K** is $(3n \times 3n)$ and the load and displacement vectors **F** and **δ** are $(3n \times 1)$ column matrices. **K** is formed from element stiffness matrices using rules (9.16) with $D_f = 3$. In the example which follows the frame would be stable even if the joint were pinned, i.e. it qualifies for the term braced frame.

Example 14.8 The bars of the frame in Figure 14.22 have $E = 210\ GN/m^2$, length 2·5 m, cross-sectional area 800 mm^2, and second moment of area 40 cm^4. Determine the displacements of node 2.

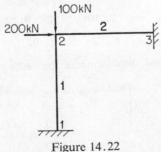

Figure 14.22

Solution Since the two bars are identical substitution in equation (9.11) gives for each in local coordinates

$$\bar{k}_j = 10^3 \begin{bmatrix} 67200 & 0 & 0 & -67200 & 0 & 0 \\ 0 & 32\cdot3 & 40\cdot3 & 0 & -32\cdot3 & 40\cdot3 \\ 0 & 40\cdot3 & 67\cdot2 & 0 & -40\cdot3 & 33\cdot6 \\ -67200 & 0 & 0 & 67200 & 0 & 0 \\ 0 & -32\cdot3 & -40\cdot3 & 0 & 32\cdot3 & -40\cdot3 \\ 0 & 40\cdot3 & 33\cdot6 & 0 & -40\cdot3 & 67\cdot2 \end{bmatrix} \quad (i)$$

Writing the coordinate transformation matrix from equation (14·12) in the partitioned form

$$\lambda = \begin{bmatrix} \lambda_0 & 0 \\ 0 & \lambda_0 \end{bmatrix}$$

we have

$$(\lambda_0)_1 = \begin{bmatrix} 0 & 1 & 0 \\ -1 & 0 & 0 \\ 0 & 0 & 1 \end{bmatrix} \quad \text{and} \quad (\lambda_0)_2 = \begin{bmatrix} 1 & 0 & 0 \\ 0 & 1 & 0 \\ 0 & 0 & 1 \end{bmatrix}$$

Performing the triple matrix product $\lambda^T \bar{k}_2 \lambda$ we see that since λ for bar **2** is a unit matrix $k_2 = \bar{k}_2$, which would be expected since the local and global coordinates are parallel. For bar **1** we find

$$k_1 = 10^3 \begin{bmatrix} 32\cdot3 & 0 & -40\cdot3 & -32\cdot3 & 0 & -40\cdot3 \\ 0 & 67200 & 0 & 0 & -67200 & 0 \\ -40\cdot3 & 0 & 67\cdot2 & 40\cdot3 & 0 & 33\cdot6 \\ -32\cdot3 & 0 & 40\cdot3 & 32\cdot3 & 0 & 40\cdot3 \\ 0 & -67200 & 0 & 0 & 67200 & 0 \\ -40\cdot3 & 0 & 33\cdot6 & 40\cdot3 & 0 & 67\cdot2 \end{bmatrix} \quad (ii)$$

The assembly stiffness matrix **K** of size (9×9) is assembled from k_1 and k_2 using rules (9.16). Since, for example, bar **2** joins nodes 2 and 3 the first, second and third rows of k_2 go into rows $((3 \times 2) - 2)$, $((3 \times 2) - 1)$, (3×2) of **K**. The load and displacement vectors are

$$F = \{X_1 \quad Y_1 \quad m_1 \quad 2 \times 10^5 \quad -10^5 \quad 0 \quad X_3 \quad Y_3 \quad m_3\}$$
$$\lambda = \{0 \quad 0 \quad 0 \quad u_2 \quad v_2 \quad \theta_2 \quad 0 \quad 0 \quad 0\}$$

Introducing the prescribed displacements the reduced stiffness equations are

$$10^5 \begin{bmatrix} 2 \\ -1 \\ 0 \end{bmatrix} = 10^3 \begin{bmatrix} 67232 & 0 & 40\cdot3 \\ 0 & 67232 & 40\cdot3 \\ 40\cdot3 & 40\cdot3 & 134\cdot4 \end{bmatrix} \begin{bmatrix} u_2 \\ v_2 \\ \theta_2 \end{bmatrix} \quad (iii)$$

Hence $u_2 = 2\cdot97$ mm; $v_2 = -1\cdot49$ mm; $\theta_2 = -0\cdot6 \times 10^{-3}$ rad.

Intermediate loading is treated by the method described in Section 7.4.7. In local coordinates equation (7.53a) becomes

$$\overline{F}_j = \overline{k}_j \overline{\delta}_j + \overline{f}_j \qquad (14.18)$$

where

$$\overline{f}_j = \{0 \quad a \quad b \quad 0 \quad c \quad d\} \qquad (14.19)$$

is a vector of element fixed end actions. The quantities a, b, c, d are taken from Figure 7.20 and the zeros allow for the extra degree of freedom. Then using the coordinate transformation relations equation (14.18) becomes

$$F_j = \lambda^T \overline{k}_j \lambda \delta_j + \lambda^T \overline{f}_j = k_j \delta_j + f_j \qquad (14.20)$$

Example 14.9 The frame of the previous example has instead a distributed load of 1·2 kN/m acting down on bar **2**. Find the displacements.

Solution Since $(\lambda)_2$ is a unit matrix, and taking from Figure 7.20 $a = \frac{1}{2}wL$, $b = wL^2/12$, $c = a$, $d = -b$, we find

$$f_2 = \lambda^T \overline{f}_2 = 10^3\{0 \quad 1·5 \quad 0·625 \quad 0 \quad 1·5 \quad -0·625\} \qquad (i)$$

The reduced stiffness equations are

$$\begin{bmatrix} 0 \\ 0 \\ 0 \end{bmatrix} = 10^3 \begin{bmatrix} 67232 & 0 & 40·3 \\ 0 & 67232 & 40·3 \\ 40·3 & 40·3 & 134·4 \end{bmatrix} \begin{bmatrix} u_2 \\ v_2 \\ \theta_2 \end{bmatrix} + 10^3 \begin{bmatrix} 0 \\ 1·5 \\ 0·625 \end{bmatrix}$$

Hence $\theta_2 = -4·65 \times 10^{-3}$; $u_2 = v_2 \doteq 0$.

Problems for Solution 14.25–14.27

14.4 References

(a) *Cited*

1. Hardy Cross, R., 'Analysis of continuous frames by distributing end moments', *Trans. A.S.C.E.*, **96**, 1 (1932).

(b) *General*

1. Matheson, J. A. L., *Hyperstatic Structures*, Butterworth, London, 1959.
2. Gere, J. M., loc. cit. Article 7.6.
3. Przemieniecki, J. S., loc. cit. Article 12.6.

14.5 Problems for Solution

(Take E (steel) = 210 GN/m^2 unless otherwise stated.)

14.1. Investigate the stability and statical determinacy of the frames shown in Figure 14.23.

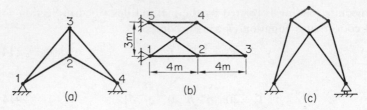

Figure 14.23

14.2. Find the direct loads in the bars of the frame in Figure 14.23b when a vertical load of 24 kN acts down (a) at joint 3, (b) at joint 2.

14.3. Repeat Problem 14.2 (b) when the bar between joints 2 and 5 in Figure 14.23b is replaced by one joining 2 and 4.

14.4. Referring to the symmetric frame in Figure 14.23a, let the supports be 4 m apart and let bars 1,2 and 1,3 make angles of 30° and 45° respectively to the horizontal. The hinged support at 4 is replaced by a roller and a force of 120 kN acts down at 2. Find the loads in the bars.

14.5. Determine whether a steel tube of 60 mm outside diameter and 10 mm wall thickness would be suitable for construction of the frame of Problem 14.4 if the permissible normal stress is 200 MN/m².

14.6. Determine the degree of statical indeterminacy for each frame in Figure 14.24.

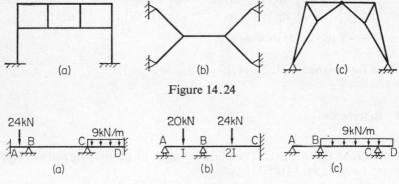

Figure 14.24

Figure 14.25 (a) AB = 2 m, BC = 6 m, CD = 4 m, the concentrated force is at the midpoint of AB, *EI* is uniform. (b) AB = 2 m, BC = 3 m, forces at midpoints, *E* uniform. (c) AB = 2 m, BC = 4 m, CD = 1 m, *EI* uniform

14.7. Carry out moment distribution for the continuous beams in Figure 14.25. (*Hint* In Figure 14.25c the moment at end C of CD is $((9 \times 1) \times \frac{1}{2})$ kNm, and $DF_{CD} = 0$ since rotation of C does not cause bending of CD.)

14.8. Find the end moments for the bars in the system of Figure 14.25a if joint B suffers a 3 mm downward displacement. $EI = 2$ MNm².

14.9. Rework part (c) of Problem 14.7 with the I value for BC twice that of AB.

14.10. Draw shearing force and bending moment diagrams for the beams of Problem 14.7.

14.11. Using the results of Problems 14.7 and 14.8 find the changes in shear force and bending moment in span AB of Figure 14.25a due to the displacement of B.

14.12. Using Problems 14.7 and 14.9 find the change of bending moment at the centre of BC (Figure 14.25c) due to doubling its I value.

14.13. Determine whether any of the cross-sections in Table 6.3 would be suitable for the beam of Figure 14.25a if the permissible normal stress is 100 MN/m².

14.14. Rework Problem 14.7 (cases b and c) using the method of Section 14.2.4.

14.15. If a fixed support is substituted at A in Figure 14.25c find the reactive moment exerted.

14.16. In Figure 7.10b $a = 0.8$ m, $b = 0.5$ m, $w = 2$ kN/m, $EI = 2.8$ kNm² and support B is 4 mm above A and C. Find the force exerted by support B.

14.17. If in Figure 14.15a C is found to be 8 mm below A and B draw the bending moment diagram, and find the vertical reaction at C. $I = 250$ cm⁴.

14.18. Referring to Figure 14.15a let all bars have the same I value and a hinge be substituted at A. Obtain the bending moment diagram and the horizontal reaction at D.

14.19. Rework Examples 14.4 and 14.5 with hinges at A and D.

14.20. Referring to Figure 14.22 let hinges be substituted at 1 and 3 and the loads be replaced by a uniformly distributed downward load of 60 kN/m on bar 2. If the bars are 4 m long and the beam has I value double that of the column find the reactions.

14.21. A portal frame ABCD (cf. Figure 14.16a) has fixed support at A and hinged supports at C and D. AB, BC and CD have lengths of 6 m, 4 m, 6 m and I values of $2I$, $3I$, I respectively. AB and BC are subject respectively to uniformly distributed loads of 10 kN/m acting to the right and 30 kN/m acting down. Find the horizontal reactive loads and the greatest bending moment for each bar.

14.22. If the frame of Figure 14.16a is unrestrained at C and has hinged supports at A and D obtain the bending moment diagram.

14.23. A portal frame ABCD has AB = BC = CD = 3 m, uniform EI, and fixed supports at A and D. A 90 kN force acts down on BC at 1 m from B. Obtain the bending moment diagram.

14.24. Find the stresses in the bars of the pin-jointed frames shown in Figure 14.26. All bars are steel and have cross-sectional areas of 800 mm².

14.25. The frame of Example 14.8 (Figure 14.22) is also subject to a uniformly distributed load of 1.2 kN/m acting to the right on bar 1, and the support at node 1 is displaced 10 mm to the right. Find the displacements of node 2 and the greatest bending moment in the frame.

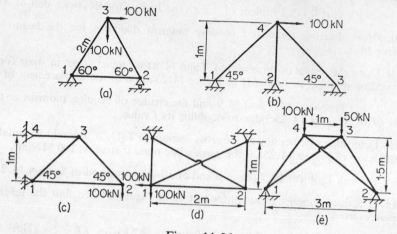

Figure 14.26

14.26. Repeat Problem 14.25 with a hinge at node 1, and $I = 80$ cm⁴ for bar **1**.

14.27. Find the reactive loads, and the direct loads and greatest bending moments in the bars of the frame of Figure 14.16a when the support at A is displaced 10 mm to the right. Take $I = 600$ cm⁴, $E = 200$ GN/m².

15

Introduction to the Bending
of Plates

15.1 Pure Bending

To introduce the problem of plate bending we refer to the results of Chapter 6
for bending of beams. The states of stress and strain were formulated in
Section 6.3.1 where it was established that $\epsilon_x = y/R$ and $\epsilon_z = -\nu y/R$ where
R is the radius of curvature in the xy plane (Figure 6.11). These strain components correspond to curvatures of $1/R$ and $-\nu/R$ in the xy and yz planes
respectively (Figure 6.11). Surfaces defined by curvatures of opposite sign

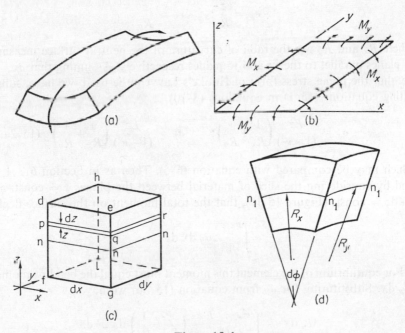

Figure 15.1

are termed *anticlastic* (Figure 15.1a) and are easily demonstrated by bending a rubber eraser.

Because of the small lateral dimensions of a beam the anticlastic curvature can occur freely but this is not so for a plate-like body extensive in two directions. A convenient coordinate system for plates (Figure 15.1b) has the xy plane as the middle surface of the plate (thickness t), and origin at one corner as shown. The plate is said to be in pure bending when uniformly distributed edge moments of intensity M_x and M_y per unit length act on the edges normal to the x and y directions respectively and there are no transverse forces (compare discussion of beams). The moments are considered positive when, as shown, their action would cause tension in the upper surface and compression in the lower surface of the plate. Then a rectangular element of the plate with lateral surfaces parallel to the edges of the plate as in Figure 15.1c is also subject to bending moments with intensities M_x and M_y.

Extending the method of Section 6.3.1 we now assume that during bending of the plate the lateral surfaces remain plane and rotate about their middle lines (Figure 15.1d). We make the further assumption that the middle surface $n_1 n_1 n_1$ is a neutral surface, i.e. one which undergoes neither extension nor contraction. Then by a discussion similar to that given for Figure 6.11b of Section 6.3.1 we can easily show that the strains of the original plane pqr distant z from the neutral surface (Figure 15.1c) are

$$\epsilon_x = \frac{z}{R_x}; \qquad \epsilon_y = \frac{z}{R_y} \qquad (15.1)$$

where R_x and R_y are the radii of curvature of the neutral surface measured in planes parallel to the xz and yz planes respectively. Assuming that $\sigma_z = 0$, i.e. that the plane stress form of Hooke's Law can be used, we have, substituting equations (15.1) in equations (3.10)

$$\sigma_x = \frac{Ez}{(1-\nu^2)} \left(\frac{1}{R_x} + \frac{\nu}{R_y} \right); \qquad \sigma_y = \frac{Ez}{(1-\nu^2)} \left(\frac{\nu}{R_x} + \frac{1}{R_y} \right) \qquad (15.2a, b)$$

which may be compared with equation (6.5). Then as in Section 6.3.1 we find by considering the slice of material between the planes $z = $ const. and $z + \mathrm{d}z = $ const. (Figure 15.1c), that the total moment on the surface degf is

$$\int_{-t/2}^{t/2} z\sigma_y \, \mathrm{d}x \, \mathrm{d}z$$

For equilibrium of the element this moment must equal the bending moment $M_y \, \mathrm{d}x$. Substituting for σ_y from equation (15.2b) we have

$$M_y \, \mathrm{d}x = \int_{-t/2}^{t/2} \frac{E}{(1-\nu^2)} \left(\frac{\nu}{R_x} + \frac{1}{R_y} \right) \mathrm{d}x \, z^2 \, \mathrm{d}z$$

i.e.

$$M_y = \frac{Et^3}{12(1-\nu^2)} \left(\frac{\nu}{R_x} + \frac{1}{R_y} \right)$$

The moment on the surface df1h is, similarly,

$$M_x \, dy = \int_{-t/2}^{t/2} z\sigma_x \, dy \, dz \qquad (15.3)$$

and is evaluated by substituting from equation (15.2a). Then writing

$$D = \frac{Et^3}{12(1-\nu^2)} \qquad (15.4)$$

we have

$$M_x = D\left(\frac{1}{R_x} + \frac{\nu}{R_y} \right); \qquad M_y = D\left(\frac{\nu}{R_x} + \frac{1}{R_y} \right) \qquad (15.5a, b)$$

The quantity D is called the flexural rigidity of the plate; it incorporates the second moment of area of a strip of unit width, i.e. $1 \times t^3/12$. Compared with a beam of unit width, D exceeds EI in the ratio $1/(1 - \nu^2)$ which measures the effect of suppressing anticlastic curvature (Figure 15.1a).

The deflexion of the plate parallel to the z axis is denoted by w and is a function of both x and y. In Section 7.1.1 it was shown that for small deflexions the curvature of a deflexion curve is (-1) times its second derivative, i.e. $1/R = -\mathrm{d}^2v/\mathrm{d}x^2$. The plate curvatures $1/R_x$ and $1/R_y$ in planes parallel to the xz and yz planes respectively, can therefore be written as

$$\frac{1}{R_x} = -\frac{\partial^2 w}{\partial x^2}; \qquad \frac{1}{R_y} = -\frac{\partial^2 w}{\partial y^2} \qquad (15.6a, b)$$

Substituting these in equations (15.5) we obtain

$$M_x = -D\left(\frac{\partial^2 w}{\partial x^2} + \nu \frac{\partial^2 w}{\partial y^2} \right); \qquad M_y = -D\left(\nu \frac{\partial^2 w}{\partial x^2} + \frac{\partial^2 w}{\partial y^2} \right) \qquad (15.7a, b)$$

These are the equivalent for plates of the beam differential equation (7.1); partial derivatives are required since the deflexion surface is a function of the two variables x and y.

The stress components σ_x and σ_y have greatest magnitudes at the upper and lower surfaces ($z = \pm\frac{1}{2}t$) and can be obtained from the bending moments by combining equations (15.2) and (15.5), giving

$$(\sigma_x)_{max} = \pm \frac{6M_x}{t^2}; \qquad (\sigma_y)_{max} = \pm \frac{6M_y}{t^2} \qquad (15.8a, b)$$

It is important to note the following restrictions on the use of equations (15.7) arising from the assumptions used in their derivation. The use of the approximate expressions (15.6) for the curvatures is valid only if the deflexions of the plate are small compared with its lateral dimensions. A more restrictive condition arises from the assumption that the middle surface of the plate is the neutral surface. It is only under certain very simple conditions of loading and support that a plate can bend in a manner which satisfies this requirement. In most cases the formation of the curved surface from a plane demands some stretching or contraction of the middle surface which is not then a neutral surface. It is found that if equations (15.7) are to give satisfactory results the deflexions must usually be small compared with the thickness of the plate.

15.2 Cylindrical Bending

A plate is said to be bent to a cylindrical surface when, as in Figure 15.2a, one curvature, here $1/R_y$, is zero. Then $(\partial^2 w/\partial y^2) = 0$ and equations (15.7) become

$$M_x = -D\frac{\mathrm{d}^2 w}{\mathrm{d}x^2}; \qquad M_y = -\nu M_x \qquad (15.9\text{a, b})$$

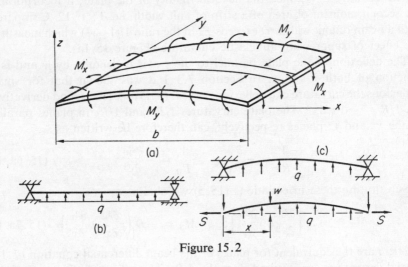

Figure 15.2

If the plate is long compared with its breadth b (Figure 15.2a) it is found in practice that even if the moments M_y are absent from the ends the surface remains cylindrical except close to the ends. Equations (15.9) can easily be integrated for pure bending. For more general loadings in which transverse forces act on the plate the bending moment is a function of x. Following the practice for beams it is assumed that equation (15.9a) is still valid. Then considering a uniformly distributed load of intensity q (force/unit area) as for example due to fluid pressure (Figure 15.2b) we have for a strip of unit length (see Figure 15.2a) and breadth b

$$M_x = \frac{qbx}{2} - \frac{qx^2}{2} = -D\frac{\mathrm{d}^2 w}{\mathrm{d}x^2}$$

Integrating subject to the support conditions $w = 0$ at $x = 0$ and $x = b$ we find that the central deflexion is $5qb^4/384D$ (cf. Figure 7.5).

Since the cylindrical surface can be derived from a plane without stretching or compression of the middle surface of the plate the above results are accurate provided deflexions are small in relation to plate width. The greatest stresses are obtained by substituting the greatest bending moments in equations (15.8). For instance if a plate 10 mm thick and 1 m wide is subject to a pressure of 40 kN/m^2 (approx. 0·4 atm) the greatest bending moment is $qb^2/8 = 5 \text{ kNm/m}$ at the centre of the plate and the greatest stress is $\pm 300 \text{ MN/m}^2$.

However, a plate is rarely in the simply supported condition of Figure 15.2b. It is more likely that the edges are not free to move together (Figure 15.2c) the supports exerting 'in-plane' forces S per unit length. The differential equation of the deflexion surface is then

$$\frac{qbx}{2} - \frac{qx^2}{2} - Sw = -D\frac{d^2w}{dx^2}$$

The solution of this equation is beyond the scope of this work (see Timoshenko[1]). The general effect is to reduce the maximum bending moment, but this is partially counteracted by the tensile stress due to the force S.

15.3 Symmetrically Loaded Circular Plates

In discussing the bending of circular plates subject to loads disposed symmetrically with respect to the plate centre it is convenient to take the latter point as origin and to use polar coordinates r, θ in the plane of the plate while retaining z for the direction normal to the plate (Figure 15.3a). Since from symmetry all diametral sections (i.e. rz planes) must be identical, the deflexion w is independent of θ. Then the slope of the deflexion surface

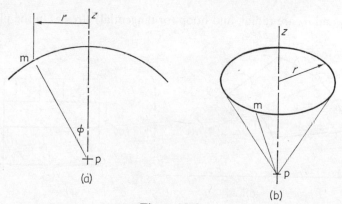

(a)

(b)

Figure 15.3

measured in the *rz* plane at any point m is (dw/dr), and is here negative because *w* decreases as *r* increases. This slope is equal in magnitude to the angle ϕ which the normal to the surface makes with the *z* axis (Figure 15.3a). The curvature of the surface in the *rz* plane is then (compare discussion of equations 15.6) since $\phi = -dw/dr$

$$\frac{1}{R_r} = \frac{d\phi}{dr} = -\frac{d^2w}{dr^2} \tag{15.10a}$$

where R_r is the radius of curvature in this plane. Figure 15.3b shows a cylindrical section of the plate obtained at radius *r*. This section is normal to the diametral section at point m and so its curvature $1/R_\theta$ together with $1/R_r$ will define the surface curvature at m. Since from symmetry the normals to the surface at any two adjacent points on the cylindrical section meet at p (Figure 15.3b) this point is the centre of curvature of the cylindrical section of the surface. The radius of curvature R_θ is then the length mp and from Figure 15.3a we have $r = R_\theta\phi$ so that

$$\frac{1}{R_\theta} = \frac{\phi}{r} = \frac{1}{r}\left(-\frac{dw}{dr}\right) \tag{15.10b}$$

Since the curvatures $1/R_r$ and $1/R_\theta$ are in planes normal to each other we can write *r* for *x* and θ for *y* in equations (15.5) and then substitute from (15.10) to obtain for pure bending

$$M_r = -D\left(\frac{d^2w}{dr^2} + \frac{\nu}{r}\frac{dw}{dr}\right); \qquad M_\theta = -D\left(\nu\frac{d^2w}{dr^2} + \frac{1}{r}\frac{dw}{dr}\right) \tag{15.11a, b}$$

where M_r and M_θ are bending moments per unit length on edges normal to the radial and circumferential directions. Similarly, equations (15.8) can be written

$$(\sigma_r)_{\max} = \frac{\pm 6M_r}{t^2}; \qquad (\sigma_\theta)_{\max} = \frac{\pm 6M_\theta}{t^2} \tag{15.12a, b}$$

where σ_r and σ_θ are radial, and hoop (or tangential) stresses in the plate.

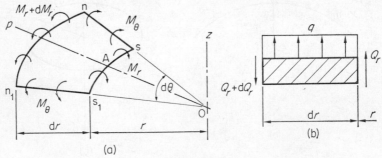

(a)   (b)

Figure 15.4

To obtain a relation between the quantities M_r and M_θ in equations (15.11) we now consider an infinitesimal element nss_1n_1 (Figure 15.4a) bounded by diametral sections ns and n_1s_1 and cylindrical sections ss_1 and nn_1. As shown, the edges ns and n_1s_1 are subject to distributed moments of intensity M_θ which from symmetry are equal in magnitude. Each of these moments has a component in the plane pOz of magnitude $(M_\theta dr)\sin\frac{1}{2}d\theta$ giving a total contribution of $M_\theta dr d\theta$. The moments on the edges ss_1 and nn_1 are respectively $M_r r d\theta$ and $(M_r+dM_r)(r+dr)d\theta$ where dM_r is the change of M_r between r and $(r+dr)$. Then the resultant moment in the pOz plane due to the edge moments is

$$(M_r+dM_r)(r+dr)d\theta - M_r r d\theta - M_\theta dr d\theta \qquad\qquad (i)$$

The presence of transverse forces acting on the plate would in general require shearing forces to act on all edges of an element such as nss_1n_1. However, in the restricted case of symmetrical loading it is clear that there are no shear forces on the edges ns and n_1s_1. The shear force intensities per unit length of the edges ss_1 and nn_1 which are normal to the r direction are denoted by Q_r and (Q_r+dQ_r) in Figure 15.4b. The intensity (force/unit area) of distributed load on the element is represented by q. Then the moment in the pOz plane about point A of the element due to the load and the shearing forces is

$$((Q_r+dQ_r)(r+dr)d\theta)dr-(q dr r d\theta)\tfrac{1}{2}dr \qquad\qquad (ii)$$

which on simplifying and neglecting higher-order terms becomes $Q_r r dr d\theta$. Adding this to (i) and dividing by $dr d\theta$ we obtain the equilibrium equation

$$r\frac{dM_r}{dr} + M_r - M_\theta + Q_r r = 0 \qquad\qquad (15.13)$$

Then substituting for M_r and M_θ from (15.11) we obtain

$$\frac{d^3w}{dr^3} + \frac{1}{r}\frac{d^2w}{dr^2} - \frac{1}{r^2}\frac{dw}{dr} = \frac{Q_r}{D} \qquad\qquad (15.14)$$

The solution of a problem in the bending of a symmetrically loaded plate is therefore reduced to that of finding a solution to (15.14) which satisfies the boundary conditions of the plate. Equations (15.11) and (15.12) would then be used to find the bending moments and stresses.

The shearing force intensity Q_r on a cylindrical section of the plate at radius r is equal to the total load on the part of the plate within a circle of this radius divided by the circumference of this circle. Here we shall consider as an illustration of the general approach the case when a plate is subject to a uniformly distributed load q and/or forces along its boundaries. Then the shearing force intensity is conveniently written in the form

$$Q_r = \frac{(q\pi r^2+V)}{2\pi r} \qquad\qquad (15.15)$$

where V will be interpreted later as required. Substituting from (15.15) in (15.14) and writing the left-hand side of the latter in a more easily integrable form, we obtain

$$\frac{d}{dr}\left(\frac{1}{r}\frac{d}{dr}\left(r\frac{dw}{dr}\right)\right) = \frac{1}{2D}\left(qr+\frac{V}{\pi r}\right) \qquad (15.16)$$

Integrating once and multiplying by r gives

$$\frac{d}{dr}\left(r\frac{dw}{dr}\right) = \frac{1}{4D}\left(qr^3+\frac{2Vr}{\pi}\ln(r)\right)+4C_1r \qquad (15.17)$$

Integrating again and dividing by r we have

$$\frac{dw}{dr} = \frac{1}{16D}\left(qr^3+\frac{2Vr}{\pi}(2\ln(r)-1)\right)+2C_1r+\frac{C_2}{r} \qquad (15.18)$$

Integrating once more we obtain finally

$$w = \frac{qr^4}{64D}+\frac{Vr^2}{8\pi D}(\ln(r)-1)+C_1r^2+C_2\ln(r)+C_3 \qquad (15.19)$$

where C_1, C_2, C_3 are constants of integration. Substituting from (15.18) and (15.19) in equations (15.11) we obtain the following general expressions for the bending moments due to loadings which obey (15.15)

$$M_r = -\frac{(3+\nu)}{16}qr^2-\frac{V}{4\pi}((1+\nu)\ln(r)+\tfrac{1}{2}(1-\nu))-2C_1D(1+\nu)+\frac{C_2D}{r^2}(1-\nu)$$

$$(15.20)$$

$$M_\theta = -\frac{(1+3\nu)}{16}qr^2-\frac{V}{4\pi}((1+\nu)\ln(r)-\tfrac{1}{2}(1-\nu))-2C_1D(1+\nu)-\frac{C_2D}{r^2}(1-\nu)$$

$$(15.21)$$

We shall now give some examples of the determination of the constants of integration for particular problems.

A. Solid Plate (Radius a)

A complete plate supported only at its outer edge (Figures 15.5a and b) has zero slope at the plate centre and so putting $(dw/dr) = 0$ at $r = 0$ in equation (15.18) we see that C_2 must be zero.

Now when the solid plate is subject only to distributed load (Figures 15.5a and b) then clearly Q_r is zero at the plate centre and therefore from (15.15) V must be zero.

(i) *Simply Supported Edge* (Figure 15.5a) The additional boundary conditions are zero bending moment M_r and zero deflexion, at the support. By

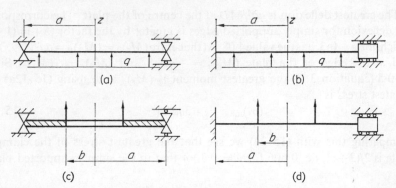

Figure 15.5 The rollers are a *convention* to signify neglect of in-plane forces

setting $M_r = 0$ at $r = a$ in equation (15.20), with C_2 and V also zero, we find C_1, and then putting $w = 0$ at $r = a$ in equation (15.19) we can obtain C_3. Thus

$$C_1 = -\frac{qa^2(3+\nu)}{8D(1+\nu)}; \qquad C_3 = \frac{qa^4(5+\nu)}{64D(1+\nu)}$$

Substituting for C_1, C_2, C_3 and V in equations (15.18–15.20) gives

$$w = \frac{q}{64D}(a^2-r^2)\left(\frac{(5+\nu)}{(1+\nu)}a^2-r^2\right)$$

$$\phi = \frac{qr}{16D}\left(\frac{(3+\nu)}{(1+\nu)}a^2-r^2\right); \qquad M_r = q(3+\nu)(a^2-r^2)/16 \tag{15.22}$$

The maximum deflexion and moment are at $r = 0$ and greatest slope is at $r = a$. Substituting in (15.21) we obtain

$$M_\theta = q((3+\nu)a^2-(1+3\nu)r^2)/16 \tag{15.23}$$

from which we see that M_θ is greatest at $r = 0$ where it equals M_r. Consequently using equations (15.12) we see that the greatest values of σ_r and σ_θ are equal and occur at the centre of the plate where

$$(\sigma_r)_{\max} = (\sigma_\theta)_{\max} = \pm 3(3+\nu)qa^2/8t^2 \tag{15.24}$$

(ii) *Clamped Edge* (Figure 15.5b) The additional boundary conditions are zero slope and deflexion at the support. Putting $(dw/dr) = 0$ at $r = a$ in equation (15.18) with C_2 and V both zero we find $C_1 = -qa^2/32D$, and then putting $w = 0$ at $r = a$ in equation (15.19) we obtain $C_3 = qa^4/64D$. Using these results in equations (15.18–15.21) we have

$$w = \frac{q}{64D}(a^2-r^2)^2; \qquad \phi = \frac{qr}{16D}(a^2-r^2)$$

$$M_r = q((1+\nu)a^2-(3+\nu)r^2)/16; \qquad M_\theta = q((1+\nu)a^2-(1+3\nu)r^2)/16 \tag{15.25}$$

The greatest deflexion is $qa^4/64D$ at the centre of the plate. The corresponding deflexion for simply supported edges is greater by the factor $(5+\nu)/(1+\nu)$ which, for $\nu = 0.3$, has the value 4.08. At the centre $(M_r)_0 = (M_\theta)_0 = qa^2(1+\nu)/16$ and at the edge of the plate $(M_r)_a = -qa^2/8$ and $(M_\theta)_a = \nu(M_r)_a$. Since $\nu \not> 0.5$ (equation 3.7) the greatest moment is $(M_r)_a$ and, using (15.12a) the greatest stress is

$$(\sigma_r)_{max} = \mp 3qa^2/4t^2 \tag{15.26}$$

Comparing this with (15.24) we see that the greatest stress in the clamped plate is $2/(3+\nu)$, i.e. 0.606 for $\nu = 0.3$, of that in the simply supported plate.

B. *Plate with Central Hole* (*Radius* b)

Due to limitations of space we shall consider only a plate with outer edge simply supported and subject to a total force P distributed evenly round the inner edge (Figure 15.5c) so that $q = 0$ and $V = P$ in equation (15.15). The bending moment M_r is then zero at both inner and outer edges. Hence putting $M_r = 0$ at $r = a$ and $r = b$ we obtain from (15.20) two equations in C_1 and C_2. After solving for these constants of integration we can write (15.20) as

$$M_r = -\frac{P(1+\nu)}{4\pi}\left(\ln\left(\frac{r}{a}\right) + \frac{b^2(a^2-r^2)}{r^2(a^2-b^2)}\ln\left(\frac{a}{b}\right)\right) \tag{15.27}$$

If C_1 and C_2 are substituted in (15.21) it is found that M_θ is non-zero at the edges of the plate.

The solution for the case of distributed load q is obtained in a similar manner except for the interpretation of the terms q and V. The total load within radius r is $q\pi(r^2 - b^2)$ and so

$$Q_r = \frac{q\pi(r^2-b^2)}{2\pi r} = \frac{qr}{2} - \frac{qb^2}{2r} \tag{15.28}$$

Comparing this with (15.15) we find $V = -qb^2/2r$. Then in equations (15.16–15.21) we require not only the terms in q but also those involving V using the value found above.

By letting b tend to zero in equation (15.27) we obtain

$$M_r = -\frac{P(1+\nu)}{4\pi}\ln\left(\frac{r}{a}\right) \tag{15.29}$$

which is the solution for a simply supported solid plate subject to a central concentrated load P. However, it should be noted that this result cannot be used close to the plate centre since there infinite moment and stress is predicted.

C. Ring Loads

As a last example we consider a solid plate with clamped edge and subject to load P distributed evenly round a circle of radius b (Figure 15.5d). In the outer portion of the plate $(r > b)$ we have $q = 0$ and $V = P$ in equation (15.15). In the inner portion $(r < b)$ there is no load within any radius r and so Q_r is zero. Therefore in equations (15.16–15.21) q is zero and V is respectively zero and P for the inner and outer portions of the plate. Then for example (15.18) is written for the two portions as

$$\frac{dw}{dr} = 2C_1'r + \frac{C_2'}{r} \tag{15.30a}$$

$$\frac{dw}{dr} = \frac{Pr}{8\pi D}(2\ln(r) - 1) + 2C_1 r + \frac{C_2}{r} \tag{15.30b}$$

where the primed constants relate to the inner portion.

Since slope is zero at the plate centre we see from (15.30a) that $C_2' = 0$. The slope is also zero at the clamped edge $(r = a)$ so that

$$\frac{Pa}{8\pi D}(2\ln(a) - 1) + 2C_1 a + \frac{C_2}{a} = 0 \tag{15.31a}$$

The slope must be continuous at $r = b$ and so from (15.30a) and (15.30b) we have

$$\frac{Pb}{8\pi D}(2\ln(b) - 1) + 2C_1 b + \frac{C_2}{b} = 2C_1' b \tag{15.31b}$$

A third equation is obtained from the condition that M_r must be continuous at $r = b$. Thus writing (15.20) for both portions and equating these expressions at $r = b$ we obtain

$$\frac{P}{4\pi}((1+\nu)\ln(b) + \tfrac{1}{2}(1-\nu)) + 2C_1 D(1+\nu) - \frac{C_2 D}{b^2}(1-\nu) = 2C_1' D(1+\nu) \tag{15.31c}$$

After solving equations (15.31) we have $(\alpha = P/16\pi D)$,

$$C_1 = -\alpha\left(1 - \frac{b^2}{a^2} - 2\ln(a)\right); \quad C_2 = 2\alpha b^2; \quad C_1' = \alpha\left(1 - \frac{b^2}{a^2} + 2\ln\left(\frac{b}{a}\right)\right)$$

Then substituting C_1 and C_2 in (15.20) we obtain

$$M_r = \frac{P}{4\pi}\left((1+\nu)\ln\left(\frac{a}{r}\right) - 1\right) + \frac{Pb^2}{8\pi}\left(\frac{1+\nu}{a^2} + \frac{1-\nu}{r^2}\right) \tag{15.32a}$$

for the outer plate. Similarly using C_1' and $C_2' = 0$ we find

$$M_r = \frac{P(1+\nu)}{8\pi}\left(\frac{b^2}{a^2} - 1 + 2\ln\left(\frac{a}{b}\right)\right) \tag{15.32b}$$

for the inner portion; this is independent of r. From equations (15.32) it is found that the greatest M_r depends on a/b.

It may be noted that if b is set zero in (15.32a) we obtain the expression for a clamped-edge solid plate with central concentrated load.

15.4 Rectangular Plates

15.4.1 *Twisting Moments*

A triangular portion of a plate subject only to edge moments M_x and M_y as shown in Figure 15.6a can only be in equilibrium if moments are exerted on the edge AC by the rest of the plate. In the most general case moments are required acting in planes parallel to both the $x'z$ and $y'z$ planes, where x' is normal, and y' parallel, to AC. These edge moments are denoted $M_{x'}$ and $M_{x'y'}$ respectively and are given positive directions as in Figure 15.6b

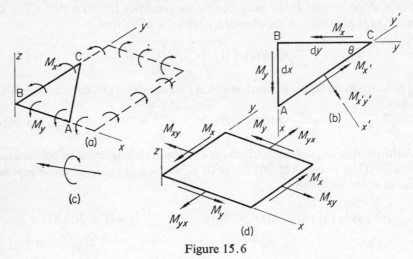

Figure 15.6

where the double arrow representation of a moment has been used for clarity. (The rotation is clockwise, as in Figure 15.6c, when looking in the direction of the arrow.) $M_{x'}$ is a bending moment, while $M_{x'y'}$ which acts in the plane of the edge could in the language of Article 8.1 be termed a twisting moment.

Resolving moments in planes parallel to the $x'z$ plane we have for equilibrium

$$M_{x'} \times \text{AC} - M_x \times \text{BC} \times \cos\theta - M_y \times \text{AB} \times \sin\theta = 0$$

Noting that $\text{AB} = \text{AC} \times \sin\theta$ and $\text{BC} = \text{AC} \times \cos\theta$ we find

$$M_{x'} = M_x \cos^2\theta + M_y \sin^2\theta \qquad (15.33)$$

Similarly, resolving in planes parallel to the $y'z$ plane

$$M_{x'y'} = -(M_x - M_y)\sin\theta\cos\theta \qquad (15.34)$$

Equations (15.33) and (15.34) have the same form as those of coordinate transformation for states of plane stress (Article 1.6) when the reference axes are the principal axes. We may therefore translate all the results of Articles 1.6–1.10 to the case of plate edge moments.

At any point of a plate there is one choice of axes in the plane of the plate (the principal axes) for which there are only bending moments on a rectangular element as in Figure 15.1b; the corresponding radii of curvature (equations 15.5) are the principal values. For any other set of axes at the point the edges of an element are subject to both bending and twisting moments (Figure 15.6d). The twisting moment on the y faces is denoted M_{yx}. It is only in the particular case $M_x = M_y$ that there are no twisting moments for any inclination of the axes. This case is called 'spherical bending' since from equations (11.5) the principal radii of curvature are equal. Following the stress analogy we expect greatest twisting moments on edges at 45° to the principal directions. A Mohr's Circle could be drawn as a graphical representation of equations (15.33) and (15.34).

Two conventions are widely used for the signs of the moments acting on an element of a plate. That in Figure 15.6d is used in references 1 and 2; an alternative form used in references 3 and 4 has one set of twisting moments represented as inward normals. The system of Figure 15.6d has been used here as it appears marginally easier to remember, with all twisting moments

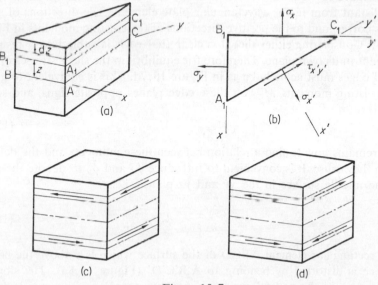

Figure 15.7

represented by outward normals. The double arrow notation makes the bending moments easy to remember, with the arrows, as it were, flowing round the element.

Figures 15.7a and b show the stresses acting on a slice above the neutral surface of the plate element of Figure 15.6a. The bending moments on the BC and AB edges cause tensile stresses σ_x and σ_y. On edge AC there must be stress components $\sigma_{x'}$, $\tau_{x'y'}$ with positive directions as shown (see Figure 1.8c). Then following equation (15.3) we can write

$$M_x = \int_{-t/2}^{t/2} \sigma_x z \, dz; \quad M_y = \int_{-t/2}^{t/2} \sigma_y z \, dz; \quad M_{x'} = \int_{-t/2}^{t/2} \sigma_{x'} z \, dz$$

$$(15.35a, b, c)$$

for the bending moment intensities on the edge planes as a whole. The twisting moment intensity on the AC edge plane is obtained in a similar manner by considering the resultant moment due to the shear stresses $\tau_{x'y'}$. Then

$$M_{x'y'} = -\int_{-t/2}^{t/2} \tau_{x'y'} z \, dz \qquad (15.35d)$$

The negative sign in this equation is required because the direction of $\tau_{x'y'}$ shown in Figure 15.7b corresponds to a negative twisting moment according to the moment sign convention (Figure 15.6b). The quantities defined by equations (15.35) are also known as *stress resultants*. Actual stress components can be obtained by use of equations of the form of (15.8).

In Figure 15.7c slices are shown above and below the neutral surface, and equidistant from it, for a rectangular plate element. The directions of shear stress correspond to the positive directions of the twisting moments in Figure 15.6d. Considering either slice it is clear that each is subject to a resultant moment in its own plane. Therefore for equilibrium the shear stresses on one set of edges must actually be as in Figure 15.7d. This is equivalent to giving the twisting moments M_{xy} on these edge planes opposite signs, and so for equilibrium

$$M_{xy} = -M_{yx} \qquad (15.36)$$

It remains now to find a relation between these moments and the deformation they cause. It is convenient to introduce ϕ_x and ϕ_y to denote the slope of the middle surface in the xy and yz planes, i.e.

$$\phi_x = \frac{\partial w}{\partial x}; \qquad \phi_y = \frac{\partial w}{\partial y} \qquad (15.37)$$

A rectangular element ABCD of the surface which is z above the neutral surface is distorted by bending to A'B'C'D' (Figure 15.8a). The slope of the line nn in the neutral surface is negative (angle positive anticlockwise)

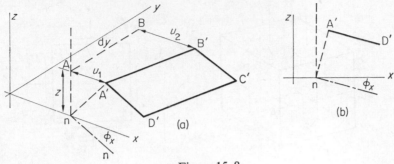

Figure 15.8

(Figure 15.8b). The line nA' has the same rotation and so point A is displaced a distance $u_1 = -\phi_x z$ in the x direction. The slope of $B'C'$ is different from that of $A'D'$ by the increment of ϕ_x due to the change in y coordinate (viz. $(\partial\phi_x/\partial y)\mathrm{d}y$). Therefore the displacement of B to B' is $u_2 = -(\phi_x+(\partial\phi_x/\partial y)\mathrm{d}y)z$. Because u_2 and u_1 are in general unequal the original right angle $D\hat{A}B$ decreases to $D'\hat{A}'B'$. This shear strain is equal to $(u_2-u_1)/\mathrm{d}y$ or $-z(\partial\phi_x/\partial y)$. A corresponding discussion of displacements of A and D in the y direction gives an additional change of angle $D\hat{A}B$ of $-z(\partial\phi_y/\partial x)$. The total shear strain is therefore

$$\gamma_{xy} = -z\left(\frac{\partial\phi_x}{\partial y} + \frac{\partial\phi_y}{\partial x}\right) \qquad (15.38)$$

Using equations (15.37) and the stress/strain equation (3.4d) we have

$$\tau_{xy} = -2Gz\,\frac{\partial^2 w}{\partial x \partial y} \qquad (15.39)$$

The quantity $(\partial^2 w/\partial x \partial y)$ is called the twist of the surface and by analogy with equations (15.6) is also written as $1/R_{xy}$. We can now obtain the twisting moment intensity as (cf. 15.35d).

$$M_{xy} = -\int_{-t/2}^{t/2} \tau_{xy}z\,\mathrm{d}z = \frac{Gt^3}{6}\frac{\partial^2 w}{\partial x \partial y} \qquad (15.40a)$$

Finally substituting for G from equation (3.8) and rearranging using (15.4) we obtain

$$M_{xy} = D(1-\nu)\frac{\partial^2 w}{\partial x \partial y} \qquad (15.40b)$$

15.4.2 *Equilibrium Equations*

An element of a plate supporting distributed transverse forces of intensity q is shown in Figure 15.9a and has shearing forces distributed over the edges,

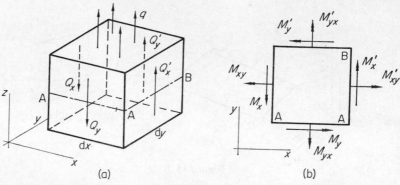

Figure 15.9

with

$$Q'_x = Q_x + \frac{\partial Q_x}{\partial x}\,dx; \qquad Q'_y = Q_y + \frac{\partial Q_y}{\partial y}\,dy \qquad (15.41a, b)$$

Figure 15.9b shows the intensities of bending and twisting moment where for example $M'_x = M_x + (\partial M_x/\partial x)\,dx$.

For equilibrium of forces in the z direction we require

$$(Q'_x - Q_x)\,dy + (Q'_y - Q_y)\,dx + q\,dx\,dy = 0$$

Substituting from (15.41) and simplifying we obtain

$$\frac{\partial Q_x}{\partial x} + \frac{\partial Q_y}{\partial y} + q = 0 \qquad (15.42)$$

Taking moments about edge AA of the neutral surface we find for equilibrium

$$(M_y - M'_y)\,dx + (M'_{xy} - M_{xy})\,dy + Q'_y\,dx\,dy + (Q'_x - Q_x)\frac{(dy)^2}{2} + q\,dx\,\frac{(dy)^2}{2} = 0$$

The last two quantities are small quantities of a higher order than the others and can be neglected. Expanding M'_y and M'_{xy} as explained above we obtain

$$\frac{\partial M_{xy}}{\partial x} - \frac{\partial M_y}{\partial y} + Q_y = 0 \qquad (15.43)$$

Similarly by taking moments about the line AB we find

$$\frac{\partial M_{yx}}{\partial y} + \frac{\partial M_x}{\partial x} - Q_x = 0 \qquad (15.44)$$

There are no net forces on the element in the x and y directions. Equilibrium of moments in the xy plane has previously been established by equation (15.36). Thus equations (15.36) and (15.42–15.44) completely specify the conditions for equilibrium of the element subject to distributed transverse forces q and stress resultants M_x, M_y, M_{xy}, M_{yx}, Q_x, Q_y.

The four equations in six variables can be reduced to one equation in terms of any three. Substituting from (15.43) and (15.44) in (15.42) and using (15.36) we obtain after simplification.

$$\frac{\partial^2 M_x}{\partial x^2} - \frac{2\partial^2 M_{xy}}{\partial x \partial y} + \frac{\partial^2 M_y}{\partial y^2} + q = 0 \qquad (15.45)$$

as the equilibrium equation in terms of moments. This equation can be expressed in terms of the deflexion of the neutral surface by substituting from the moment curvature relations (15.7) and (15.40b). Hence

$$\frac{\partial^4 w}{\partial x^4} + 2\frac{\partial^4 w}{\partial x^2 \partial y^2} + \frac{\partial^4 w}{\partial y^4} = \frac{q}{D} \qquad (15.46)$$

The solution to a plate bending problem involves finding the function $w(x, y)$ which satisfies (15.46) and the prescribed boundary conditions. It has to be emphasized that the deflexions must be small in relation to the thickness in order that the middle surface is a neutral surface (see Article 15.1). Once $w(x, y)$ has been determined equations (15.7) and (15.40b) can be used to find M_x, M_y, M_{xy} from which in turn the greatest values of σ_x, σ_y, τ_{xy} are found using equations (15.8) and

$$(\tau_{xy})_{\text{max}} = \mp 6\frac{M_{xy}}{t^2} \qquad (15.47)$$

This equation results from the combination of (15.39) and (15.40b) when $z = \pm t/2$. The stress resultants Q_x, Q_y are obtained from M_x, M_y, M_{xy} using equations (15.43) and (15.44) together with (15.36). Then assuming that Q_x, Q_y are the resultants of stress distributions τ_{xz} and τ_{yz} respectively, distributed parabolically as in Section 6.4.1, we have

$$(\tau_{xz})_{\text{max}} = 1\cdot5\frac{Q_x}{t}; \qquad (\tau_{yz})_{\text{max}} = 1\cdot5\frac{Q_y}{t} \qquad (15.48a, b)$$

The study of the known analytical solutions of equation (15.46) is beyond the scope of an introductory text (see for example reference 1). Recently the solutions of plate problems have been obtained mainly by Finite Element Methods. The material given above is sufficient background for descriptions of this method such as in reference 5.

15.5 References

1. Timoshenko, S. P. and Woinowsky-Krieger, S., *Theory of Plates and Shells*, 2nd ed., McGraw-Hill, New York, 1959.
2. Love, A. E. H., *A Treatise on the Mathematical Theory of Elasticity*, 4th ed., Dover, New York, 1944.
3. Flugge, W., *Stresses in Shells*, Springer, Berlin, 1960
4. Turner, C. E., *Introduction to Plate and Shell Theory*, Longmans, London, 1965.
5. Zienkiewicz, O. C., loc. cit. Reference 1, Article 13.6.

Appendix I—
Equilibrium Conditions

In accordance with Newton's first law a body remains in a state of rest or of uniform motion unless obliged to change that state by external forces. In the case of a particle, i.e. a body of negligible volume but significant mass, the condition for equilibrium can be stated conveniently as 'The sum of the resolved components of the forces in any three mutually perpendicular directions must be zero'. Formally these conditions may be expressed as

$$\Sigma X = 0; \quad \Sigma Y = 0; \quad \Sigma Z = 0 \qquad (I.1)$$

In the case of an extensive body, i.e. one containing many particles and having significant volume and mass, the above conditions are necessary, but not sufficient, for equilibrium. Since the forces on such a body are not in general concurrent, the moments of the forces about any point in the body must also be zero. This additional condition can be conveniently expressed as 'The sum of the moments of the forces about any three mutually perpendicular directions must be zero'. Thus, in addition to (I.1) we require

$$\Sigma m_x = 0; \quad \Sigma m_y = 0; \quad \Sigma m_z = 0 \qquad (I.2)$$

The equilibrium conditions stated above can be expressed in an alternative manner by introducing the concept of a virtual displacement. Such a displacement is one during which the forces acting on the system do not change. For example consider a particle subject to forces $F_1, F_2 \ldots F_n$ (as in Figure I.1a) which, by successive application of the Parallelogram of Forces, can always be replaced by a single resultant R as in Figure I.1b. If the particle is given a virtual displacement δ_1^* in the direction of the resultant then the work done is the product $R\delta_1^*$ since the force R is constant during the virtual displacement. This quantity is known as the virtual work and is zero if the particle is in equilibrium since the resultant is then zero. However, the particle is not necessarily in equilibrium if the virtual work is zero during a single virtual displacement, as is easily shown by considering a virtual displacement δ_2^* at right angles to the line of action of the resultant (Figure I.1b). The virtual work performed is then zero whether or not the resultant is zero. Thus the particle is only in equilibrium if the virtual work performed during every

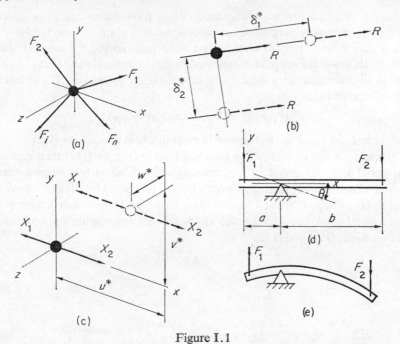

Figure I.1

virtual displacement is zero. All possibilities can be tested by virtual displacements in three mutually perpendicular directions.

For example let the particle in Figure I.1c be given virtual displacements u^*, v^*, w^* in the x, y, z directions respectively. Then for equilibrium of the particle, we require, *inter alia*, that

$$X_1 u^* + X_2 u^* = 0 \quad \text{i.e.} \quad -X_1 + X_2 = 0$$

which is the same as the first of equations (I.1). In the y and z directions it is obvious that no virtual work is done during the virtual displacements v^* and w^*. Thus the particle is in equilibrium if X_1 and X_2 are equal in magnitude.

It may seem that nothing is gained by this procedure but this is not so as will be indicated later. For an extensive body to be in equilibrium it will be necessary for the virtual work performed by the forces on all the particles to be zero for every possible virtual displacement. This can only be properly explored by imposing three arbitrary displacements for each particle. For any body of finite size this is evidently a very large number and is clearly impracticable. Fortunately considerable simplification is possible when there are restrictions on the possible sets of displacements. In particular, in the case of a rigid body the distances between the particles are considered to remain constant and it is possible to dispense with any set of virtual displacements which does not satisfy this requirement. Equilibrium can then be tested in

14

the most general case by imposing three virtual translations and three virtual rotations in mutually perpendicular directions and planes, respectively.

As a simple and classical example we need only consider a lever (Figure I.1d). In this case the support imposes additional constraints so that the only possible displacement as a rigid body is a rotation as shown. If the bar is to be in equilibrium

$$F_1 a\theta - F_2 b\theta = 0 \qquad \text{i.e. } F_1 a = F_2 b$$

which is just the same as the third of the equilibrium conditions (I.2).

If the lever is not considered to be a rigid body (Figure I.1e) then equilibrium could only be tested by considering also those virtual displacement patterns which permit relative movements of the particles. The method of carrying this out is discussed in Section 12.5.1 and utilized in Chapter 13 where virtual work provides a much more convenient equilibrium condition than equations (I.1) and (I.2).

Appendix II—
Matrix Algebra

II.1 Introduction

A summary of the matrix properties and algebra used in the text will now be given. For general theory see references 1 and 2, for example.

A matrix is a rectangular array of numeric or symbolic quantities enclosed by square brackets, e.g.

$$\begin{bmatrix} 1 & 0 \\ 6 & -5 \\ 1 & 1 \end{bmatrix}; \begin{bmatrix} X \\ Y \\ Z \end{bmatrix}; \begin{bmatrix} a_{11} & a_{12} & a_{13} \\ a_{21} & a_{22} & a_{23} \\ a_{31} & a_{32} & a_{33} \end{bmatrix}; [P \quad Q] \qquad \text{(II.1a, b, c, d)}$$

If all the quantities are zero the array is called a *null* matrix. The order of a matrix is specified by the number of rows and columns so that the above examples are, respectively, (3×2), (3×1), (3×3), (1×2); the second example is called a *column* matrix, and the fourth a *row* matrix. The quantities comprising a matrix need not be functionally related or have the same physical dimensions. A matrix can be represented by a single symbol (here printed in bold type). For example the third of matrices (II.1) might be denoted a. A physical example is provided by the second of matrices (II.1) which could be interpreted as comprising the components of a force in the x, y and z directions. Denoting the force by $\mathbf{F}$ we have

$$\mathbf{F} = \{X \quad Y \quad Z\} \qquad \text{(II.2)}$$

where the braces $\{\}$ indicate a column matrix written as a row *in order to save space*. Since expression (II.2) defines the magnitude and direction of the force relative to a particular coordinate system it follows that $\mathbf{F}$ can be referred to as a vector quantity.

The advantage of using matrices lies in this ability to represent concisely large amounts of information. When operations involving two or more matrices are defined this economy is even more apparent. For instance, as will be explained later, the set of simultaneous equations

$$X = a_{11}u + a_{12}v + a_{13}w$$
$$Y = a_{21}u + a_{22}v + a_{23}w \qquad \text{(II.3a)}$$
$$Z = a_{31}u + a_{32}v + a_{33}w$$

may be written in matrix form as

$$\begin{bmatrix} X \\ Y \\ Z \end{bmatrix} = \begin{bmatrix} a_{11} & a_{12} & a_{13} \\ a_{21} & a_{22} & a_{23} \\ a_{31} & a_{32} & a_{33} \end{bmatrix} \begin{bmatrix} u \\ v \\ w \end{bmatrix} \qquad (II.3b)$$

and in symbolic form as

$$\mathbf{F} = \mathbf{a}\boldsymbol{\delta} \qquad (II.3c)$$

where $\mathbf{F}$ and $\mathbf{a}$ are as defined by (II.2) and (II.1c) and

$$\boldsymbol{\delta} = \{u \quad v \quad w\}$$

If u, v, w are components of displacement in the x, y and z directions then equations (II.3) can be interpreted as relating force and displacement at a point on a body. Matrix $\mathbf{a}$ then consists of coefficients dependent on the properties of the body. A further advantage of the form (II.3c) is that it is valid regardless of the actual number of quantities required to define a vector. For example if forces and displacements are confined to the xy plane we have $\mathbf{F} = \{X \quad Y\}$ and $\mathbf{a}$ is a (2×2) matrix.

II.2. Square Matrices

A *square* matrix has an equal number of rows and columns (e.g. II.1c). The principal diagonal is that running from the upper left-hand corner to the lower right. If all quantities outwith the principal diagonal are zero this is a *diagonal* matrix. A *unit* matrix is a diagonal matrix in which all quantities in the principal diagonal are unity.

e.g.
$$\mathbf{I} = \begin{bmatrix} 1 & 0 & 0 \\ 0 & 1 & 0 \\ 0 & 0 & 1 \end{bmatrix} \qquad (II.4)$$

A *band* matrix has its non-zero quantities in a band around the principal diagonal, e.g.

$$\begin{bmatrix} 4 & -2 & 0 & 0 & \cdots \\ -2 & 2 & 3 & 0 & \\ 0 & 3 & 1 & -3 & \\ 0 & 0 & -3 & 5 & \\ \cdot & \cdot & \cdot & \cdot & \\ \cdot & \cdot & \cdot & \cdot & \end{bmatrix} \qquad (II.5)$$

If the quantities in a square matrix are symmetrical about the principal diagonal the matrix is said to be *symmetric*. For instance if in (II.1c) $a_{21} = a_{12}$, $a_{31} = a_{13}$ and $a_{32} = a_{23}$ the matrix is symmetric; matrix (II.5) is symmetric. In general we require $a_{ij} = a_{ji}$ for $i \neq j$.

II.3 Equality; Addition; Multiplication

(a) *Equality*

Two matrices of the same order are equal if the corresponding elements are equal, i.e. for $\mathbf{a} = \mathbf{b}$ we require $a_{ij} = b_{ij}$ for all i, j.

(b) *Addition*

The sum of two matrices of the same order is obtained by adding corresponding elements, i.e. taking $a_{ij} + b_{ij} = c_{ij}$ for all i, j we have

$$\mathbf{a} + \mathbf{b} = \mathbf{c} \tag{II.6}$$

e.g.
$$\begin{bmatrix} 4 & 3 & 1 \\ -6 & 2 & 0 \end{bmatrix} + \begin{bmatrix} -2 & 1 & -4 \\ 1 & 0 & 3 \end{bmatrix} = \begin{bmatrix} 2 & 4 & -3 \\ -5 & 2 & 3 \end{bmatrix}$$

The commutative and associative rules apply, i.e. $\mathbf{a} + \mathbf{b} = \mathbf{b} + \mathbf{a}$ and $(\mathbf{a} + \mathbf{b}) + \mathbf{c} = \mathbf{a} + (\mathbf{b} + \mathbf{c})$.

(c) *Subtraction*

For matrices of the same order, if $c_{ij} = a_{ij} - b_{ij}$ for all i, j we have $\mathbf{c} = \mathbf{a} - \mathbf{b}$.

(d) *Multiplication*

The matrix product
$$\mathbf{ab} = \mathbf{c} \tag{II.7}$$

exists only if the number of columns in $\mathbf{a}$ is equal to the number of rows in $\mathbf{b}$. Then if $\mathbf{a}$ and $\mathbf{b}$ are of order $(m \times q)$ and $(q \times n)$ respectively they are said to be *conformable* and their product is a matrix $\mathbf{c}$ of order $(m \times n)$. A typical element c_{ij} of the product matrix is equal to the sum of the products of each element of the ith row of $\mathbf{a}$ with the corresponding element in the jth row of $\mathbf{b}$

e.g.
$$\begin{bmatrix} 2 & -4 & 1 \\ 0 & 6 & -2 \end{bmatrix} \begin{bmatrix} 3 & 0 \\ 1 & -3 \\ 4 & -2 \end{bmatrix}$$

$$= \begin{bmatrix} ((2 \times 3) + (-4 \times 1) + (1 \times 4)) & ((2 \times 0) + (-4 \times -3) + (1 \times -2)) \\ ((0 \times 3) + (6 \times 1) + (-2 \times 4)) & ((0 \times 0) + (6 \times -3) + (-2 \times -2)) \end{bmatrix}$$

$$= \begin{bmatrix} 6 & 10 \\ -2 & -14 \end{bmatrix}$$

The operation may be expressed formally as

$$c_{ij} = \sum_{r=1}^{q} a_{ir}b_{rj} \tag{II.8}$$

In (II.7) $\mathbf{b}$ is premultiplied by $\mathbf{a}$ and $\mathbf{a}$ is postmultiplied by $\mathbf{b}$. It is important to note that, even if both products exist, $\mathbf{ab}$ is not in general equal to $\mathbf{ba}$. The associative and distributive laws apply provided the order of multiplication is not reversed, i.e.

$$\mathbf{a(bc)} = \mathbf{(ab)c} \quad \text{and} \quad \mathbf{a(b+c)} = \mathbf{ab+ac} \tag{II.9}$$

II.4   Scalar Product; Transpose; Inverse

(a) *Scalar Product*

The product of matrix $\mathbf{a}$ by a scalar b is obtained by multiplying all elements of $\mathbf{a}$ by b, e.g. ba_{ij}.

(b) *Transpose*

The transpose of a matrix $\mathbf{a}$ is formed by interchanging each row with the corresponding column, and is denoted $\mathbf{a}^T$, e.g. the transpose of (II.1a) is

$$\begin{bmatrix} 1 & 6 & 1 \\ 0 & -5 & 1 \end{bmatrix}$$

Clearly $(\mathbf{a}^T)^T = \mathbf{a}$, and the transpose of a symmetric matrix is equal to the original matrix. The transpose of a product of any number of matrices consists of their transposes with the multiplication sequence reversed,

e.g.   if $\mathbf{p} = \mathbf{abcd}$ then $\mathbf{p}^T = \mathbf{d}^T\mathbf{c}^T\mathbf{b}^T\mathbf{a}^T$

(c) *Inverse*

We have previously noted how a set of linear equations (II.3a) could be represented in matrix symbolic form as (II.3c)

i.e. $\mathbf{F} = \mathbf{a\delta}$ (II.3c)

If a matrix denoted $\mathbf{a}^{-1}$ could be found such that

$$\mathbf{a}^{-1}\mathbf{a} = \mathbf{I} \tag{II.10}$$

where $\mathbf{I}$ is a unit matrix, then premultiplying both sides of (II.3c) by $\mathbf{a}^{-1}$ we have
$$\mathbf{a}^{-1}\mathbf{F} = \mathbf{a}^{-1}\mathbf{a\delta} = \mathbf{I\delta} = \mathbf{\delta}$$

Thus if the matrix $\mathbf{a}^{-1}$ exists it permits the inversion of the equations; it is therefore called the *inverse* or *reciprocal* matrix. Only a square matrix with non-zero determinant can have an inverse. Various methods have been devised for obtaining the inverse especially with the aid of a digital computer. In the next article one method will be described.

II.5 Inversion (Adjoint Matrix)

It can be shown that

$$\mathbf{a} \, (\text{adj } \mathbf{a}) = |\mathbf{a}| \, \mathbf{I} \tag{II.11}$$

where $|\mathbf{a}|$ is the determinant of the matrix $\mathbf{a}$ and adj $\mathbf{a}$, called the *adjoint* matrix, is the transpose of the matrix of cofactors of the determinant. Comparing (II.10) and (II.11) we see that

$$\mathbf{a}^{-1} = \frac{\text{adj } \mathbf{a}}{|\mathbf{a}|} \tag{II.12}$$

from which it is clear that the inverse does not exist when $|\mathbf{a}|$ is zero, in which case $\mathbf{a}$ is said to be *singular*.

To illustrate the method we shall determine the inverse of the matrix

$$\mathbf{H} = \begin{bmatrix} 1 & x_i & y_i \\ 1 & x_j & y_j \\ 1 & x_m & y_m \end{bmatrix} \tag{II.13}$$

If we delete the pth row and qth column from the determinant of the matrix we obtain the minor H'_{pq}, e.g. deleting row 3 and column 1 we have

$$H'_{31} = \begin{vmatrix} x_i & y_i \\ x_j & y_j \end{vmatrix} \tag{II.14}$$

The cofactor $\bar{H}_{pq}$ is the product of the minor and $(-1)^{(p+q)}$. When the cofactors are written as a matrix and then transposed we have the adjoint matrix

$$\text{adj } \mathbf{H} = \begin{bmatrix} \begin{vmatrix} x_j & y_j \\ x_m & y_m \end{vmatrix} & -\begin{vmatrix} x_i & y_i \\ x_m & y_m \end{vmatrix} & \begin{vmatrix} x_i & y_i \\ x_j & y_j \end{vmatrix} \\ -\begin{vmatrix} 1 & y_j \\ 1 & y_m \end{vmatrix} & \begin{vmatrix} 1 & y_i \\ 1 & y_m \end{vmatrix} & -\begin{vmatrix} 1 & y_i \\ 1 & y_j \end{vmatrix} \\ \begin{vmatrix} 1 & x_j \\ 1 & x_m \end{vmatrix} & -\begin{vmatrix} 1 & x_i \\ 1 & x_m \end{vmatrix} & \begin{vmatrix} 1 & x_i \\ 1 & x_j \end{vmatrix} \end{bmatrix} \tag{II.15}$$

For example H'_{31} of (II.14) is transposed to row 1 column 3. Expanding the determinants we have

$$\text{adj } \mathbf{H} = \begin{bmatrix} (x_j y_m - x_m y_j) & -(x_i y_m - x_m y_i) & (x_i y_j - x_j y_i) \\ -(y_m - y_j) & (y_m - y_i) & -(y_j - y_i) \\ (x_m - x_j) & -(x_m - x_i) & (x_j - x_i) \end{bmatrix} \qquad (\text{II}.16)$$

The inverse is obtained by dividing adj $\mathbf{H}$ by the determinant of $\mathbf{H}$.

II.6 Characteristic Equation

A set of homogeneous equations

$$(a_{11} - \lambda)x_1 + a_{12}x_2 + \ldots + a_{1n}x_n = 0$$

$$a_{21}x_1 + (a_{22} - \lambda)x_2 + \ldots$$

$$\ldots \qquad \ldots \qquad + (a_{nn} - \lambda)x_n = 0$$

has a non-trivial solution if

$$\begin{vmatrix} (a_{11} - \lambda) & a_{12} & \ldots \\ a_{21} & (a_{22} - \lambda) & \ldots \\ \ldots & \ldots & \end{vmatrix} = 0$$

The expansion of this determinant is a polynomial in λ of degree n known as the *characteristic equation* of the matrix

$$\mathbf{a} = \begin{bmatrix} a_{11} & a_{12} & \ldots & a_{1n} \\ a_{21} & a_{22} & \ldots & \\ \cdot & & & \\ \cdot & & & \\ \cdot & & & a_{nn} \end{bmatrix}$$

This equation has roots $\lambda_1, \lambda_2, \ldots \lambda_n$ called the *characteristic*, or *eigen-*, *values* of the matrix $\mathbf{a}$.

II.7 Partitioning

A matrix may be subdivided into a number of *sub-matrices* by means of horizontal and vertical lines drawn through the array of elements

e.g.
$$\mathbf{a} = \begin{bmatrix} a_{11} & a_{12} & a_{13} \\ a_{21} & a_{22} & a_{23} \\ a_{31} & a_{32} & a_{33} \end{bmatrix} = \begin{bmatrix} \mathbf{a}_{11} & \mathbf{a}_{12} \\ \mathbf{a}_{21} & \mathbf{a}_{22} \end{bmatrix}$$

where $\mathbf{a}_{22} = [a_{32} \quad a_{33}]$ etc.

Bold type has been used for sub-matrices since they may represent a number of scalar quantities. Provided the appropriate rules are followed the algebraic operations previously described can be applied to sub-matrices as if they were elements of an ordinary matrix. In particular in multiplication of conformable matrices it is necessary that the columns of the premultiplier are grouped in the same manner as the rows of the postmultiplier.

e.g.

$$\left[\begin{array}{cc|c} a_{11} & a_{12} & a_{13} \\ \hline a_{21} & a_{22} & a_{23} \\ a_{31} & a_{32} & a_{33} \end{array}\right] \left[\begin{array}{cc} b_{11} & b_{12} \\ b_{21} & b_{22} \\ \hline b_{31} & b_{32} \end{array}\right] = \left[\begin{array}{cc} \mathbf{a}_{11} & \mathbf{a}_{12} \\ \mathbf{a}_{21} & \mathbf{a}_{22} \end{array}\right] \left[\begin{array}{c} \mathbf{b}_1 \\ \mathbf{b}_2 \end{array}\right]$$

$$= \left[\begin{array}{c} \mathbf{a}_{11}\mathbf{b}_1 + \mathbf{a}_{12}\mathbf{b}_2 \\ \mathbf{a}_{21}\mathbf{b}_1 + \mathbf{a}_{22}\mathbf{b}_2 \end{array}\right]$$

II.8   References

1. Aitken, A. C., *Determinants and Matrices*, Oliver and Boyd, Edinburgh, 1967,
2. Gere, J. M. and Weaver, W., *Matrix Algebra for Engineers*, Van Nostrand. Princeton, N.J., 1967.

Appendix III—
Answers to Problems

1.1. See Figure III.1

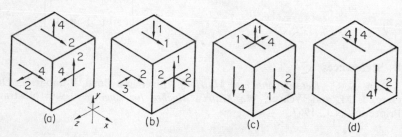

Figure III.I

1.2(a). $\begin{pmatrix} -1 & -1 & 1 \\ -1 & 0 & 0 \\ 1 & 0 & 1 \end{pmatrix}$ (b) $\begin{pmatrix} 0 & 1 & -1 \\ 1 & -1 & 1 \\ -1 & 1 & 0 \end{pmatrix}$ (c) $\begin{pmatrix} -1 & 0 & 0 \\ 0 & -1 & 0 \\ 0 & 0 & -1 \end{pmatrix}$

1.3. $\begin{pmatrix} 2 & 2 & -2 \\ 2 & 8 & -2 \\ -2 & -2 & 2 \end{pmatrix}$ 1.4. $\begin{pmatrix} -3 & 2 & -2 \\ 2 & 3 & -2 \\ -2 & -2 & 0 \end{pmatrix}$ 1.5(a). $\begin{pmatrix} 0 & -100 & 0 \\ -100 & 0 & 0 \\ 0 & 0 & 50 \end{pmatrix}$

1.5(b). $\begin{pmatrix} 50 & -50 & 0 \\ -50 & 50 & 0 \\ 0 & 0 & 50 \end{pmatrix}$ (c) $\begin{pmatrix} -32\cdot3 & 33\cdot3 & 0 \\ 33\cdot3 & 52\cdot3 & 0 \\ 0 & 0 & -50 \end{pmatrix}$

1.6. $\begin{pmatrix} -43\cdot9 & 0 & 0 \\ 0 & 63\cdot9 & 0 \\ 0 & 0 & -50 \end{pmatrix}$

$\theta_0 = 10°54'$ 1.8. 72·4, 361, 50 MN/m²   1.9. σ_1, σ_2, σ_3 (80, 40, 0 MN/m²)
$\theta_0 = -45°$ 1.10. σ_x, σ_y, τ_{xy}, $(-123\cdot2, -76\cdot8, 359$ MN/m²); unchanged

1.11. Yes, 2·76; no; yes, 4   1.12. 125, 16 MN/m²   1.13. 160 kN   1.14. p_x, τ_{max}; $-400, 400$; $0, 200$; $400, 200$ 1.15. 20, 25 mm.

2.1(a). $-22°30'$; $(483, -83, -400; 283) \times 10^{-6}$ (b) $22°30'$; $(-59, -341, 0; 171)$ $\times 10^{-6}$ 2.1(c). $0°$; $(100, 100, 0; 50) \times 10^{-6}$ 2.2. $(0, -400, 200) \times 10^{-6}$ 2.3(a) $36°55'$; $(583, -617) \times 10^{-6}$, ϵ_z (b) $45°$, $(200, -200) \times 10^{-6}$, ϵ_z.

3.1. ϵ_x, ϵ_y, ϵ_z, $(1, -0\cdot25, -0\cdot25 \times 10^{-3})$; $\frac{1}{2}\gamma_{xy} = \frac{1}{2}\gamma_{yz} = \frac{1}{2}\gamma_{xz} = 0$; $0\cdot5 \times 10^{-3}$, 20 mm^3 3.2 ϵ_1, ϵ_2, ϵ_3 $(562, -429, -57 \times 10^{-6})$: 76×10^{-6}; 0 3.3. $-0\cdot2$ mm, 40 MN/m^2; $-0\cdot24$ mm 3.4. ϵ_x, ϵ_y, ϵ_z $(875, 250, -375 \times 10^{-6})$, $\frac{1}{2}\gamma_{xy} = 625 \times 10^{-6}$; 262, 38, 0 MN/m^2; 1263, -138, -375×10^{-6}; 131 MN/m^2 3.5. ϵ_x, ϵ_y, ϵ_z $(1000, 375, -875 \times 10^{-6})$, $\frac{1}{2}\gamma_{xy} = 625 \times 10^{-6}$; 262, 38, -100 MN/m^2; 1388, -13, -875×10^{-6}; 181 MN/m^2 3.6. $1\cdot31$, $1\cdot05$; $99\cdot5 \times 10^{-6}$, 0 3.7. Principal strains, rod $(952, -286, -286 \times 10^{-6})$, tube $(-476, 143, 143 \times 10^{-6})$ 3.8. σ_x, σ_y, τ_{xy} $(23\cdot0, -32\cdot1, 92\cdot8)$ MN/m^2.

4.1. $79\cdot2$ Nm 4.2. 209 Nm 4.3. $57\cdot7$ MN/m^2 4.4. $Ab^2L^3/6E$ 4.5. $22\cdot6$, 890, $33\cdot3$ kNm/m^3 4.6. 295 Nm 4.8. $13\cdot5$ mm 4.9. 90 MN/m^2 4.10. 357 MN/m^2 4.11. $0\cdot8$ 4.12. $L(P+Q)/3AE$ 4.13. $PL/4AE$.

5.1. $0\cdot714$ mm, $2\cdot4$ 5.2. $1\cdot19$ mm, $1\cdot6$ 5.4. K (row by row); 420, -420, 0; -420, 630, -210; 0, -210, 210; all $\times 10^{-6}$; 140, -140, 0; -140, 350, -210; 0, -210, 210; all $\times 10^{-6}$. 5.5. -42, 98, -56 kN; 105, -70 MN/m^2 5.6. $1\cdot11$ MN 5.7. $1\cdot43$ mm, 150 MN/m^2 5.8. $u_2 = u_3 = -0\cdot238$ mm; ± 50 MN/m^2 5.9. 280 kN 5.10(a). $\mathbf{F} = \{-28, 28, -42, 42\}$ kN, $\boldsymbol{\delta} = \{0, 1, 0, 0\cdot5\}$ mm, (b) $\mathbf{F} = \{-34\cdot3, 34\cdot3, -51\cdot4, 51\cdot4\}$ kN, $\boldsymbol{\delta} = \{0, 1\cdot22, 0, 0\cdot61\}$ mm 5.11. $-46\cdot2$, $34\cdot7$ MN/m^2 5.12. $23\cdot7$, $-63\cdot2°C$ 5.13. $-106\cdot6$, $-53\cdot3$ MN/m^2 5.14. 12 mm 5.15. $14\cdot56$ kN 5.16. $48\cdot3$, $153\cdot3$ MN/m^2 5.17. 284 kN 5.18. 96 kN 5.19. 168 kN 5.20. -78, -39 MN/m^2 5.21. 80, -40 MN/m^2.

6.1. Shearing forces at A, B, C, D (kN); 30, 30/10, 10/-50, -50; 50, 50/-50, -50; 0, 0/-8, -8, -8. Bending moments at A, B, C, D (kNm); 0, 30, 50, 0; -50, 50, 0 16, 16, 8, 0. Contraflexure at $x = 1$ m in Figure 6.29b. 6.2. 16 kN. 6.3 Shearing forces at A, B, C, D, E (kN); -10, -10, -2, -2; -5, $-1/1$, 1, 1/-2, -2, $(-3$ midpoint AB); -60, $-60/0$, 60/-60, 0, (30, -30 midpoints of BC and CD). Bending moments at A, B, C, D, E (kNm); 24, 14, 2, 0; 0, -12, $-8/2$, 6, 0, (-8 mid point AB); 60, $-60/0$, 60, 0 (15 midpoints BC, CD). Contraflexures; none; C; $x = 2$ m, 4 m. 6.4. Shearing forces at A, B, C, D (kN); 0, 20/0, 40/-50, -50, (10, 20 midpoints AB, BC); -10, 30/-70, -50, (10, -60 midpoints AB, BC). Bending moments at A, B, C, D (kNm); 0, 10, 50, 0 ($2\cdot5$, 20 midpoints AB, BC; 40, 60, 0 (40, $27\cdot5$ midpoints AB, BC). 6.5. A, midpoint, B; shearing forces, 0, $w_0L/8$, $w_0L/2$; bending moments, 0, $w_0L^2/48$, $w_0L^2/6$. 6.6. >16 kNm anticlockwise. 6.7. A, B, C; shearing forces (xy) $-5\sqrt{3}$, $-5\sqrt{3}$, $-5\sqrt{3}$ (xz) -15, $-15/-5$, 0; bending moments (xy) $10\sqrt{3}$, $5\sqrt{3}$, 0 (xz) 20, 5, 0. 6.8 $0\cdot0446$ m^4. 6.9. $1\cdot113 \times 10^{-4}$ m^4. 6.10. $hb^3/12$; $(\times 10^{-4}$ m^4), $0\cdot267$, $0\cdot278$, 1567. 6.11. 491×10^{-8} m^4; $133\cdot2$, 121 mm. 6.12. $22\cdot4$, $12\cdot8$ MN/m^2 6.13. Yes 6.14. 285 kN 6.15. 277 kN 6.16. $10\cdot43$, $4\cdot23$ m 6.17. $5\cdot22$, $2\cdot12$ m 6.18. $-21\cdot9$, $45\cdot4$ MN/m^2 6.19. $6\cdot6$ kN/m 6.20. $5\cdot04$ m 6.21. $47\cdot5$ kN 6.22. $8\cdot8$ kN/m 6.23. No 6.24. E 6.25. E 6.26. E, F. 6.27. $L/4$ from free end 6.28. $6PL/bh^2$ 6.29. $2PL/bh^2$ 6.30. $\pm 10\cdot2$ MN/m^2 6.31. $4\cdot5$ mm 6.32. $42\cdot7$ MN/m^2 6.33. $10\cdot1(\tau_{av} = 10)$ MN/m^2 6.34. $83/77\cdot7$ MN/m^2 (top flange/web) 6.35. 400 N/m 6.36. 136, $-152\cdot5$ MN/m^2 6.37. $22\cdot9$ kNm 6.38. $-14°21'$; 2392, 252 cm^4; $-14°13'$; 6900, 1520 cm^4 6.40. $119\cdot7$, $-174\cdot8$ MN/m^2; $y = -0\cdot322z$.

7.1. $mL/EI - wL^3/6EI$; $mL^2/2EI - wL^4/8EI$ 7.2. $mL/3EI$, $-mL/6EI$; $v_{\max} = 0\cdot0641$ mL^2/EI, ratio $= 0\cdot975$ 7.3. At A $-12EI/L^3$, $-6EI/L^2$, at B $12EI/L^3$, $-6EI/L^2$; shearing force at A and B $12EI/L^3$; bending moment at A, B $-6EI/L^2$, $+6EI/L^2$ 7.4. $-2\cdot25 Pa^3/EI$; $-15Pa^2/8EI$, $13Pa^2/8EI$; $x = 1\cdot92a$, $-2\cdot4Pa^3/EI$

7.5. $-57wa^4/24EI$, ∓ 11 $wa^3/6EI$ 7.6 Ends $-1/900$, centre $+1/450$ 7.7. $-12\cdot7$
mm, $-7\cdot93 \times 10^{-3}$; 4630 cm^4 7.8. $-8\cdot92$ mm   7.9. $8Pb/21$; shearing force (at
A, B, C) $-P$, $-P/0$, 0; bending moment (at A, B, C) $13Pb/21$, $-8Pb/21$, $-8Pb/21$
7.11. $3wL/8$; $-3EI/L^3 + 3wL/8$ 7.12. $\frac{1}{2}wL$; $-wL^2/12$ 7.13. $m^2L/2EI$; $w^2L^5/40EI$
7.14. 250 Nm   7.15. mL/EI 7.16. $Pb^2c^2/3EIL$ 7.17. $wL^4/8EI$, $wL^3/6EI$ 7.18.
$w_0L^4/30EI$, $w_0L^3/24EI$ 7.19. $PLH^2/2EI$, $PH^3/3EI$, $PH^2/2EI$ 7.20. $\pi Pr^3/4EI$,
$Pr^3(1+(\pi/4)),/EI)$ Pr^2/EI 7.21. $Pr^3/2EI$.   7.22. $2Pr^3/EI$, $\pi Pr^3/2EI$, $3\cdot52$ mm,
$8\cdot55$ N   7.23. Moments at A, B $\pm wL^2/12$, forces at A, B $wL/2$.   7.24. $0\cdot0909$ m side
7.25. $13P/32$, $11P/16$, $-3P/32$: bending moments at A, B, C 0, $3PL/64$, 0,
under load $-13PL/128$ 7.26. See Example 7.15 and Problem 7.23   7.27. $Y_A =$
$P - Y_C$, $m_A = \frac{1}{2}PL - Y_CL$, $Y_C = (-3EI\delta/L^3) + 5P/16$ 7.28. $3wL^4k/(24EI + 8kL^3)$
7.29. 77$\cdot$8 MN/m^2, $-3\cdot53$ mm   7.30. Horizontal P/π (inwards), vertica
$\frac{1}{2}P$ 7.31. $10^3\{-12$ -36 $+12$ $-12\}$ 7.32. $\{1160$ 644 -1160 $516\}$
7.33. $10^3\{-15$ -15 60 30 -45 $45\}$; -15, $15/45$, -45 kNm   7.34.
$10^4\{3$ 4 9 32 -12 $24\}$; 4, $-8/+24$, -24 kNm   7.35. X_1, X_3 are $6\cdot87$, $3\cdot13$ kN;
$m_1 = 15$ kNm; $v_2 = -5\cdot83$ mm; θ_2, θ_3 are $-0\cdot625$, $2\cdot5 \times 10^{-3}$ rad   7.36. X_1, X_2
are $6\cdot59$, $3\cdot41$ kN; $m_1 = 6\cdot36$ kNm; $v_2 = -3\cdot94$ mm; θ_2, θ_3 are $0\cdot455$, $2\cdot73 \times 10^{-3}$
rad   7.37. 0, $-3/+9$, 0, kNm   7.38. $v_3 = -51\cdot7$ mm; θ_2, θ_3 are -31, -62×10^{-3}
rad; -2, 4, 0, kNm   7.40. $-12\cdot7$ mm, $-7\cdot93 \times 10^{-3}$ rad   7.41. Shape factors,
$1\cdot144$, $1\cdot34$, $1\cdot53$, 2, $1\cdot8$; moments, $25\cdot5$ MNm, $90\cdot2$, $51\cdot1$, $0\cdot471$, $22\cdot7$ kNm   7.42.
Stresses $87\cdot5$, $98\cdot75$, 200 MN/m^2 for $y = 0\cdot2$, $0\cdot19$, 0 m.   7.43. $-4\hat{M}/3$ 7.44.
$\hat{M}/b$, $9\hat{M}/L$.

8.1. Torsion moments at A, B, C (kNm) (a) $0/-1\cdot6$, $-1\cdot6/-3\cdot6$, $-3\cdot6/0$ (b) 0, $0/2$,
$2/0$ 8.2. A, B, C (kNm) $0/1\cdot91$, $1\cdot91/3\cdot82$, $3\cdot82/0$ 8.3. 547 kW, $5\cdot47$ MW, $0\cdot0375$
rad   8.4. $54\cdot4$ mm, $0\cdot046$ rad   8.5. $1\cdot57$ kNm   8.6. $34\cdot4 \leqslant d \leqslant 38\cdot6$, $34\cdot4$ mm,
$0\cdot187$ rad   8.7. 78, $70\cdot2$ mm   8.8. $43\cdot3$ mm, $0\cdot058$ rad   8.9. $0\cdot0176$ rad, $35\cdot2$,
$5\cdot54$ MN/m^2 8.10. 'C'   8.11. 202 MN/m^2 (in steel)   8.12. $99\cdot5$ MN/m^2 8.13.
$0\cdot0766$ rad   8.15. $87\cdot5$ mm   8.16. 209 Nm   8.17. $58\cdot5$ Mm   8.18. 32 mm,
$0\cdot109$ m   8.19. 16 : 27   8.20. $(L-5q)\delta_B/4qL$ (N.B. $\delta_B = f(q)$)   8.21. $56\cdot1$ kNm,
$0\cdot076 \times 10^{-3}$ m^{-1} 8.22. $0\cdot0304$ rad, $55\cdot6$ MN/m^2 8.23. $8\cdot40$ kNm^2, 300 Nm;
179 kNm^2, $3\cdot2$ kNm   8.24. 75, 1/15   8.26. 183 kNm; $37\cdot9$, $68\cdot95$ MN/m^2 at outer
and inner radii respectively   8.27. $1\cdot90$.

9.1. $8\cdot2$, $-34\cdot8$ MN/m^2; $19\cdot1$ kN.   9.2. σ_{max} exceeds P/A by $120y_e$ (y_e in m)
9.3. $22\cdot9$, $17\cdot8$ kN/m   9.4. $2\cdot54$ kN/m   9.5. $\pm 0\cdot0678$ m   9.6. $8\cdot73$, $6\cdot79$ kN/m
9.7. Limits on axes ± 5, ± 10 mm, joined by straight lines   9.8. 1500 kW   9.9. $65\cdot2$
MN/m^2 9.10. 207 mm   9.11. $3\cdot98$ mm   9.14. u_2, v_2 are 5×10^{-5}, $-0\cdot712 \times 10^{-3}$ m
$\psi_2 = 0$; $\theta_2 = 0\cdot267 \times 10^{-3}$ rad.

10.1. 10 mm   10.2. $18\cdot4$ kNm   10.3. Zero   10.4. $2\cdot36 \leqslant P \leqslant 3\cdot93$ MN   10.5. 10
mm; $1\cdot81 \times 10^{-3}$; $1\cdot5 \times 10^{-3}$ m^3 10.6. $0\cdot607$, $0\cdot56 \times 10^{-3}$ 10.7. $0\cdot34 \times 10^{-3}$
10.8. $0\cdot25$ mm; $31\cdot4$ kN   10.9. 10 mm; $13\cdot1 \times 10^3$ mm^3 10.10. $46\cdot3$ mm   10.11.
$3\cdot67$ km   10.12. 75, $187\cdot5$ MN/m^2 10.13. $201 \leqslant P \leqslant 552$ kN   10.14. $1\cdot32 \times 10^{-3}$;
$2\cdot55 \times 10^{-3}$ m^3 10.15. 212 kN   10.16. 118 MN/m^2 10.17. σ_x, σ_θ, σ_r, are $63\cdot5$,
$135\cdot5$, -150 MN/m^2 10.18. $52\cdot5$, 16 MN/m^2; 202 kN   10.19. $2\cdot18$ 10.20. $83\cdot4$
MN/m^2 10.21. $0\cdot5$, $0\cdot1$, $0\cdot2$ m (radii); $0\cdot715 \times 10^{-3}$ interference   10.22. 75 MN/m^2;
111 mm; 139 MN/m^2 10.23. 118, -115 MN/m^2 10.24. 12,700, 9220, r.p.m.

11.2. $(k_1 + k_2)h$; k/a, $k/2a$; $2k_2$ 11.3. $0\cdot423$ kL, $1\cdot577$ kL 11.4. $19\cdot15$ kN, xz;
$2\cdot46$ MN, xy 11.5. $47\cdot0$ mm, $17\cdot3$ kN   11.6. $3\cdot69$ MN, $3\cdot69$ MN   11.7. $28\cdot8$ kN
11.8. C, E   11.9. All   11.10. $2\cdot05$ MN.

12.1. Plate loaded as in Figure 1.15   12.2. Beam of rectangular cross-section subject to linearly varying bending moment   12.3. No   12.4. Plate as in Figure 1.15; $A/2$; A 12.5. σ_x, σ_y, τ_{xy} are $8Cxy/h^2$; $C(1-(4y^2/h^2))$, 0; cantilever with force at $x = 0$, and support at $x = L$ 12.6/12.7. Compare with Figure 7.5.

13.1. 0, 64·7 kN; $\sigma_x = \sigma_y = 0$, $\tau_{xy} = 12\cdot9$ MN/m²   13.3.(b). Upper symmetric half, each row starts with coefficient on diagonal (0·35, 0, −0·35, −0·7, 0, 0·7; 1, −0·6, −1, 0·6, 0; 4·35, 1·3, −4, −0·7; 2·4, −0·6, −1·4; 4, 0; 1·4; all times 550×10^6 N/m)   13.4. $u_2 = -u_4 = -0\cdot14$ mm; σ_x, σ_y, τ_{xy} are 56, 19·5, 0 MN/m²   13.5. $10\cdot4 \times 10^{-6}$ m, 0; 0, 0, 40, MN/m²   13.6. $u_2 = u_4 = 1\cdot245 \times 10^{-3}$ m.

14.1(a). Stable, indet. (b) Stable, det. (c) Unstable   14.2. Bars 1, 2; 2,3; 3,4; 2,5; 1,4; 4,5.   Loads kN (tensile positive) (a) − 32, − 32, 40, 0, − 40, 64 (b) − 32, 0, 0, 40, 0, 0   14.3. Bars 1,2; 2,3; 3,4; 2,4; 1,4; 4,5; loads kN 0, 0, 0, 24, − 40, 32   14.4. Bars 1,2; 2,3; 1,3; loads kN 164, 284, − 201   14.5. No. Bars 1,3 and 3,4 would buckle   14.6(a). 12 (b) 5 (c) 12   14.7. Distribution Table column moments kNm (a) 9·23, ±0·46, ∓4·15, − 15·92 (b) 0, ∓8·04, − 9·48 (c) 0, ∓10·5, ∓4·5   14.8. 15·08, ±3·15, ∓5·39, − 15·31 kNm   14.9. 0, ∓7·875, ∓4·5 kNm   14.10. Checks (mid-span values, units kN or kNm, shear first) (a) − 16·85/7·15, 0·77, 2·94; − 7·62, 1·85, − 7·96 (b) − 5·98/14·02, − 11·52/12·48; − 5·98, − 9·24 (c) 5·25, − 1·5, − 4·5; 5·25, − 10·5, 1·13.   14.11. S.F. and B.M. increased in magnitude by 25 % and 63 % respectively   14.12. 13·5 % increase in magnitude   14.13. B, C, D, E, F   14.15. 5·73 kNm clockwise   14.16. 1·97 kN   14.17. Mid-span values 1·76, 12·71, 1·0, kNm; 15·36 kN   14.18. Mid-span values 2·7, 11·7, 1·8 kNm; 2·7 kN   14.19. (positive to right) − 5·48, 2·41, kN   14.20. 10, 130; − 10, 110; kN   14.21. − 27·6, − 1·2, − 31·2, kN; 39·8, 39·8, 7·3, kNm   14.22. Bar end moments 0, ±8·46, + 7·16, 0, kNm   14.23. Bar end moments − 8·6, +21·4, ∓18·6, 111·4, kNm.

Index

432 — Index

Strain—cont.

- load, 308
- magnitude of, 33
- measurement, 41
- normal, 36
- principal, *see* Principal strains
- rosette, 41–44
- state of, *see* State of strain
- thermal origin, 93, 308
- total, 308, 342
- volume, 41

Strain/amplitude relations, 323, 346
Strain compatibility condition, 302, 305
Strain/displacement relations, 300–302, 308, 323, 342, 346, 350
Strain energy, 60–67, 73
- and complementary energy, 70
- density, 61, 62, 64, 66, 73, 223
  - general expression, 64
  - in terms of strain, 64
  - in terms of stress, 64
  - positive quantity, 64
  - virtual, 310
- dilatational, 73, 100
- distortional, 72, 73, 99, 100
- in
  - beam, 162, 163
  - curved beam, 167
  - direct loading, 60–63, 72, 283
  - discretized plane body, 327
  - simple shear, 63
  - spring, 220
  - torsion, 218, 219
  - uniaxial tension, 60–62, 74
- matrix forms, 66
- methods in structural analysis, 71–72, 162–169, 173, 180, 285, 292, 295, 311–316
- non-equilibrium loading, 67–69
- non-linear material, 61, 71, 72
- per unit volume, 62
- virtual, 309, 310, 327, 347

Strain hardening, 47, 99, 276
Strain transformation equations, 38
Strength, *see* Failure conditions
Stress, 2–5
- array, *see also* State of stress, 4, 9, 12, 51, 79, 120, 132, 232
  - polar co-ordinates, 210, 237, 252
  - principal, *see* Principal stress array
- boundary condition, 302–305, 308
- circle, *see* Mohr circle

components, 2, 4
- polar co-ordinates, 210, 252
concentration, 225, 226, 267, 280
direct, 2
failure, *see also* Failure conditions, 22–24
fields, 303–305
in
- bars of frame, 378
- beam, 118–138, 141–143
- combined bending and direct load, 232–237
- combined bending and torsion, 237–243
- compound cylinder, 269, 270
- composite beam, 138–140
- compression, 79, 289, 294
- cylinders, 251–276
- direct loading, 79, 94, 98
- discretized plane body, 325, 345, 347
- disk, 278–280
- gravitational loading, 305
- impact, 69
- leaf spring, 161
- non-equilibrium loading, 69
- plane body, 298
- plate, 386, 387, 390, 399, 401
- ring, 277
- rod, 212, 237–243, 278
- rotating body, 277–280
- self-weight, 305
- shaft, 212, 237–243
- spherical vessel, 258, 259
- tension, 25, 26, 48, 60, 69, 74, 79
- thick cylinder, 260–264, 266, 267, 273–276
- thin cylinder, 252–255
- torsion, 212, 219, 220, 222, 223, 225
- tube, 79, 118, 212, 222, 223, 232–243, 252, 266
initial, 97, 105, 269, 270
on a surface, 1
plane, *see also* Plane elasticity, 8
principal, *see* Principal stresses
resultants, 400
residual, *see* Elastic/plastic analysis
shrink fit, *see* Cylinders, compound
sign convention, 5
space, 100
thermal, 93–95, 308, 342–345
Stress compatibility equation 306